Marketing Management:
Fundamentals and Practices

Marketing Management:
Fundamentals and Practices

S.K. Kapur
Formerly Senior Specialist
State Council of Educational
Research and Training, Haryana

Oxford & IBH Publishing Co. Pvt. Ltd.
New Delhi

Oxford & IBH Publishing Company Pvt. Ltd.
113-B Shahpur Jat, *Asian Village Side*
New Delhi 110 049, India

Fax: (011) 4151 7559
Email: oxford@oxford-ibh.in

Printed at Chaman Enterprises

ISBN 978-81-204-1738-0

Preface

It is being realised now that modern business cannot be run efficiently and successfully unless it is managed professionally. Efficient marketing is the backbone of a business enterprise and plays a pivotal and critical role in its success. In fact, it contributes directly to keep the wheels of the business organisation moving on the path to progress and prosperity. Therefore, the need for an efficient and effective system of marketing management in improving the functioning of an enterprise or organisation, especially in these days of intense global competition and globalisation of the Indian economy, cannot be overemphasised.

Understanding of the principles, concepts and techniques of marketing management and acquisition of distinctive abilities and skills to apply them in day-to-day operations are the basic requirements for those who seek careers in sales and marketing management.

The aim of this book is to give a simple and lucid exposition of basic principles, concepts and techniques of marketing management. The book, especially written in the "Indian Context" and the "Indian scenario", provides a comprehensive treatment of all aspects of marketing management. Alongwith traditional and established concepts of marketing management, the book presents the accomplishments of more recent thinking and research in the field so as to instill a global outlook.

The book is primarily intended for use as a standard textbook for B.Com. (Hons.), M.Com., MBA, PGDBM, and other students studying marketing management in universities and institutes of management. One important feature of the book is that it presents the principles and basics of marketing management as a conceptual framework. In this manner it can realistically serve the needs of students, professors and practitioners including marketing professionals in India and abroad. It can be used as reading and teaching tool by all levels of management at management training courses/workshops.

The book contains 20 cases/case studies and numerous examples from the Indian marketing environment to provide a deeper insight into contemporary marketing practices and to improve the analytical skills of the reader. One unique feature of the book is that the cases, case studies and examples from the Indian marketing scene have been so integrated in the text that they befittingly cover, reinforce and correspond to the key concepts in marketing theory and processes discussed in various chapters.

In writing this text, I have drawn freely from the works of traditional theorists, from writings and research done by scholars in the area of marketing management and behavioural science, and from the experience of many eminent practitioners. Some of them have been referred to in the text and bibliography. I feel indebted to them all.

I wish to express my special thanks to Dr. D.R. Aggarwal (Academic Head, Management Programme, Institute of Technology and Management, Gurgaon), Professor S.K. Sachdeva (IMT, Ghaziabad), Mr. J.S. Arora, and Mr. Rajnish Sharma for their useful suggestions and candid advice. Mrs. Ritu Sharma, PGDBM, Marketing Executive, presently working in Malaysia, helped me in writing the chapter on Analysing Markets and Buyer Behaviour. I wish to express my deep gratitude to her for her contribution.

I wish to acknowledge the excellent assistance of my sons Man Mohan, Sushil and Sanjeev who have given me some useful suggestions based upon their personal experiences.

I shall be indebted to any reader who draws my attention to mistakes and errors. Criticisms and suggestions will be welcomed.

Sudarshan Kumar Kapur

Acknowledgements

I acknowledge with appreciation the cooperation accorded to me by the following companies and organisations and am indebted to them for providing me with much needed information for use in the book and sharing their marketing experiences with me.

1. Hindustan Lever Limited (now Hindustan Unilever)
2. Marico Industries Limited
3. Ashok Leyland
4. Colgate Palmolive (India) Limited
5. Brooke Bond Lipton Limited
6. Raymond Limited
7. Dabur India Ltd.
8. American Center Library, Delhi
9. British Council Division, Delhi

I also benefitted from some excellent writings and advertisements published from time to time in The Statesman, The Hindustan Times, The Pioneer, The Times of India, The Economic Times and The Business Today. I acknowledge with thanks their important influence on my thinking and their contribution in making this text really authentic and instructive.

Sudarshan Kumar Kapur

Contents

1

Marketing Management: Nature and Significance

1.1 INTRODUCTION

Food, clothing and shelter are basic human needs. Besides these, there are other human wants like education, health, transport, entertainment, electricity, light etc., which promote human efficiency and make human life comfortable. For satisfying their economic and social needs, people need numerous goods and services. These goods and services are produced in production units, namely fields, factories, public undertakings, schools, colleges, professional and training institutes, hospitals, hotels, shops, ships etc. A large number of producers — business enterprises, organisations and undertakings are engaged in producing goods and services i.e. products which satisfy the needs and wants of consumers or customers. The producers have to make it sure that the products they produce cater to the wants and needs of the people who have to buy them. They are of fairly good quality, are supplied unadulterated, and at reasonable prices as and whenever people need them if they have to earn a permanent goodwill. And herein lies the importance of marketing and marketing management. Marketing is, in fact, the culmination or ultimate activity of production. According to American Marketing Association, *"marketing is the performance of business activities that direct the flow of goods and services from the producer to the consumer or user"*. The function of marketing is to meet the demand of people and make available to them the goods and services which they feel they want.

It must be remembered that organisations and undertakings are set up to earn profit. Only then they can survive and grow. A manager owes responsibility first to the organisation that employs him and he must make it sure that the organisation earns reasonable profit. This would necessitate taking steps to reduce costs besides increasing yield. It should also be remembered that it is the customers and consumers who provide market and an organisation or firm owes a lot to them. In fact its survival and growth depend on them. It is the responsibility of marketing manager to see that goods are produced and provided per demands of the customers and, therefore, marketing management involves many other things besides providing goods at reasonable prices per demands of the customers, such as ensuring their quality in terms of their packaging, reliability and safety, low consumption of fuel or energy, low cost of maintenance and use, machinery for after-sales service, prompt attention to customers' grievances, fair advertising and fair trade practices etc.

However, before going into the study of various aspects of marketing and marketing management, we must understand the meanings of some important terms and concepts referred to in above paragraphs such as basic human needs, business activities, goods and services etc.

1.2 SOME IMPORTANT TERMS/CONCEPTS

1.2.1 Human Needs

Maslow classified human needs into five groups or classes *viz.*

1. **Physiological Needs:** These are the basic needs of the organism such as food, water, oxygen, sleep, sex or activity.

2. **Safety or Security Needs:** These are needs for ordered existence in a stable environment such as need for safety and freedom from fear or threat, need for security in one's position.

3. **Love or Social Needs:** These are needs for good relations with other individuals, need to be accepted by one's peers, need for belongingness.

4. **Ego or Esteem Needs:** These are needs for self-respect, self-esteem, recognition, prestige in one's position within the organisation and outside.

5. **Self-fulfilment or Self-actualisation Needs:** These are needs for self-fulfilment including need to achieve one's full capacity for doing, opportunity for personal growth and development, feeling of worthwhile accomplishment.

The importance of Maslow's classification is that it forms a "*Hierarchy of Needs"* i.e. it starts from very basic needs (group 1 and 2) to a cluster of higher social needs (group 3, 4 and 5).

Though marketing helps mainly in satisfying basic human needs yet it has assumed a crucial role lately to help in the satisfaction of higher needs up the hierarchy as well.

1.2.2 Business Activities

Henry Fayol observed that work done in a business enterprise usually included the following activities:

 (i) **Technical activities:** These activities involve production, manufacture and adaptation.

 (ii) **Commercial activities:** These include buying, selling, exchange.

(iii) **Financial activities:** The work involves search for and optimum use of capital i.e. finance

(iv) **Security activities:** These mean measures adopted for protection of property and persons.

 (v) **Accounting activities:** These include stock taking, preparing balance sheet, accounting costs, statistics.

(vi) **Managerial activities:** Managerial work includes planning, organisation, command, coordination and control (of all the above activities).

1.2.3 Goods and Services (Products)

A. Definition: In economics, anything which satisfies human want is called a good. A good is simply something that somebody wants. Anything material or non-material which has utility i.e. which has a quality of satisfying some want is called a good such as pen, book, table, television, water, soap, shampoo, services of a doctor, lawyer or barber. In this sense, a good is tangible whereas service is intangible. Service is referred to an act performed by one person for the benefit of another. Service can yield income. As it provides utility to receiver, we can say that service is a non-material good. For example, the acts of a lawyer, musician, doctor or consultancy provided by a consultant are non-material goods or services. Again production and consumption of services takes place simultaneously and there is no time lag between their production and consumption.

B. Classification of goods: The goods are classified as:

 (a) (i) *Free goods:* These are those goods which are free gifts of nature and have no market price that may be determined by forces of demand and supply. Their supply is unlimited. Benham has defined free good as something of which more is not wanted e.g. water, air, sunshine etc. They are also called *non-economic goods*.

(ii) *Economic-goods* are those goods for which price has to be paid such as foodgrains, house, book, TV. These goods are scarce and limited. Benham has defined economic good as something of which more is wanted. *Marketing is concerned with economic goods only*. Free and non-economic goods are not the subject matter of economics or marketing.

(b) (i) *Material goods:* These are goods that can be seen and touched by us and have a shape. These can be transferred from one place to another.

(ii) *Non-material goods:* These are those goods or services which have no shape. They cannot be seen, touched or transferred e.g. services of a doctor, business goodwill, expertise in music.

(c) (i) *Consumer goods or consumption goods:* These goods include all those things that directly satisfy people's wants. Some consumer goods like food, fruit, milk, bread and butter are consumed instantly or at once in a single act of consumption. These are called single-user consumer goods. Some consumption goods like furniture goods, car, TV, washing machine are used slowly over a time. These can be used repeatedly. They are *called durable use consumer goods or durables.*

(ii) *Producers' goods or Production goods:* These are the goods which are not directly used by individuals for themselves but help in production of other consumption goods i.e. they satisfy wants indirectly e.g. factories, machinery, equipment, raw material, supplies, means of transport etc. These are called *capital goods or investment goods.*

C. The goods and services which directly satisfy human wants i.e. consumption goods can be divided in three classes *viz.,* necessaries, comforts and luxuries:

(i) *Necessaries*: These are those things or goods which are necessary to sustain life and for maintaining efficiency. In other words, these are essential and must be consumed. These include goods desired intensely per habits and customs.

(ii) *Comforts*: These are those goods and services beyond the necessaries level which make living easy. They are not essential but add to efficiency.

(iii) *Luxuries*: These are goods which do not add to efficiency. They may satisfy some superfluous wants and in the process may sometimes harm efficiency.

D. Keeping consumer behaviour in view, consumer goods have been grouped into four categories *viz.* convenience goods, shopping goods, specialty goods and unsought goods:

(i) *Convenience goods* are those products which consumers purchase without taking much time such as food, medicines, soaps, grocery goods, ice cream, locks etc. They do not cost much.

(ii) *Shopping goods* are goods that are a bit costly and need thorough examination and evaluation with competing goods. The customer spends enough time on purchasing them e.g. colour television, refrigerators, air conditioners, furniture, interior decoration etc.

(iii) *Specialty goods* are goods for which the customer has special liking and is willing to make special effort to get them. They may or may not cost much. Customers will not go in for substitutes. Quality and brand items like Adidas shirts, Lotus shoes, Forhan's toothpaste.

(iv) *Unsought goods* are goods which customers usually do not want or are not familiar with. For example, in India, people are not yet familiar with time sharing at holiday resorts. Insurance, selling of plantations 20 or 30 years hence also come under this category.

Anything that can be offered to a market for attention, acquisition, use or consumption so as to satisfy a want or need is called a product.

These goods or products include services, sometimes known as service products such as hair cuts, concerts, consultancy services etc. Marketing is concerned with sale and promotion of such products.

1.2.4 Demand

The demand for a commodity is the quantity of commodity that a consumer is ready to purchase at a given price at a given time. Demand implies three things:

(i) effective desire to possess or buy a thing,

(ii) ability to pay for it or means of purchasing it, and

(iii) the willingness to use these means for purchasing it.

1.3 BARTER VERSUS MARKET ECONOMY

Barter means the direct exchange of commodities for commodities. It refers to the direct exchange of one good or service for another without the mediation of money. In mid-eighteenth century, prior to the advent of British rule in India, our villages had a self-sufficient and independent agrarian economy. People produced goods for their own consumption and not for sale. In villages, goods and services were exchanged for goods and services and barter system prevailed in the countryside.

Need for a common medium of exchange, commercialisation of agriculture, establishment of factories and industrialisation gave rise to a new economic system known as "market economy" in place of traditional barter system.

Market economy is that economy in which goods and services are produced for sale in the market and the produce is sold at a price determined by the market. In market economy, goods are exchanged through the medium of money, therefore market economy is also known as *monetary economy*. The main features of a market economy are:

(i) goods or services are produced for sale in the market.

(ii) prices of commodities are determined by demand and supply conditions.

(iii) money is used as a medium of exchange.

(iv) there is competition among buyers, sellers and labourers.

(v) the factors of production are bought and sold and payment to them is made in terms of money.

(vi) the production in market economy is carried on to earn profit, the main aim being to earn maximum profit.

1.4   MARKETS: DEFINITIONS AND CLASSIFICATION

As stated above a market economy operates through a network of markets and necessitates existence and development of markets with the dual task of facilitating the sale of goods and services produced by various producers to those who want them and ensuring the satisfaction of needs of people and organisations at large. A brief explanation of what is meant by "a market" and "what types of markets" one usually comes across in a marketing economy will be helpful in our study.

Market (Economics Context)

As seen above, the term *"market" refers to a group of buyers and sellers exchanging a commodity.* Benham has defined market as "any area over which buyers or sellers are in such close touch with one another either directly or through dealers that the prices obtainable in one part of the market affect the prices paid in other parts." The result of this free intercourse is that the price of the same good tends to equality easily and quickly. Therefore market is not a particular place. It means dealings or transactions in any product or factor of production between sellers and buyers. This definition suggests the following essentials of a market:

(i) a commodity or service to be bought or sold.

(ii) existence of buyers and sellers willing for exchange.

(iii) place, be it a certain area, region, a country or entire world.

(iv) free intercourse and contact between buyers and sellers.

(v) prevalence of one price for the same commodity at the same time in the entire region or area, and

(vi) complete knowledge and information about the market.

Based on the above concepts of market, markets are classified in a number of ways:

1. According to area covered, a market can be local, regional, national and international market.

2. According to the period or time of transaction, a market may be ready market or futures market.

3. On the basis of commercial commodities, a market may be a fruit market, crop market, produce exchange or stock exchange.

4. According the quantities of the commodities involved in transactions, market may be a retail market or wholesale market.

5. Markets are also classified on the basis of some conditions as perfect markets and imperfect markets.

6. On the basis of presence or absence of competition, markets are classified as perfect competition, monopoly, oligopoly, duopoly, monopolistic competition etc.

In this context, a marketer is a person who is willing to engage in a transaction and can be a seller or buyer.

Markets (Marketing Context)

In context of marketing of goods and services, we are more concerned with the satisfaction of wants and needs of the buyers, consumers or customers, i.e. with the demand market and the term "market" needs to be defined with emphasis on this aspect.

A market consists of all the potential customers sharing a particular need or want who might be willing and able to engage in exchange to satisfy that need or want.

In short, the term market is used to mean people with wants and purchasing power to satisfy them i.e. all the potential buyers, individuals or organisations.

By type of institution, buyer markets are identified as:

1. *Consumer Market:* It consists of all individuals and households who buy goods and services for personal consumption. It must be kept in view that different groups of consumers acquire different types of products and services depending on their needs and wants.

2. *Industrial Market:* This market comprises individuals and organisations that buy and produce goods and services *viz.* raw materials and intermediary goods and services for use in production of other goods and services that are sold to others. This market is also known as *business market or producer market.*

3. *Reseller Market:* This market comprises all individuals and organisations that procure goods for reselling to others at a profit. This includes whole-salers, supermarkets and traders.

4. *Government Market:* Government and its agencies at central, state or local levels spend vast sums on the purchase of goods and services to fulfil their administrative and other functions and to provide welfare services like education, health etc. This constitutes government market.

5. *Institutional Market:* As the name suggests, this market consists of institutions like schools, colleges, hospitals, nursing homes, charitable institutions that procure goods and services for providing them to people i.e. for redistribution. This includes firms and companies that procure them for providing them to their employees.

1.5 MARKETING AND ITS IMPORTANCE

Marketing: Definition

It must be noted that it is the need or want of a thing that gives rise to its demand and it is to cater to the needs of the people at large that goods and services are produced. The above study suggests that a proper definition of marketing should at least encompass the following aspects:

  (i) flow of goods from producers to customers or consumers in exchange for money i.e. through value exchange.

  (ii) satisfaction of wants and needs of the customers or consumers.

(iii) accomplishment of objective for which the firm or organisation is set up i.e. of making profits.

Marketing has been variously defined by different practitioners and thinkers over the time. In simple words, marketing is referred to activities that are udertaken in relation to markets. According to the American Marketing Association,

Marketing is the performance of business activities that direct the flow of goods and services from the producer to the consumer or user.

This definition explains only the distribution aspect of marketing and does not reflect any concern for satisfaction of wants and needs of the consumers.

E.G. McCarthy improved upon this definition by incorporating above missing aspects in his explanation. According to him,

"Marketing is the performance of business activities that direct the flow of goods and services from the producer to the consumer or user in order to satisfy customers and accomplish firm's objectives."

Satisfaction of wants and needs of a particular group of customers necessitates designing, development and manufacture of products in accordance with their specific requirements. Marketing therefore involves manufacturing (product designing and developing), financing and selling.

Edward W. Cundiff refers marketing to as "managerial process by which products are matched with market and through which transfers of ownership are affected". In this sense, marketing is the process by which an organisation relates creatively, productively and profitably to the market place.

Marketing is thus the sum total of all business activities which deal with the movement of goods and services from producer to customer in order to satisfy his needs. The required goods and services may include raw materials, supplies, semi-finished goods and finished goods as well as those services that are necessary to keep consumers using and enjoying them. Marketing in fact is concerned with satisfaction of customers' wants and needs through markets at a profit.

Importance of Marketing

Marketing is an important and essential function of business and has a significant role in conducting business, in helping society and in the economic development of the country. Its importance can be seen as:

1. Marketing helps the firm in planning and making decisions of business.
2. It establishes balance in the demand and supply and helps in earning maximum profit for the firm besides providing satisfaction to the customers or consumers.
3. Marketing establishes a communication system between the firm (i.e. producer) and the consumer (i.e. customers).
4. Marketing helps in ensuring maximum production at minimum cost.
5. It increases employment opportunities and provides employment to a large number of people.
6. Efficient marketing makes the goods available to the buyers at competitive prices.
7. It assists in raising the standard of living of the people and helps in improving their living conditions.
8. It helps in increasing demand of products and hence in checking the recession in the economy.
9. Marketing helps in increased industrialisation by catering to the demand of vast population by ensuring large scale of production of existing and new products caused by increased standards of living of the people and thus has an important role in the economic development of a country.

1.6 MARKETING FUNCTIONS OR ACTIVITIES

Marketing is a process of bringing together the producer and the buyer. It encompasses all those activities that represent working through markets. Marketing involves movement of goods and services from the producer to the consumer with a view to satisfying the latter's wants and needs. It necessitates undertaking many activities called marketing functions. Marketing functions can be categorised as:

1. *Exchange functions* comprising buying, assembling and selling.
2. *Physical distribution* or *supply functions* comprising activities including materials procurement, packaging, transportation, shipping, storage and warehousing.
3. *Facilitating Functions* i.e. activities aiding the performance of the first two categories of functions and comprising financing, pricing, grading and standardisation, advertising and sales promotion, market information, research and development (R & D).

Marketing is thus a total system of interacting business activities designed to plan, price, promote and distribute goods and services. *These activities form operative functions of marketing management and will be the main subject matter of our study in this book.*

1.7 MARKETING MANAGEMENT: DEFINITIONS AND NATURE

1.7.1 Marketing Management

Broadly speaking, marketing management means management of marketing activities i.e. marketing management is concerned with the planning, organising, directing and controlling of the activities pertaining to marketing of goods and services to satisfy customers' wants and needs.

According to American Marketing Association, *"marketing (management) is the process of planning and executing the conception, pricing, promotion and distribution of ideas, goods and services to create exchanges that satisfy individual and organisational objectives."*

This work is to be carried out by a number of personnel employed in the marketing department of an organisation holding positions right from marketing-vice president to sales people down the hierarchy. It must be noted that marketing is not only confined to the movement of goods and services from the producer to the consumer, it has to see that the consumer is satisfied. Therefore marketing management transcends manufacturing of product, financing and after-sales service also.

From the above definition, we can highlight the nature and basic features of marketing management.

(i) Marketing management is a process involving planning, analysis, execution and control. It requires overall coordination and cooperation of all to achieve organisation goals. This process consists of:

1. analysing marketing opportunities.
2. researching and selecting marketing programmes.
3. designing marketing programmes and organising, implementing and controlling the marketing effort.

(ii) It covers flow of ideas, goods and services. Marketing of ideas includes marketing of patents and franchise.

(iii) It rests on the idea of exchange and it has to actualise potential exchanges. This involves engaging in transactions and relationship building.

(iv) It is aimed at ensuring the satisfaction of customers/consumers and achieving the organisational goals.

Marketing management is thus extension of general management, that of motivating and inspiring every member to contribute his or her best to accomplish organisation objectives which in this case includes satisfaction of the consumers also.

1.7.2 Marketing Management—A Field of Knowledge and Study

Marketing is the art of creating and satisfying customers at a profit. The main objective of marketing management is to satisfy the wants and needs of people through exchange of goods and services. Marketing management is the study of how to "get the right goods and services to the right people at the right places at the right time at the right-price with the right communications and promotion". In this sense, marketing management is a branch of knowledge and study. It involves study of market behaviours i.e. buyer behaviour, market trends, marketing strategies, theories of motivation to achieve various goals. Marketing managers are required to perform well-defined task and responsibility and make decisions pertaining to product developing, pricing, marketing channels, promotion and physical distribution of goods and services and they need to acquire and master several positive attitudes to be an effective and successful marketing manager. One needs specialised knowledge, training and experience to be effective in the market place.

Marketing management is a distinct discipline having a well-defined body of knowledge with formalised methods of acquiring training and experience. This has its theories, procedures and strategies that can be applied to practical situations. Due to the popularity of this field, a number of institutions have been set up in the country to impart formal training in marketing management and professional marketing managers are replacing traditional managers.

1.7.3 Marketing Management: Functional Definition

The above study suggests that besides managerial functions, a marketing manager has to perform and execute a number of specific tasks and duties himself, such as collecting and analysing market information, forecasting and setting marketing objectives, designing, developing and launching new products, pricing, advertising and promoting them, ensuring post-sale services, recruiting, selecting and training sales workforce etc. Therefore, a definition of marketing management should cover both the basic managerial functions and operative functions of a marketing manager. Henri Fayol has listed planning, organising, commanding, coordinating and controlling as five basic managerial functions. Others have suggested staffing, directing, actuating, innovating, communicating etc. as managerial functions. For the purpose of simplicity, modern writers take planning, organising, staffing, directing and controlling as five functions of general management. To cover both the managerial as well as the operative functions, we can express the above definition of marketing management in the following way

Marketing management is the process of planning, organising, directing and controlling the conception, pricing, promotion and distribution of ideas, goods and services to create exchanges that satisfy individual and organisational objectives."

The definition of marketing management as given above is comprehensive one and points out the three dimensions of marketing management *viz.* management functions, operative functions and the objectives of marketing management. A brief explanation of these components of the definition is given here.

A. Management Functions Involved in Marketing Management

All those who are involved in marketing management have to perform basic managerial functions, namely planning, organising, directing and controlling of operations involved in the process. In other words, marketing management involves planning, organising, direction and control of marketing activities or functions required to be performed to provide satisfaction to the customers and to attain the objectives for which the organisation has been set up.

(i) **Planning:** Planning involves determining objectives, formulating policies, programmes and procedures in advance and selecting the most appropriate and suitable course of action. It is essentially a decision-making process leading to determination of action in future. In context of marketing management, the process consists of analysing information about marketing opportunities, selecting target markets, determining marketing objectives, designing marketing strategies, policies and procedures and planning marketing programmes for achieving organisational objectives. Use of computers and electronic data processing (EDP) can be of great help in planning marketing policies and programmes.

(ii) **Organising:** Organising involves identifying and assigning activities to be performed to individuals or groups and establishing relationships among jobs, persons, departments. Marketing organisation involves determining marketing functions and designing the structure of relationships among jobs and personnel. In a small firm, marketing functions may be entrusted to one person only. In case of big organisation, it would mean establishment of a marketing department. A marketing department in a large establishment is usually headed by Marketing Vice-President. Jobs in marketing department may be designated as marketing vice-president, market managers, product managers, advertising managers, marketing researchers, market trainers, sales·managers, sales supervisors, sales persons etc. The marketing Vice-President has the dual task of guiding and coordinating the work of all persons engaged in marketing department and working in close coordination and cooperating with Vice-Presidents or Heads of other departments such as finance, manufacturing, R&D, purchasing, with a view to satisfy customers.

(iii) Directing: Directing is the process of guiding the subordinates towards achieving the organisational goals. It means guiding and leading to better work performance. It is motivating and getting the people to do the work willingly and enthusiastically. In our present context, it involves guiding and issuing directions to the subordinates to put the marketing strategies and programmes into action i.e. initiate action in case of planned marketing activities. It is leading them to implement marketing programmes as planned.

(iv) Controlling: Controlling involves monitoring and measuring the actual performance, comparing it with planned goals and taking steps to rectify distortions, if any. Controlling a marketing programme or marketing department means monitoring and evaluation of marketing activities and providing feedback. It will include a systematic and comprehensive audit of marketing activities, periodical reassessment of marketing effectiveness through instruments of marketing audit, marketing probability analysis and marketing efficiency studies.

The importance of the marketing department in the realisation of overall objectives of the organisation cannot be overlooked, therefore it will be advisable on the part of top management to ensure active participation of the marketing manager in the determination of company's overall objectives, policies, programmes and strategies.

B. Operative Functions of Marketing Management

The above functions *viz.* planning, organising, directing and controlling are basic functions to be performed by every manager including a manager in marketing department. Besides these functions, marketing management includes execution or performance of some operative functions. It is, in fact, for performance of these very functions that marketing departments are established. We have already referred to some important marketing functions in Section 1.6 of this chapter. In the above definition, these operative functions or marketing activities have been termed as *conception, pricing, promotion and distribution of ideas, goods and services.*

These activities include segmenting the markets, formulating marketing strategy, selecting target market, studying buyer behaviour, sales forecasting, deciding the product policy, designing and developing products, branding and packaging, pricing the product, transportation, warehousing, deciding market channels, managing advertising, building brand image, organising sales promotion campaigns and managing sales, organising marketing information and research.

Though planning, organising, directing and controlling are basic functions of all levels and types of managers but the operative functions as referred to above form the core marketing activities. Obviously these operative functions of marketing management will be the main subject matter of our study in this book. We shall discuss them in details in later chapters.

C. Individual and Organisational Objectives

The third component of above definition suggests that marketing management seeks to satisfy individual and organisational objectives. A marketing manager is not only to safeguard the interests of the organisation but also has to see that individual and societal objectives are also satisfied in the process. We have already referred to consumer markets, industrial markets, governmental markets, institutional markets, reseller markets that constitute potential buyers or customers. A market manager must see that the interests of consumers, customers and society at large are reconciled with the objectives for which organisation has been established.

The needs of individuals i.e. consumers and customers will be well-satisfied if they get the right goods and services at the right places at the right time at the right price. A marketing manager has to see that goods are produced and provided per demands of the customers. He has to make it sure that goods produced are of a fairly good quality and are supplied unadulterated at reasonable prices, as and whenever they need. He has also to ensure that products are reliable, safe and give satisfactory service. It is also to be ensured that unfair trade practices are not resorted to. This would involve questions of morals, values and ethics. At the same time, organisational objectives also need to be ful-filled. According to the American Management Association, "the purpose of a business enterprise is the profitable production of goods and services." A market manager has to make it sure that the organisation earns reasonable profit so as to survive and grow.

It is clear from the above study that marketing management is an integrating process in the sense that it effects a reasonable reconciliation of individual, organisational and societal interests. In fact, individual, organisational and societal objectives are *"beacons that should guide day-to-day activities of marketing management"*. We shall study some marketing philosophies in the next chapter.

1.8 MARKETING ORGANISATION

In small enterprises, marketing activities are usually performed by the owner or the manager. As the enterprise grows a bit larger, the marketing functions are entrusted to some members of the organisation or are performed by the management in consultation with others in the field. In large establishments where the interests of individuals, organisation and society compete with each other and where work is based on division of labour and specialisation, the task of performing marketing activities becomes complex and specialised. In a large establishment, a separate marketing unit or department is set up to perform marketing functions.

Marketing department or unit may have a *supervisory role* with the authority to command and direct or mere an *advisory role* or a combination of both. Thus marketing department can have line organisational structure, staff organisational structure, or line and staff organisation. In management jargon, *line* refers to those positions which have the direct authority of exercising supervision over subordinates, have right to command and communicate and are accountable for the tasks assigned to them. *Staff* refers to those positions or elements that exist primarily for the purpose of providing advice and service to the line and other departments in the attainment of the objectives. *Line and staff* organisation is an organisation in which both line and staff relationships exits together. It is an organisation in which line organisation has the support of the staff to give it specialised advice and service to achieve the organisational goals.

The main purpose of marketing management is to determine the needs, wants and interests of customers (consumers or target markets) and to deliver the desired satisfactions more effectively and efficiently than competitors in a way that preserves or enhances the consumer's and the society's wellbeing. We have already referred to various marketing functions that need to be performed to achieve this objective. Therefore a marketing department is usually organised keeping in view the various functions to be performed by it. It can be structured into specific function based sub-departments or divisions with jobs or positions pertaining to the functions. For example, a large marketing department may be organised with a hierarchy of jobs or positions such as Vice-President or Marketing Director as the Head of Marketing Department, responsible to the President or to the Board, managers responsible for sales, advertising, promotion, physical distribution, customer support service, marketing research, product designing and development and the subordinate executives and employees at lower and entry level. Chart 1.1 illustrates the hierarchy of jobs in a large marketing management department.

Chart 1.1 shows hierarchy of positions within a large marketing department and structure is based on functions and activities that are usually performed by the marketing department. It is evident that the department has line and the staff relationships within the department and the organisation. Various job positions in the hierarchy in a marketing department responsible for the performance of their respective functions report to the top position in the department. Again middle level managers, handling various marketing functions, have their subordinates, executives and supervisors, to assist them in their functions. Besides, marketing department has to advise other departments such as manufacturing department, materials department etc., in matters concerning needs and demand of people.

Marketing departments in a number of establishments are organised on *geographical (regional) basis or on basis of marketed products.* The positions in this case could be regional managers, product group managers, brand managers etc. It is not necessary that they may form a separate organisational structure. Product management organisation may serve as another layer of management within the functional structure.

1.9 MARKETING MANAGEMENT—A SYSTEM OF INTERDEPENDENT ACTIVITIES

In order to fulfil its functional, societal and organisational objectives, marketing management involves performance of a number of interconnected activities, each activity being directly related to the other activity or affecting performance of other activities. Marketing management activities comprise a system with a number of sub-systems of activities functioning together to meet the specified objectives. As W.G. Stanton has observed— "Marketing is a total system of interacting business activities designed to plan, price, promote and distribute want satisfying products and services to present and potential customers". For example, it is in the light of requirements of the organisation and marketing environment, both internal and external, that the planning unit of marketing department plans its marketing strategy. This has a bearing

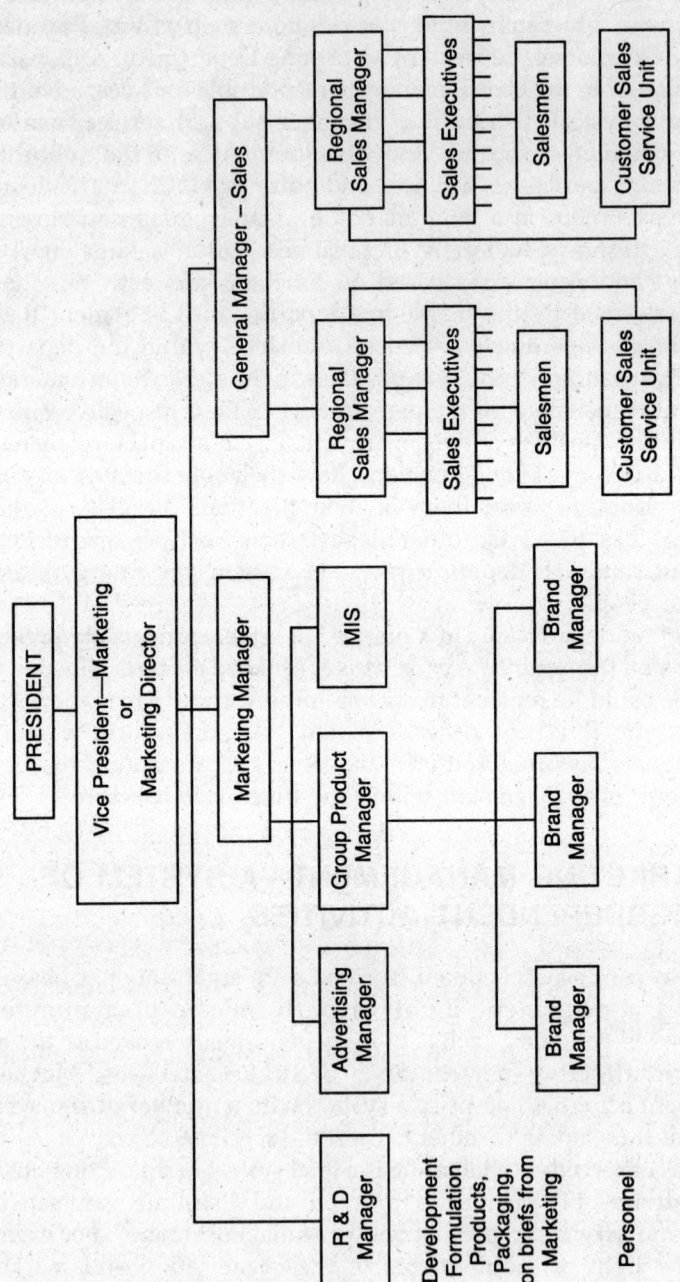

Chart 1.1 Organisation of a Marketing Management Department.

on the devising of sales, advertising, promotion, transportation, warehousing, customer support and public relations programmes of the organisation. Thus marketing management functions form an integrated system consisting of a number of subsystems with well-defined limits of their operations, each affecting the others and contributing to the accomplishment of marketing management objectives. The system is influenced by marketing approach or philosophy of the organisation and business or marketing environment, internal or external, micro or macro.

We have already referred to various activities that are required to be performed by the marketing departments of the organisations in earlier sections of this chapter, namely planning, selling, promotion, distribution (transportation, warehousing), packing and packaging, product launching, customer service, evaluation and feedback etc. All these sub-systems have their well defined boundaries. Chart 1.2 illustrates the marketing management system with its sub-systems.

We shall discuss the various approaches to marketing management (Marketing philosophies or concepts) and the environmental challenges faced by marketing management and the role of marketing managers in the next two chapters.

1.10 IMPORTANCE OF MARKETING MANAGEMENT

It is apparent from the above discussion that marketing management plays an important and vital role in the success of a business enterprise. Efficient marketing is the back bone of a business enterprise. Marketing management is primarily concerned with movement of goods and services from the producer to the consumers or customers in order to satisfy their needs. Obviously it contributes directly to keep the wheels of the organisation moving on the path to progress and prosperity. Therefore the need for an efficient and effective system of marketing management in improving the functioning of an enterprise or organisation cannot be overlooked. The importance of management can be summed up as follows:

1. Marketing management helps in the realisation of the objectives for which the organisation has been set up. It helps in the task of running the establishment smoothly and effectively. In fact, an effective marketing management is essential for the survival and growth of the organisation.

2. It helps the community to satisfy their economic and social needs and thus raise their standard of living. Marketing management is also defined as *"the creation and delivery of standard of living to the society"*. It ensures better deal and services for the consumers. It helps the enterprise to fulfil its social responsibilities.

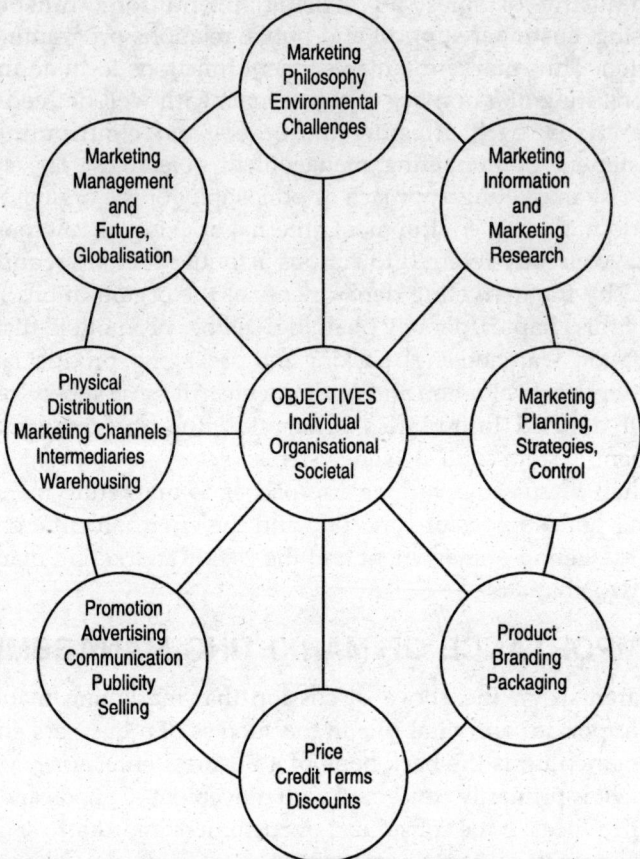

Chart 1.2 Marketing Management—An Integrated System.

3. It helps in producing those products that are needed by the consumers and community at large. It activates the production-consumption chain. Thus it helps in an efficient and productive utilisation of resources, both human and materials, eliminating wastages.
4. It helps the enterprise to adapt to the changing conditions and circumstances.
5. It provides guidance to the organisation on the innovations to be adopted, enabling it to face competition more squarely.
6. It helps the enterprise in achieving the maximum efficiency, productivity and profitability with the minimum of effort and cost.

7. Effective marketing management ensures effective performance of other functional departments and hence better and effective results. The effectiveness of other departments depends on effectiveness of marketing management.
8. It makes organisational planning more meaningful and relevant.
9. It helps in creating efficient entrepreneurs and managers.
10. It ensures the economic growth of the enterprises which results in growth and economic development of the country.

EXERCISES

1. What do you mean by the term 'market'? Explain briefly the main types of markets.
2. Define the term "Marketing". What are its main functions?
3. What is meant by marketing management? Explain its nature and scope.
4. "Marketing management is an integrated system." Comment on this statement.
5. Explain the set up of a marketing management department in a modern enterprise.
6. Write short notes on the following
   (a) Marketing functions.
   (b) Operative functions of marketing management.
   (c) Importance of marketing management.
   (d) Business market and reseller market.
7. "Marketing management is a decision making process which is primarily useful for firms involved in the manufacture of consumer goods". Comment.
8. Discuss the similarities and differences in the marketing strategies adopted by companies such as Hindustan Unilever producing soaps and detergents and Maruti, the car manufacturer.

2

Marketing Philosophy and Various Approaches

Marketing management is referred to as a process by which an organisation relates creatively, productively and profitably to the market place. It is the management of marketing activities to satisfy both individual and organisational objectives. Its main objective is to achieve exchange relations that may help in the realisation of the objectives for which the organisation has been set up. Obviously these exchange relationships between the organisation and customers are determined and governed by organisation objectives, norms, criteria, beliefs and values held by the management as to the behaviour of customers, their needs and desires, and other forces, both internal and external, such as societal, cultural and legal factors. In short, these relations are affected by what is called *marketing philosophy*, which is based on values, beliefs and economic and social objectives of the organisation. We shall briefly trace some concepts of marketing philosophy and the nature of change that has occurred in it over the time.

2.1 MARKETING PHILOSOPHY AND VALUES

A. Marketing Philosophy

The term "philosophy" is referred to as a science which deals with the ultimate reality, cause and principles of things. It establishes fundamental criteria and principles for determining and shaping action and provides a framework of thought and attitude for managing certain affairs. It is the values, beliefs, standards of conduct and the like that shape a philosophy. Philosophy is dynamic in nature. It changes as human values and beliefs are prone to change over a time. A philosophy thus provides foundations for making decisions and determining policies and programmes concerning particular tasks in hand.

Marketing philosophy means the philosophy that is related to marketing management. It is the philosophy that provides the sound foundations for a marketing management policy or programme. It is an integrated and consistent body of organisational objectives and beliefs as to the conduct pertaining to the management of marketing activities. Marketing philosophy represents the beliefs and views of management, organisation, and the ideas and principles that guide its marketing efforts.

B.   Values and Marketing Philosophy

It is evident from the above study that it is the values or ideas held by the management and customers that underline the objectives of the organisation and shape its marketing philosophy. Both these sets of values are key consideration in framing a sound marketing philosophy and policy of marketing management. The organisation as well as its customers are constituents of a system called *Society*. Therefore values cherished by the society and its constituents have an important bearing on the functioning, stability and effectiveness of an organisation.

As stated earlier, values underlie the objectives of management and of customers. Values are referred to as needs, interests, wants, desires of individuals, groups, organisation or demands of society. Sometimes customs, mores and folkways become values. For example, values may include maximisation of profit, concern for production, concern for the sale, concern for the customer, mutual trust, intergrity, equity. All these expressions are value-judgements. Different organisations may have different value systems.

In this context, it must be noted that *individual values* are the values which are cherished by individuals i.e. customers and guide and control their conduct and action. They are based on subjective judgements and may exhort a person to pursue personal desires rather than group welfare or company welfare. On the other hand, *organisational values* are values which are held by an organisation or a company as a whole. These are derived from its objectives and affect its structure and policies. These values may be the logical conclusions of *"what is good for the company"*. Regarding these values, individual or personal preferences are irrelevant.

The conflict between values

The basic values of the organisation may be different from the value system of the customers. This conflict between values of the organisation and customers is a critical factor that can adversely affect the working of an organisation. For example, management is more concerned with maximum production and profit whereas customers may be more interested in acquiring the products at lower prices resulting in reduced profits. The divergence of values of management and individuals leads to conflict. Moreover environmental values i.e. values of various other contributors such as suppliers, competitors, government and interest groups are also involved. In context of philosophy, a brief study of the Blake and Mouton value theory of organisation will be helpful. This theory throws light not only on marketing philosophy but on managing or selling styles as well.

C.   Blake and Mouton Managerial (Selling) Theory

According the R.R. Blake and Jane Mouton, two major dimensions (values) of managerial/selling style are:

(1) concern for the people i.e. concern for customers' needs and benefits expected by them from the product and hence their satisfaction and

(2) concern for production or the sales.

In their view, these two concerns can be combined in different degrees. They suggest that an optimal managerial or selling style involves maximising both dimensions *viz. concern for sales* and *concern for the customers.*

Again, fulfilment of social responsibility may require a marketing manager to inculcate new values or a value system that may guide and direct towards accomplishment of societal needs as well as organisational objectives.

It is evident form the above discussion that values are a key consideration in matters concerning organisation - customer relations. In fact, it is the values that shape marketing philosophy and give rise to concepts that affect exchange relationships. Again values to customer may become values to management when they are perceived as being relevant to customers' wants. Values of management and customers thus need to be balanced and matched and these must be reflected in the philosophy. To illustrate the values, we are giving a case study at the end of the chapter. See Case 2.3 i.e. Appendix to Chapter 2

2.2 ROLE OF PHILOSOPHY

The philosophy as related to marketing management thus presents a framework of thought and attitude of the administration and establishes fundamental criteria for determining and shaping action and how resources, both material and human, are to be managed and marketing problems to be tackled. A proper knowledge of the philosophy of marketing management and its various concepts helps the marketing managers in:

   (i) understanding views and feelings of individuals and groups,

  (ii) in establishing relationship between objectives and accomplishments resulting from various actions,

 (iii) in understanding the complete nature of marketing activities, and

 (iv) in establishing relationship between value-related decision and work.

Values are a part of total complex and they change over a long time under economic, social, cultural and political forces. The marketing philosophy also needs to be reviewed and modified in the light of shifting values of both organisation and individuals. The present day marketing philosophy of most of the organisations is quite different from the marketing philosophy that was being followed by the organisations a few decades back or a century earlier. In the following pages we shall study briefly how various concepts or approaches of philosophy pertaining to marketing management have evolved over the period.

2.3 MARKETING PHILOSOPHY—VARIOUS CONCEPTS AND APPROACHES

Studies have revealed that there are five distinct concepts of business philosophy under which marketing activities are carried out. They are— the production concept, the product concept, the selling concept, the marketing concept, and the societal marketing concept.

The Production Concept (Production-oriented Philosophy)

This is the oldest concept of business philosophy and as the name suggests the emphasis is on increasing production. According to this philosophy, *consumers will usually buy those products that are widely available and have low cost.* Organisations holding this concept strive to increase the production efficiency and improve technology to have higher and higher production and to take measures that aim at reducing per unit cost. This approach is also called *manufacturing-oriented philosophy.* It also strives to make the product easily and widely available through a wide distribution network.

This concept is usually followed in underdeveloped countries where demand for a product exceeds it supply. The concept also suggests that high price of a product can be reduced only by increasing the productive éfficiency or the productivity of the enterprise. Obviously more production would need expansion of the market so as to achieve its disposal and higher profits. Organisations, government or non-governmental, like employment exchanges, hospitals, charitable institutions that organise and provide mass facilities such as employment services, eye-operations, dental care etc., follow production concept as they provide these facilities to a large number of customers widely spread at a very low price. Here consumers or beneficiaries are more interested in getting the product or service rather than the quality of the product or service. Obviously this concept does not help where customers are more interested in the quality and performance of the product and considerations other than wide availability and low price.

The Product Concept
(Product-oriented Philosophy)

The business philosophy known as *"the product concept"* is based on the notion or perception that *the consumers or customers are interested more in the quality and performance of the product or the service rather than its volume of production or price.* They prefer to purchase products with higher quality and performance. Obviously in product-oriented approach, the sellers concentrate on producing goods of higher quality and performance and on improving upon them continuously over the time through innovations and inventions. This approach is *technology-oriented.* Volume of production or lower cost per unit or prices are not the primary considerations with the seller or producer. The emphasis here is on the improved quality and excellent performance of the product. Organisations following this approach believe that they can accomplish the same results by offering good quality products as are expected from producing goods on mass scale and supplying them to consumers at lower prices and the consumers would not mind paying higher prices for quality products.

There are several Indian companies which can be said to be following the product-oriented approach. To quote a few examples, one can name Godrej and Boyce (manufacturing refrigerators, washing machines, safe vaults, office furniture, store wells, typewriters), Crompton Grieves (manufacturing fans), Kirloskar (makers of electric motors).

As stated above, product-oriented approach necessitates extensive research and development (R&D) programme to make improvements in existing products and to design and develop new products of better quality. The product concept lays emphasis on product excellence and quality assurance. The products are regarded *"good buy"* for the money.

However, one serious draw back of the product concept is in the case of an overemphasis on the quality of product. The producers, in their zeal to produce goods of higher and higher quality to achieve excellence, may overlook or ignore the actual needs of the customer or consumer or market trends and land themselves in the red. This approach thus sometimes leads to business short-sightedness or what Prof. Theodore Levitt called *"a marketing myopia"*.

> *Marketing myopia means excessive emphasis or concentration of attention on achieving quality of the product, ignoring the actual needs of the customer and growing market challenges in the process.*

This obsession with the product and its quality leads to a wrong and inadequate perception and understanding of the market, resulting in failure or adversely affecting the business.

The Sales Concept (The Selling Concept)

Another business philosophy that most of the companies and organisations pursue is known as the sales concept or the selling approach. *The selling concept is based on the assumption that the customers, on their own, will not buy adequate quantities of products produced by the company and they have to take strenuous and aggressive measures to sell them and persuade or rather coax the customers to buy them.* In this approach, the emphasis is on hard selling, particularly so, when the company is to deal with a buyers market i.e. a market wherein buyers are dominant and there are a few of them willing to buy the products of the company particularly when the products or services produced by the company remain unsought.

Hard selling envisages a vigorous and heavy advertising campaign through print and audiovisual media such as TV commercials and consistent sale and promotion effort to boost sales. Publicity, exhibition-sales, price discounts, personal selling, public relations and a whole battery of other selling and promotion tools are employed to push the sales.

Insurance, holiday time sharing in holiday resorts at tourist centres, long term bonds like gift bonds, education bonds, retirement bonds of finance companies, courses offered in computer software by various computer training institutes, management and other professional courses offered by various foreign colleges and universities to Indian students, automobiles and other consumer goods which attract strong competition are a few examples which require hard selling. Most of the companies in our country are practising the selling concept. Their main aim is to sell their wares anyhow whether they are needed or not. And as has been aptly pointed out by Philips Kotler in American context, that *"their main aim is to sell what they make rather than make what they can sell"*, is true in case of India also. No wonder that in the wake of liberalisation of the last two decades, Indian market is being flooded with household products unheard and unsought so far. Thanks to aggressive sales campaigns of the companies over television and in news papers, people are tempted to purchase even those goods which normally they are not inclined to buy.

As stated above, the selling concept lays emphasis on selling the product, notwithstanding the post-sale disappointment or dissatisfaction of the customer. The seller is interested in serving his needs and interests and satisfaction of the consumer or buyer is not his concern. An approach which serves the interests of only the seller and does not cater for needs, interests and satisfaction of the consumer can bring the seller only bad name and cannot help the organisation to achieve its long term objectives. Thus this approach also suffers from that flaw of overlooking the needs and interests of the customers and leads to *"marketing myopia"*. This approach also leads to a wrong and inadequate perception and understanding of business. An adequate and proper concept or approach of marketing must encompass satisfaction of needs and interest of the buyers. Such a concept demands that the firm should produce those value satisfying products and services that the customers need and would themselves like to buy.

It is clear from the above discussion that selling concept does not connote and represent the whole concept of marketing, it is just a part of it. Marketing is much more than mere selling. We shall discuss the marketing concept or philosophy of business in the following pages.

The Marketing Concept

It is evident from our study that the three marketing philosophies discussed above *viz.* the production concept, the product concept and the selling concept suffer from the defect that they, by and large, ignore the needs and interests of the customers and overlook all important aspect of customers' satisfaction as a marketing objective. In other words, the traditional approaches lack a market focus and customer perspective which must form two important ingredients of a business or marketing philosophy. The business philosophy, which is named as the *marketing concept* meets these requirements. This approach also helps in achieving individual and organisational objectives. The concept, as it obtains presently, was developed in 1950s.

Again, it is obvious that different groups of consumers or customers have different needs and wants depending upon their socio-economic conditions and a firm or organisation cannot satisfy the needs and wants of all types of customers simultaneously. However, it can meet the demands of a particular group of customers who want a product of a specific quality. In other words, it can satisfy the needs of a target market. Determination of target market should thus be an essential element of an effective marketing or business philosophy.

The marketing concept is based on the assumption that once a firm determines and knows the needs and wants of consumers i.e. the target-market, designs and develops products to satisfy those needs, make them available to consumers at reasonable prices, they will be ready to buy them on their own and this would be the best course to achieve individual as well as organisational objectives.

> *Marketing concept is a business philosophy that believes in achieving organisational objectives including profits by determining needs and wants of target markets and satisfying them through goods and services wanted by them, more effectively and efficiently than competitors.* This philosophy is amply reflected in the marketing maxim: *Know your target market and know how to satisfy it.*

It is clear from the above study that the marketing concept has four central tenets, *viz.* a market focus, consumer or customer orientation, integrated and coordinated marketing, and profitability or accomplishment of organisational goals. A brief explanation of these tenets is given below.

(i) **Market Focus:** A firm or organisation cannot satisfy every need of a customer nor all the customers or consumers would need the same type of a product. For example, while a large number of consumers may be using a popular or ordinary brand of toilet soap (say Lux, Liril, Cinthol, Margo), others may be interested in buying a brand of higher quality (say Dove, Camay or Pond's Cleansing Bar). It is for the firm to decide what type of the consumers or segment of the market it is to serve. It has to determine this segment of the market as a target and keep this market and its needs in focus and at the centre of its marketing actions.

> *Target market can be defined as market that is to be served by the business enterprise. It is a group of existing or potential customers of a product toward whom a firm's marketing efforts are directed.*

They are also called *'served markets'*. These markets comprise different sales territories, customer groups or other market segments. The marketing concept advocates that a firm should first determine the needs and wants of the target markets and then produce those products that may satisfy those needs.

(ii) **Customer-Orientation:** It is an approach which regards customer or consumer as the king, is oriented towards what the customer wants and satisfaction of his needs. Determination of customers' needs involves market research and market forecasting. In this approach, product-design, features, quality, price, production schedules, advertising, distribution channels, promotion of the product and personal selling are all aimed at satisfying the customer. The main purpose is not only to attract new customers but also to retain the existing customers. Satisfaction of customer must be the sole motto of the organisation. Customer - orientation approach requires the gauging of needs of customers from customers' point of view and not from the view of marketing personnel. This concept has been aptly focused in the following statement:

> *The end objective of Hindustan Lever is to delight consumers and customers by satisfying their needs and aspirations and the company has gone well beyond standard market research methodology into habit surveys, qualitative research and intensive consumer contacts in its quest for creating superior consumer value at affordable prices.*

June 28, 1996   S.M. Datta, Chairman, Hindustan Lever.

*Hindustan Lever now is known as Hindustan Unilever

(iii) **Coordinated and Integrated Marketing:** Another salient feature of the marketing concept is the emphasis on a coordinated and integrated programme of marketing. It requires coordination and cooperation among all concerned within the department as well as outside it in the organisation. Satisfaction of the customer should be the main concern of the organisation and all efforts should be geared to this end. This implies that all marketing functions *viz.* marketing research, product, designing and developing, advertising, pricing, promotion and distribution need to be integrated and coordinated in a planned manner to satisfy customer needs. Again there should be coordination amongst various departments within the organisation as well. It should be always kept in view that an organisation owes its existence and growth to the customers. The management must ensure that all departments *viz.* finance, manufacturing, personnel, materials and marketing work for a common purpose, namely the satisfaction of the customer and growth of the organisation. Marketing department cannot do its task effectively and efficiently if manufacturing department does not produce goods of the required quality at a feasible cost and financing department does not provide necessary finances. The interdependence of various departments in an organisation cannot be overlooked. Again, it is very essential that internal marketing is integrated with external marketing.

Internal marketing will imply the task of proper hiring, training, developing and motivating marketing personnel of the organisation so that they can contribute their best towards meeting the need of customers and achieving the organisational objectives.

External marketing is the task done by the organisation to plan, price, promote and distribute the product or service to the customers.

These need to be fully integrated if the company has to realise its objectives.

(iv) **Profitability (Accomplishment of Organisational Objectives):** Business organisations are set up to earn profit. Huge funds are invested in them, some of which are borrowed and are to be repaid. Therefore a business firm must get a reasonable return on the investment. Profit is one of the main objectives of an organisation. Therefore notion of satisfaction of customer needs, which is the most important tenet of the marketing concept, does include the prospects of earning reasonable profits. The marketing concept advocates that the efforts to satisfy customers needs should be so integrated as may enable the organisation to earn reasonable profits as a by-product out of the process. Though a firm should not indulge in what is called profiteering, it must earn due profit if it is to survive in the market. It is for this reason that normal profit is considered as a part of the cost of production in economics. Even non-profit organisations and the public sector agencies need funds to perform their work and for their smooth running. Therefore the marketing concept holds that the firm should seek profit making opportunities and strive to find profitable ways to satisfy customer needs. Only then, it can be successful in achieving organisational as well as individual objectives.

Difference between Selling and Marketing

The above study of the sales concept and the marketing concept as business philosophies helps in appreciating better the difference between the selling and the marketing. It will be useful here to specifically spell out the difference between the two.

Theodore Levitt has given a distinction between the selling and marketing concepts. His views are summed up in the following chart.

The Societal Marketing Concept

With rapid advancement in the field of science and technology during last half century coupled with new innovations and inventions, not only

Chart 2.1:   Distinction between Selling and Marketing

	Selling	Marketing
1.	Selling focuses on serving the needs and interests of the seller.	Marketing is concerned with satisfaction of wants and needs of the buyer. Its main task is to generate customer satisfaction.
2.	This lays emphasis on selling what a firm makes.	The emphasis here is on making what a firm can sell i.e. what a customer wants.
3.	This believes in producing goods without identifying marketing opportunities.	This lays stress on identification of marketing opportunities and target market before goods are produced.
4.	It is concerned with converting the product into cash notwithstanding the post-sale disappointment or dissatisfaction of the buyer. There is absence of customer consciousness.	It lays emphasis on offering value-satisfying goods and services. It ensures post-sale satisfaction of the customer. Customer's consciousness pervades all organisational activities.
5.	A seller's basic skill lies in coaxing the buyer to buy the product or service anyhow.	The marketer's basic skill lies in "influencing the level, timing, and composition of demand for a product or service."
6.	The seller determines what product is to be produced. The product precedes the selling effort.	The buyer determines what product he needs and should be produced by the producer. Here, marketing effort leads to production of goods and services.

the nature of production has changed considerably but the society's needs also changed dramatically. For example, factories have started producing many hazardous goods like guns, high-speed automobiles, junk food, mobiles, plastic goods, that have harmful effects on health, ecology and environment. Health, education, welfare, environment, pollution etc. have become important issues of society's concern.

One drawback of the marketing concept is that this business philosophy does no reflect the concern for welfare and wellbeing of the society at large, though it betrays enough concern for interest of individual customers or buyers i.e. target markets. Therefore, in this sense, it cannot be regarded as proper or adequate perception or business philosophy. Therefore, besides caring for interests of the organisation and the customers, a third dimension *viz.* society's welfare or wellbeing needs to be introduced in the business philosophy. It should have a human angle. *"The societal marketing concept"* holds that the organisation's task is to determine the needs, wants and interests of target markets and to deliver the desired satisfaction more effectively and efficiently than competitors in a way that preserves or enhances the consumers and the society's wellbeing.

This approach postulates that the managements should take into consideration the social implications of their decisions and ensure that they contribute to the welfare of the society at large. That will require a strong commitment to the ideal of social responsibility and matters of social concern. We shall further discuss the concept of social responsibility in Chapter 4.

The societal marketing concept enjoins on the marketers to maintain a proper balance between individual, organisational and societal objectives. The societal marketing concept lays emphasis on showing equal concern for all the three dimensions, namely concern for the customer, concern for the organisation and concern for society's wellbeing and public interest i.e. concern for social issues.

2.4 HURDLES IN IMPLEMENTATION OF MARKETING CONCEPT

Modern marketing management is founded on the marketing concept. It is a customer-oriented philosophy, integrated throughout the organisation, striving to serve customers better than competitors in order to achieve desired goals. It calls for close coordination and cooperation among all departments of the company and a corporate team effort aimed at providing customer satisfaction.

Explaining the marketing concept may look simple but implementing it is said easily than done. Effecting coordination among various departments or functional areas and grasping the principle and spirit of the marketing concept are not so simple. There are many hurdles in the way of converting an organisation into a really market-oriented one. Many a successful multinational companies which are regarded as master practitioners of the marketing concept have floundered, especially when they have expanded their marketing activities in developing countries. The main hurdles in its implementation are—conflict among functional departments, slow learning of marketing principles and tendency of forgetting them fast.

(i)   **Conflict among functional departments:** Despite various measures undertaken to encourage cooperation and coordination among functional areas *viz.* marketing, production, finance, R&D, conflicts do sometimes occur. Though knowing fully well that successful marketing is in fact the lifeline of the organisation, they may tend to think that the management is building up the marketing department at their cost. There can be various areas of conflict among functional areas such as:

(a)   competition for scarce resources like finance, personnel etc.,

(b)   difference in goal priorities of various departments e.g. while production department would like to reduce costs, the marketing department may need high quality product.

(c)   power struggles among managers in different functional departments who are competing for top management positions in the organisation. They resist marketing in becoming a major function and do not like that the head of marketing department may command a central position or position of power in the organisation. Sometimes this tussle may arise out of subjective perceptions of those in the highest hierarchy about the role of marketing department in the organisation or as a result of power game among the top management or what we may call a *"clash of personalties"*, which may have nothing to do with the marketing principles.

(ii) **Slow learning:** It takes time to understand the marketing concept, learn and grasp its principles and apply them in actual practice. Inspite of the best organisation inputs in terms of its predominance *vis-à-vis* other functional areas, job position, higher allocations, marketing enlightenment comes slowly. It involves advertising, sales promotion, innovating, image making, analysing, planning and controlling market and above all creating favourable and friendly atmosphere to retain existing customers and attract new-ones. All this cannot be learnt and implemented rapidly.

(iii) **Forgetting fast the marketing principles:** Even those organisations that have been known for their commitment to the marketing concept are likely to forget the basic marketing principles. This tendency seems to be conspicuous in case of many big multinational companies which have expanded their marketing activities into new found markets in Asia and Africa, thanks to liberalisation and globalisation of economies therein. They have overlooked the socio-economic variations prevailing in these regions. They believe that the products meant for consumers in America or European countries suit Asian and African people as well and the needs, wants and interests of the people all over the world are the same. They have started manufacturing and providing harmful products like junk food and gadgets and creating narrowly defined customer groups as target markets through excessive advertising, generally unethnical and fleecing them in the process. The result is that they have started in indulging in exploitation of consumers which is against the basic tenet of marketing philosophy. Thus, there is a strong tendency on the part of even those well-reputated organisations that are regarded to have implemented the marketing concept, to overlook the interests and satisfaction of the customers.

CASE 2.1: A NEWS REPORT*

Builder made to pay for 10-year delay

The National Consumer Disputes Redressal Commission on Monday ruled that builders have to pay the compensation fixed in the agreement with buyers if they failed to hand over possession of flats within a stipulated time.

The ruling came on a complaint filed by two buyers, Kunj Behari Mehta and another person, who sought direction to Ansal Properties and Industries Ltd., to deliver possession of an apartment in Celebrity Homes, Palam Vihar.

They had demanded 24 per cent per per annum interest on Rs. 26,26,790—the amount they had deposited with the builders from October 10,1998 till the delivery of possession. They demanded interest at least at the rate of 17 per cent per annum as agreed by the Ansals in August 1996.

The complainants had prayed that if Ansals were unable to deliver the possession of the apartment, the builders should be directed to refund the amount of Rs. 26,26,790 with an interest of 24 per cent per annum from the date of payment of the installment till the date of actual realization from Ansals.

Allowing Mehta's complaint, the commission directed the Ansals to

---

*(For professional students)

pay interest at the rate of 12 per cent per annum from November one, 1998 till December one, 2007 on the actual amount deposited with the builder i.e. 25,29,770. Interestingly, the commission said in its ruling that the compensation rate cannot be lowered.

It also ordered the builder to pay a compensation of Rs. 50,000 for rough behaviour.

The complainants and Ansals executed an agreement on March 2, 1995 regarding a flat. The complainants made the payment in advance before October 1997 and Ansals were supposed to deliver the possession of the flat to the complainants on or before October 1998. But the actual delivery of the apartment was made on December 20, 2007 only after the commission issued a direction.

The Ansals opposed the complainants' plea on ground that they had already delivered the flat's possession and the value of the said flat had increased.

But the commission said the builder's contention as "totally misconceived"

Source: Hindustan Times, May 13, 2008

CASE 2.2: A NEWS REPORT*

'Unethical' scheme cost Vodafone 50 lakh

The Delhi Consumer Commission has imposed punitive damages of Rs. 50 lakh on telecom major Vodafone for resorting to "unethical" and "deceptive" trade practice by offering prize to subscribers calling for more than 20 minutes in a day.

"We hold the telecom company guilty for indulging in highly unfair trade practice affecting crores of subscribers by adopting unfair and deceptive method to promote its business interest directly as well as indirectly", Commission President Justice JD Kapoor said.

Vodafone had launched in July last year a scheme 'Baaton se banaiye sona, bees minute mein', under which it offered 10 gold coins and one bumper prize of Maruti SX4 through a lucky draw to subscribers whose talk time was more than 20 minutes in a day. The Commission held that the company, by putting a condition of calling for at least 20 minutes in a day, had promoted its interest unethically and to the "detriment" of crores of consumers by encouraging them to make unnecessary calls.

"There is no consumer interest involved. Interest of consumers means that benefit should reach every consumer and not to few or 10 or 11 of them," it said. It said Vodafone had made roughly Rs. two crore during

---

*(For professional students)

the promotion period through unnecessary calls by the subscribers to enable them to participate in the contest and out of it, only a few lakh rupees were given in prizes.

The company on the other hand, contended that the aim of the scheme was not to generate profit, but to give benefits to its high-end users and to honour their loyalty. Vodafone claimed: "The scheme is perfectly legitimate and legal and cannot be termed as an unfair trade practice as no extra money was charged from the subscribers".

Source: Sunday Pioneer, May 18, 2008

Despite these inadequacies and aberrations, more and more Indian companies are appreciating the need for adopting this approach, if they have to meet the challenges of their competitors:

2.5 USE OF THE MARKETING CONCEPT

Contemporary marketing management is founded on the marketing concept and a large number of organisations in business sector, both national and international, are increasingly adopting marketing approach. Indian organisations like Nirma, Godrej, Sail, Reliance, ACC, Tata, Hindustan Unilever, DCM, Birla etc. are adopting the marketing approach and pledging *better products, better value, better living.*

Besides, a number of non-profit and service organisations are also using the marketing approach. Schools, colleges, universities, hospitals, theatre companies are offering their services through advertisements. Reputed universities in USA, England and Australia are offering courses in business administration, medicine, engineering etc. through advertisements in Indian Newspapers to attract Indian students and addressing to new target markets. No wonder, reputed managements institutes like IIMs, MDI, IMM, Faculties of Management Studies in universities etc. in the country are organising short duration training and development programmes and workshops to satisfy the specific needs of in-service management personnel of all levels and marketing their services. Some of them charge 5-star fees and offer 5-star facilities to the participants who are usually sponsored by the corporate and public sector companies. This shows that all types of organisations within and outside the business sector are trying to adopt the marketing concept now-a-days.

APPENDIX TO CHAPTER 2*
*(Meant for students in Professional Colleges)

CASE 2.3: VALUES—PRODUCT VALUES, PROFIT VALUES AND PEOPLE VALUES

Marico Industries Limited, the marketers of Parachute Coconut Oil and Saffola Refined Oil have set the following values for themselves.

Product Values

A. Product Image:

1. We are committed to serving our consumer — the Indian household.

2. We believe that a consumer-oriented company like ours has a responsibility to offer to the consumer, confidence through differentiated, branded products of superior quality.

3. We should provide to our consumers satisfaction in several ways; namely intrinsic utility, product improvements, easy availability, consumer friendly packaging and value for money.

4. Our products should be popular national brands, with a market leadership position of number one or two in their respective segments.

B. New Products:

1. We will look for altogether new products to meet the emerging needs of the Indian household, where, we can make a contribution synergistic with our strengths.

2. We wish to create winners with a distinctive edge based on excellent homework in market research, product and technology development, packaging, productionisation and delivery systems, consistent with market opportunities.

C. Quality:

1. We believe in total quality—in inputs, processes, systems and people—resulting in assured product quality for the consumer.

2. All our products will meet and surpass legal standards, with comfortable margin.

3. We must aim for rising standards of quality, superior to the competition.
4. We are committed and shall get our vendors committed to the principles of GMP (Good Manufacturing Practices).

Profit Values

A. Earning:

1. We are interested not just in short-term profits but a balance of short-term profits and long-term profitability.
2. We shall optimise profits, through productivity improvements, automation, economies, of scale, control of wasteful expenditure and the effective combination of resources.
3. We aim for profit based on all-round efficiencies, increased volumes and deserved price premiums.
4. We view such earned, relative profitability as reward and recognition for competitive successes.

B. Use:

1. We see profit not as final end in itself, but as a resource for further contributions to our Mission.
2. We have trustee obligations to set aside a part of the profits for continous reinvestments in people, products, processes and plants.
3. We are committed to rewarding our shareholders with handsome returns in the short run and with wealth maximisation, in the long run.
4. We would generate, mobilise and use funds in a manner that will give us returns on investments comparable with the best managed companies in the Indian capital market, irrespective of differences in product portfolios.

People Values

A. Our People:

1. We offer our existing and new entrants not just a job but-a career and even beyond—membership of our organisation community.
2. We believe people have a variety of untapped potentials and we shall provide them the opportunities to harness this potential.
3. We regard the competence and achievements of a person as more important than age, qualification and experience. Competent and talented people will be nurtured through varied role responsibilities

and attractive compensation based on intrinsic worth of the person and job.

4. We seek to provide our people not just autonomy and job satisfaction but continuous enrichment of their task roles and beyond that by empowering and inviting people to take on and perform organisational roles.

5. We believe that a flatter and leaner organisation is more conducive to speed, effectiveness, personal growth and contribution.

6. Alongside the above values for quality of worklife, we believe in firm action to curb lapses in integrity, system discipline and non-performance despite feedback and opportunities to measure up.

B. The way we Work:

1. We believe people perform better with goal clarity and result orientation. We will invest in continuously upgrading the goal setting and accomplishment process.

2. We will support experimentation, calculated risk-taking and innovative approaches. If per chance there are some failures, we shall focus on learning from these.

3. We are convinced of the power of participation. We shall involve people in most of the things we do. We believe that truly involved people have a sense of ownership and can achieve almost anything.

4. We value openness and critique in our everyday interactions. We prefer constructive "no-men" to spontaneous "yes-men"

5. We realise the motivational power of praise and public acknowledgment of genuine and ever rising standards of high performance.

6. We prefer to foster unilateral trust and its reciprocation to an excess of rules, regulations, checks and balances.

7. We believe that organisational and interpersonal care and concern make people want to give off their best.

C. Business Associates:

1. We regard our business associates as our "partners in progress". We believe that an enduring long-term relationship, meeting each other's needs, will be mutually most beneficial.

2. All such partnerships should be based on merit. We will select and retain associates on merit.

EXERCISES

1. What do you mean by the concept "marketing philosophy"? What part do the values play in the formulation of a marketing philosophy?
2. Enlist briefly various concepts of marketing philosophy.
3. What is meant by marketing concept? Explain its basic tenets.
4. Differentiate between marketing and selling.
5. What do you mean by the term "values"? Give specifically some sets of values in context of marketing management.
6. Write short notes on the following
   (a) Product concept of marketing
   (b) Societal marketing concept
   (c) Customer orientation in the marketing
7. "The present marketing is customer-oriented rather than product oriented". Discuss.
8. "Marketing is wider term than selling". Discuss the statement.
9. "Marketing is the science of actualizing the buying potential of a market for specific product". Discuss.

The Marketing Environment and Challenges

In our earlier discussion, we have referred to marketing management as an integrated system of interacting marketing activities and functions which are largely influenced by the business philosophy of the organisation and the marketing environment or the environmental changes. In the previous chapter, we studied various approaches or concepts (i.e. business philosophies) that are pursued by the organisations in their marketing efforts. In this chapter, we shall explore the nature of environment and its challenges to modern marketing.

3.1 BUSINESS ENVIRONMENT

Broadly speaking, every organisation is set up with a specific purpose or objective. A business organisation generally produces goods and services needed by the people or society. It arranges for various factors of production and inputs like land, labour, capital, raw materials etc. and organises them into producing the required goods and services. It receives these inputs from the society, transforms them through various production processes into outputs i.e. into finished goods and services, and gives them back to the society which pays price or money to the organisation in exchange of these goods which is invested for further production. Figure 3.1 illustrates this simple business environment.

We can easily guess the responsibilities of a manager in this context. Obviously his responsibilities in this immediate context will include matters like meeting needs of people for a specific product or service,

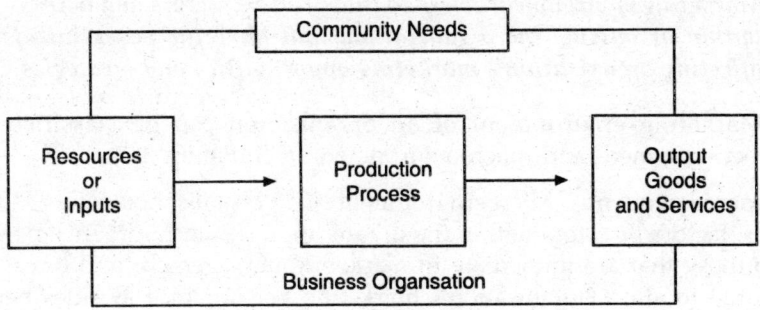

Fig. 3.1 Business Environment.

finding necessary resources for marketing activities, employing competent staff to meet the needs of the community. This is not an easy task. Even procurement of financial resources or finding competent persons to man jobs in the department or organisation, or producing goods of required quality may pose problems. There may be competitors in the market who may be vying for these very customers. Again, there may be persons or other factors affecting his functioning or that of his organisation.

In short, a business organisation or a manager does not work in a vacuum. They function in an environment, both within the organisation and outside it, comprising factors and forces which are continually affecting and influencing their functioning and the manager must be conversant with all facets of this environment, if he is to be effective and successful.

A business environment is thus a setting in which a business organisation operates — government, customers, workforce, suppliers, competitors, publics, communities etc.

We can define *environment as "the whole array of conditions, factors and forces surrounding the business organisation that influence the functioning or the success of the organisation or a manager.* It may consist of organisational as well as social, technological, economic, political and legal forces, factors or actors.

3.2 MARKETING ENVIRONMENT AND ITS FACETS

Marketing environment is thus a complex of factors and conditions that affect the marketing functions of the organisation. These factors and forces are generally intertwined. Specifically,

> *Marketing environment refers to those forces, factors and actors, within or outside the organisation, that have the potential of affecting organisation's marketing opportunities and strategies.*

Marketing environment of an organisation can be classified as microenvironment and macroenvironment, (See Figure 3.2).

Microenvironment: Microenvironment of an organisation refers to the forces, factors or actors within the organisation (i.e. intra-organisational) or to those that are immediate to marketing management and have the potential to affect organisation's marketing performance. Besides being closer to the management, these are the factors with which the management has to frequently deal with. Microenvironment thus represents two layers *viz.* intraorganisational and more-immediate or what is called *'task environment'*.

(a) **Intraorganisational environment** represents the factors and actors within the organisation and consists of marketing department that formulates and implements the marketing plans, top management, and non-marketing departments like finance department, R&D department, production department, accounting department and their personnel. All these constitute the internal microenvironment of the company and have an impact on the plans and functioning of the marketing department. The marketing department has to work in close cooperation with these departments if it has to achieve organisational goals. The interests of marketing department may clash with the interests of other functional departments. There is need for striking a healthy and fruitful balance. The marketing management has to take into consideration these forces and conditions in its internal microenvironment. An important aspect of intraorganisa-tional environment is the organisation's business philosophy or culture itself i.e. beliefs, expectations and values cherished by the members of the organisation. We have already discussed various business philosophies in Chapter 2.

(b) **The immediate microenvironment or task environment** represents factors or actors that are immediate or closer to the marketing management. They comprise suppliers, marketing intermediaries, customers, competitors, publics, i.e. groups that have interest and stake in company's performance. Suppliers, marketing intermediaries and customers form what is known as *"the core marketing system"* and immediate to this system are other constituents of microenvironment, namely competitors and/or other groups. A brief introduction of these constituents of microenvironment is given here.

    (i) *Suppliers:*   They include business firms, organisations and individuals who provide raw materials, equipment, tools, labour, energy (fuel and electricity) and other factors of production. Quality product can be produced with quality raw materials and reliable equipment only. Changes in prices, capacity and supply schedules can have effect on costs, production capacity and marketing plans of the organisation. Again, the marketing department is the direct purchaser of certain services from other agencies to implement its marketing strategies such as advertising, publicity, marketing consultancy, training packages for marketing personnel, marketing research. Marketing department is required to make decisions regarding the selection of agencies and firms to get the best efficient services and full value of money.

(ii) *Marketing Intermediaries or Task environment:* These are the firms that help the marketing department in promoting, selling and distributing goods to the customers or final consumers. They comprise distributors, wholesalers, retailers, agents, brokers or sales representatives, transportation firms, warehousing firms, cold storages, advertising and consultancy agencies and financial institutions like banks, insurance companies, credit companies. All these form part of immediate environment. Their cooperation is necessary to achieve marketing goals.

(iii) *Customers/Markets:* Customers provide market and firm's survival and growth depend on them. They comprise the target markets. We have already referred to various types of markets such as consumer markets, industrial markets, reseller markets, institutional markets, etc. in Section 1.4, Chapter 1.

The relationships of the company with its customers are dynamic. The varying consumer demands and desires would necessitate a change in the marketing plans and in quality, specifications and delivery schedule of raw materials and other supplies. At the same time, marketing department has to ensure many things like quality and reliable product, safe product with low energy consumption, fair price, fair trade practices, fair advertising, attending to customers grievances and efficient post-sale services etc.

(iv) *Competitors:* Competitors also influence the functioning of marketing department. Competitors include those firms or individuals who are seeking to satisfy the same customers and customer needs and offer similar products or services, close substitutes or product brands to them. The major task of the marketing management is to identify the competitors and to know their strengths and weaknesses. It would involve collection of full and reliable information about competitors and on their strategies and goals. This intelligence will help the manager in drawing up his own marketing plans.

(v) *Publics or Interest Groups:*   Besides competitors, there are many special interest and pressure groups which have the potential of affecting the performance of the marketing department. These groups are called publics. *A public, in present context, is a special interest group with a potential of affecting the marketing functions of an organisation.* These groups include financial institutions like banks, insurance companies, investors, mass media (newspapers, radio & TV), citizen action groups, unions, etc.

All the above forces and actors as explained under intraorganisational and task environments constitute the organisation's microenvironment.

(c) **Macroenvironment:** Macroenvironment is an environment, external to the organisation and is filled with factors and variables over which the organisation has little influence or control. This is a wider environment consisting of factors and conditions surrounding the organisation. These factors and conditions include economic, technological, political, legal, governmental, socio-cultural and international developments. These factors affect the functioning of a business organisation and put tremendous pressures on a marketing manager. The job of a marketing manager here is to keep in view these factors, anticipate changes in these variables, and plan accordingly. Thus macroenvironment is a wider environment consisting of socio-cultural, economic, technological, political and legal, natural and international environments. These are studied here briefly.

(i) *Socio-Cultural Environment:*   As the name suggests, socio-cultural environment encompasses people's characteristics, their culture and values and their specific social concerns. Social environment may include total social forces within which an organisation operates. According to Louis Allen, "cultural environment refers to the traditions, laws, rules, conventions and beliefs". Thus socio-cultural environment comprises demographic factors and developments like population trends, birth and death rates, life expectancy, occupational and ethnic distribution, education levels, income levels, their life styles, mobility of labour, behavioural patterns of people, their background etc. This would include socialite attitudes towards business values, customs, traditions, class structure, sex structure, participation of women, social issues like quality of life, social responsibility, consumerism, ethics etc. These factors greatly affect consumer markets and pose a great challenge to marketing managers in their task.

(ii) *Economic Environment:* The economic environment is filled with factors like market pressures, trade cycles, boom or recession, market turbulence, general economic conditions, needs and wants of customers and clients, bargaining power of suppliers and customers, economic system, government's fiscal and monetary policies, degree of liberalisation and globalisation, financial and credit facilities or financial constraints, economic growth and development. These factors are of critical importance to markets as they affect income, savings and hence consumers' purchasing power and expenditure patterns and create new challenges for marketing department.

(iii) *Technological Environment:* Technological environment consists of technological factors *viz.* technology, manufacturing processes, innovations and inventions, technological changes and developments in the country and abroad i.e. application of new ways of transforming resources into products and services, computerisation and automation. Technological changes and developments influence the functioning of the organisation in many ways. These factors change the nature of production and productivity levels. Change in technology may affect the growth of entire industries. Technological changes representing changes in equipment, methods, techniques result in production of new products, products at cheaper cost and more speedily and competitively, and open up new marketing opportunities for the marketing department. Computers and other electronic devices have revolutionised information gathering and communication systems and help marketing management in many ways.

Technological environment has become a critical factor effecting management in general and marketing management in particular. Moreover, technology affects social and economic environment also. For example, new electronic gadgets have changed the life styles of a large section of the society. Technological developments thus pose a great challenge to marketing managers now-a-days.

(iv) *Political and Legal Environment:* This environment consists of political environment, government agencies, laws and regulations and public interest groups who act as watch-dogs of the interests of their members. These have a great impact on marketing management.

   (a) *Political Environment:* It includes political atmosphere and stability, political parties and their philosophies, political influence of trade unions, government administration and policies concerning business and international policies. Political environment and changes therein affect marketing decisions and offer many challenges to marketing management.

(b) *Government—Business Relations:*   The macro-economic policies of government influence the flexibility of products, labour and financial markets. For example, high structural budgetary deficits and hikes in administrative prices of petroleum products and the like pre-empt savings, increase prices and adversely affect consumers' demand, business confidence and consequently the marketing strategies. On the other hand, a stable macro-economic environment with a consistent budgetery mix conducive to price stability and building business confidence poses less problems for marketing department. Again elimination or curtailment of regulatory barriers or liberalisation of economy will help restructuring of business enterprises and marketing programmes.

(c) *Legal Environment:*   Legal environment covers laws, both central and state and law-enforcement mechanisms which have direct and immediate impact on the business system, organisations and producer-market relationships. The government enforces laws to regulate business-consumer relations. These laws have direct and immediate impact on marketing function. Some of the laws are addressed to specific consumer problems. They usually aim at — (i) maintaining healthy and workable competition within the business system and (ii) preventing deceptive, fraudulent or unsafe marketing practices and curbing those practices which are unfair to consumers, (iii) safeguarding consumers' interests. For example, we in India have Monopolies and Restricted Trade Practices (MRTP) Act, Companies Act, Essential Commodities Act, Agricultural Produces (Grading & Marketing) Act, Prevention of Food Adulteration Act, Drugs Control Act, Standard Weights and Measures Act, Trade Marks and Merchandise Marks Act, Display of Price Order, Packaged Commodities (Regulation) Order etc. All of these laws affect marketing management in one way or the other and a manager needs to have a thorough knowledge of these factors. (See Appendix to this Chapter for Legal Measures in India : Case Study 3.2).

(d) *Public Interest Groups:* These groups are watchdogs of the interests of their members and cater to the interests of people like senior citizens, women, minorities or other special groups. They include consumerists. *Consumerism is an organised movement of citizens and government to protect and strengthen the rights of consumers in relation to producers or sellers.* It is a movement organised by vigilant consumers in the market place to safeguard their interests and rights and get a fair deal. The consumers' rights include the *right to information, the right to satisfaction, the right to choose, the right to be heard, the right to redress, the right to consumer education* and the *right to a safe and healthy environment.*

(v) *Natural Environment:* Natural environment encom-passes natural resources, renewable (forests / food) and nonrenewable (coal, oil, zinc), climate, physical terrain, varying seasons, air and water pollution. They greatly affect marketing decisions. The problems like shortage of raw materials and paucity of energy, air and water pollution, chemical pollution in food and produce affect marketing decisions and marketing opportunities. Not only the nature of production has changed considerably but the society's needs have also changed dramatically. Health, education, pollution, maintenance of ecological balance have become issues of society's concern.

All these environments, namely socio-cultural environment, economic environment, technological environment, political and legal environment and natural environment constitute organisations' macroenvironment.

Characteristics of the Macroenvironment

As is evident from above study, macroenvironment comprises factors and conditions which are continually changing. Some important environmental changes are regarded as *"megatrends'*. These have the following characteristics:

(1) These changes and trends are uncertain and are largely uncontrollable

(2) Some of these forces are potentially relevant to the organisation's marketing management function i.e. marketing decision making.

(3) Different types of organisations have not the same factors relevant to them. Therefore, a given firm should scan that environment i.e. identify and monitor those changes which are potentially relevant to the organisation's marketing.

(4) The forces and factors represented by marketing environment pose many challenges to the marketing management and are constraining in nature. They need to be tackled professionally.

Role of the Marketing Manager

Marketing environment, especially external environment, comprises factors and conditions which are continually changing and changing scene of this wider environment can have an impact on any one or all the present and planned activities of a marketing manager. There are two choices open to him. First that he should wait for the variables to change and react, and secondly that he may anticipate the changes that will take place and plan and act accordingly.

The above study amply points out that a manager has to live, work and survive in three concurrent environments—internal, immediately surrounding and external or macro, each with its challenges and changing pressures. Figure 3.2 illustrates the layers and components of these environments within and surrounding an organisation and a marketing manager has a pivotal role to play in order to achieve organisational objectives.

Chart 3.2:   Marketing Environment

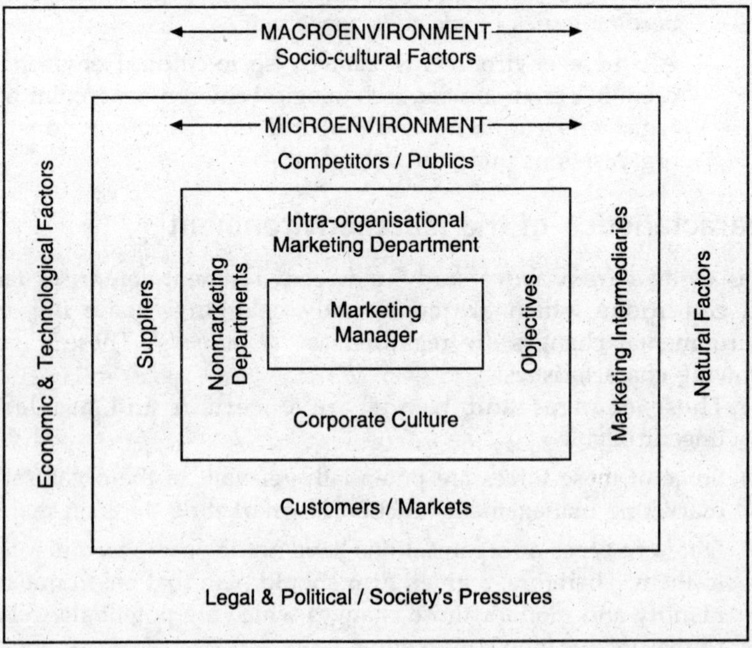

While knowledge of internal environment helps a manager in deciding "what he can do?", understanding of macroenvironment i.e. external factors would help him to determine 'what he might do?' to achieve organisational goals. We shall further discuss the tasks, responsibilities and obligations of a marketing manager in next chapter.

3.3 CHALLENGES TO MODERN MARKETING MANAGEMENT

Challenges to marketing management both arise from managerial practices (i.e. from within the organisation) and from the changing external environment. Requirements and needs of professionals also need to be tackled. Generally, these challenges are intertwined.

I. Organisational Challenges

Although there have been rapid improvements in managerial practices in recent decades, improvements are still needed to meet these challenges. Most of the challenges start from within, due to demands for an expanded contribution, modern information data and professionalism.

Organisational Objectives: The traditional concerns of a manager have been control, improvement of efficiency, performance and profitability, compliance with legal requirements, growth of business and fair return on capital. With human relation approaches in management and change in value-system as reflected in ideas like *concern for people, customer satisfaction, quality assurance, social responsibility,* the objectives of management and organisation have undergone a sea change. More and more emphasis on respect for the customer and his emerging needs, wants and expectations has put added pressures on the organisation and management. Managers are required to work so as to achieve maximising of both dimensions *viz.* concern for production and concern for the people, and the management will have to adjust their objectives in the light of these changes. It may necessitate change in current marketing practices. The marketing department will have to see how organisational objectives can be achieved while serving the interests of the customers and how furthering the cause of customers can be helpful in achieving organisational objectives. Reconciling the interests of the organisation and customers is a great challenge for marketing department.

II. Marketing Information System (MIS)

A manager is required to make a number of decisions while performing various managerial functions. These decisions are made on the basis of data and information available with the decision maker. Good decision making necessitates availability of adequate authentic and timely relevant information, processed and presented in usable form. An organisation, therefore, must build some management information system that may provide accurate, adequate and relevant information at the time it is needed. Walter Kennevans defines management information system as *"an organised method of providing past, present and projection information relating to internal operations and external intelligence. It supports the planning, controlling and operational functions of an organisation by furnishing uniform information in the proper time frame to assist the decision making process"*. Thus MIS is designed to provide needed information to each manager at the right time in right form so as to enable him to make proper decision.

Marketing information system is a continuing and interacting structure of men, equipment and procedures to collect, analyse, evaluate and distribute pertinent, timely and accurate information to those who make marketing decisions so as to improve their marketing planning, implementation and control.

Besides information about various aspects of operations within the organisation, a marketing manager needs information about developments in the marketing environment i.e. information pertaining to environment-changes, markets, market channels, competitors, public interest parties and other aspects of macroenvironment. It would involve analysis of marketing records and reports, marketing research, gathering of marketing intelligence etc.

In order to provide regular flow of authentic and relevant information for managerial decision making and control, huge amount of data are required to be gathered, stored and processed. This poses a great challenge to marketing management. In modern business, it is done with the help of electronic data processing (EDP) devices or computers. Use of electronic devices or computers has made the collection, processing and supply of information economical. It helps in developing an integrated MIS in the organisation but computerisation involves substantial cost. Such an integrated system is based on the premise that the same basic data can be used for several purposes.

In modern business, effective marketing management is not possible without MIS. So it is very important that marketing information system is designed properly and administered effectively. It must cater to needs and objectives of the organisation.

III.   External Challenges

Today, most of the pressures for change come from external environment. The ability to keep up with social, technological, economic and political/

legal forces will largely determine how effective management will be. Changes in society also affect management. Many of these changes will challenge marketing management during this decade and beyond. These challenges are likely to stimulate many innovations in marketing management in future.

In the previous section, we have discussed various components of marketing macroenvironment and the challenges they pose to the modern marketing management. Changes in society and business practices coupled with environmental developments imply that there are many other challenges facing marketing management. Each of these holds unseen implications. As the list grows, the role of marketing management will grow and become more dynamic. The need for innovation will be the greatest challenge of all.

IV. Professionalism (Professional Challenges)

In order to cope with micro and macroenvironmental challenges effectively, a marketing expert or manager is required to develop a professional approach or what is called *professionalism* in management. Professionalism is a major challenge. Marketing functions or activities are too important to be done by unqualified and untrained persons. The need for education, professional training and development is being felt and felt more and more. With the explosion of knowledge and new developments in the fields of marketing management and behavioural sciences, marketing manager is required to stay informed about advances in these fields. This requires self-renewal on his part through reading, seminars, workshops with special association of experts. He needs to learn continuously and to strive hard to acquire marketing and managerial skills for achieving the high standards expected of professionals. Possession of sound knowledge about marketing environment and technical ability, proficiency in management and resource utilisation, and acquisition of interpersonal skills go a long way in enhancing effectiveness of a market manager in to-day's environment.

A proactive approach on the part of a marketing manager requires increased *professionalism*. Otherwise he will not be able to meet the challenges of changing demographic factors, values, third party pressures and other internal and external challenges. There is an urgent need for a statutory body or of marketing management association to look after matters concerned with it.

V. Globalisation of the Economy and Internationalisation of Management

As the economy of a country opens up to the world outside and invites foreign investments for development and other projects, it creates new pressures on marketing management. In this context, even in our own country, winds of change are evident. As we enter into an era of liberalisation, privatisation and dynamic business activity, the future offers both opportunities and challenges. Some multinational companies (MNCs) have started overtaking national companies. Some national companies are merging to gain strength. Globalisation is becoming almost a reality. No county or business enterprise can be sheltered from the winds of change and the business organisations will have to cope with these challenges and adjust accordingly.

Globalisation of the economy, and entry of multinational corporations pose great challenges to marketing management and domestic companies especially in case of the Third World. Products developed in one country are being pushed into the other country without regard to the needs of people therein. With their vast resources and consistent and aggressive advertising campaigns, they are trying hard to get those products accepted by people in the new country while some other products are finding easy acceptance. In some countries, this has resulted in change in the life style of some sections of society and a consumerist trend is discernible. Domestic companies are finding new powerful competitors and many of the domestic companies and brands have been swallowed by the multinationals. Globalisation has considerably affected exports and imports.

Another consequence of entry of multinational companies and internationalisation of management has been that now MNCs and some Indian big private companies are offering fabulous salaries and perks to higher level and middle level managers and professionals — salaries unheard and unimaginable only a few years back. It can cause much heart burning. Doubtlessly, the domestic small scale industries are really finding it hard to function smoothly in such an environment. Marketing management has to devise means and strategies to face these pressures successfully and effectively.

A marketing manager has to take all these challenges into consideration and plan and act in the best interest of the origanisation, customers and the society at large. Thus a manager has a social role also.

We shall discuss the tasks, responsibilities and obligations of the marketing manager besides his social role and the distinctive abilities and skills required for a marketing manager, in the next chapter.

APPENDIX TO CHAPTER 3

CASE 3.1:
DEMOGRAPHIC SCENE IN INDIA
(2001 CENSUS)

(These data throw a light on market potential in the country).

1. **Total population**   102.9 Crores (2001)
                                      114.2 Crores (2008-09)
2. **Birth rate (per 1000) in 2006**   **23.5**
3. **Death rate (per 1000) in 2006**   **7.5**
4. **Life Expectancy (2001-2005)**   **63.2 years**
5. **Urban Population**   **27.8%**
6. **Rural Population**   **72.2%**
7. **Literacy (Male)**   **75.26%**
8. **Literacy (Female)**   **53.67%**
9. **Distribution of villages according to population**   **593643**

Less than 1000	364482
1000-1999	129979
2000-9999	95220
Above 10000	3962

10. **Occupational Distribution**

Primary Sector (Percentage)		**52.7%**
Agriculturalists	31.2%	
Labourers	20.9%	
Others	0.6%	
Secondary Sector (Percentage)		**19.0%**
Manufacturing	12.9%	
Construction	5.6%	
Electricity, water	0.5%	
Tertiary Sector (Percentage)		**28.3%**
Trade and Commerce	12.6%	
Transport and Communication	4.6%	
Other services	11.1%	

11. **Per capita income (2007-08)**   **Rs. 29786**
    (At 1999-2000 prices)
12. **People living below poverty line**   **27.5%**
13. **Income Distribution (2003)**

Category	%age of income
Poorest 20 Per cent	11.6%
Richest 20 Per cent	46.0%
Middle 60 Per cent	42.4%

14. **Distribution of Consumption Expenditure (Household)**

Category		Rural	Urban
Poorest (Lowest)	30 per cent	15.0	13.6
Middle	40 per cent	33.1	32.4
Richest (Highest)	30 per cent	51.9	54.0

15. **Appropriation of National Disposable Income (2006-07, at current prices)**

    Government final consumption expenditure

                    Rs. 427007 crores

    Private final consumption expenditure

                    Rs. 2324109 crores

    Net Domestic Savings   **Rs. 1006956 crores**

16. **Private Final Consumption Expenditure in the Domestic Market By Object (%age)**

(at 1999-2000 prices)	2006-2007 (%)
(i) Food, Beverages and Tobacco	42.1
(ii) Clothing and footwear	5.1
(iii) Gross Rent, Fuel and Power	10.0
(iv) Furniture, Furnishing, Appliances and Services	4.0
(v) Medical care and Health Services	4.4
(vi) Transport and Communication	18.4
(vii) Recreation, Education and Cultural Services	4.8
(viii) Miscellaneous GooCis and Services (Including Personal Care and Effects)	11.2
Total (Rs. 1833673 crore)	100.0

17. **Sex Ratio (2001)**   933 females per 1000 males.

CASE 3.2:
IMPORTANT INDIAN LEGISLATION
AFFECTING MARKETING

1. Monopolies and Restrictive Trade Practices Act, 1969

This act came into force from June 1, 1970. The main objectives of this act are to control monopolies and restrictive trade practices and to prevent concentration of economic power. The act was to ensure that the operation of economic system does not result in the concentration of economic power to the common detriment. The act is empowered to regulate expansion, amalgamation and merger of companies. A Monopolies and Restrictive Trade Practices Commission (MRTPC) has been set up to implement the act. This commission processes the inquiry into unfair trade practices being indulged by companies, on a reference made by government, or on receipt of a complaint from consumers, or on an application from Director General or on its own. The Commission was repealed and has been replaced by *Competition Commission of India.*

2. Competition Commission of India (CCI)

The Government of India has established a new regulatory institute 'the Competition Commission of India (CCI) by repealing MRTPC to prevent anti-competitive practises, promote and sustain competition in markets in India and to ensure freedom of trade carried on by participants in market in India.

3. Essential Commodities Act, 1955 and Essential Commodities (special provisions) Act 1981

Essential Commodities Act, 1955 is the main legislation governing the production, procurement, price control and distribution of essential commodities. This act aims at safeguarding the interests of the consumers and restricting the activities of the anti-social elements indulging in black marketing, hoarding etc. The act was amended in 1981 and as a result was enacted the Essential Commodities (Special Provisions) Act, 1981 which made the penal provisions of the act more stringent and included provisions for summary trial of the offenders. The enforcement implementation of the provision of the act lies in the State Governments.

4.   Prevention of Blackmarketing and Maintenance of Supplies of Essential Commodities Act, 1980

This act aims to check the trend of increasing prices and artificial scarcity of commodities sought to be created by unscrupulous traders and provides for preventive detention of persons with a view to deter them from acting in any manner prejudicial to the maintenance of supplies of commodities essential to the community.

5.   The Prevention of Food Adulteration Act, 1954

This act has been in force sine 1 June, 1955 with the objective of ensuring that food articles sold to the consumers are pure and whole some. It also aims at preventing fraud or deception and encouraging fair trade practices. The act has been amended from time to time to plug loopholes. The administration of this act is the primary responsibility of the State Governments and Union Territory Administrations.

6.   The Drugs and Cosmetics Act, 1940

As amended from time to time, this act regulates the import, manufacture, sale and distribution of drugs and cosmetics in the country. Under the act, the import, manufacture and sale of sub-standard, spurious, adulterated or misbranded drugs are prohibited. The government is empowered to check the quality of imported drugs.

7.   Standards of Weights and Measures Act, 1940

This act contains provisions for effective legal control on weights, measures and weighing/measuring instruments used in commercial transactions, industrial production and protection of public health and human safety.

Standards of Weights and Measures (Packaged Commodities) Rules 1977

These rules are in force since September 1977. According to these provisions every package intended for retail sale is required to carry information as regards name of the commodity, name and address of manufacturer, packer, net contents, month and year of manufacturing/ packing and sale price (the maximum retail price inclusive of all taxes). Mandatory declaration of sale price on package prohibits retail dealer from charging more than the declared price and protects consumer against arbitrary charging.

Consumer Protection Legislation

A. Consumer Protection Act 1986

The government has enacted a comprehensive consumer protection legislation, namely, the Consumer Protection Act, 1986, to safe-guard consumer interests. All the provisions of the act came into force since July, 1987. The act applies to all goods and services unless specifically exempted by the central government. It covers public, joint, private and cooperative sectors. The act provides for setting up consumer protection councils in the centre and states, the main object being to promote and protect consumers' right. It enshrines rights of consumers such as right to safety, right to be informed, right to consumer education, etc. The act also provides for setting up of three tier Quasi-judicial Machinery at the District, State and National Level to provide speedy and inexpensive redresses for consumers' grievances, namely the Consumer Disputes Redressal Forum (District Forum), the State Consumer Disputes Redressal Commission (State Commission), the National Disputes Redressal Commission (National Commission). The central government has also constituted state-level consumer protection councils.

B. In December 1986, seven major consumer protection legislations, namely the Essential Commodities Act, 1955, the Prevention of Food Adulteration Act, 1956, the Drugs and Cosmetics Act, 1940, the Monopolies and Restrictive Trade Practices (MRTP) Act, 1969, the Agricultural Produce (grading and marking) Act 1975, the Standard of Weights and Measures (enforcement) Act 1985, were amended to empower consumers and registered consumers organisations to file complaints in court, which hitherto were vested with government officials only.

EXERCISES

1. What is meant by marketing environment? Explain its important facets.
2. Describe some major challenges to modern marketing management.
3. Describe important Indian legislation that has a bearing on marketing management.
4. What do you mean by marketing microenvironment? Explain its main dimensions.
5. What is meant by marketing macroenvironment? Explain its facets.

6. Write short notes on
   (a)   Marketing Information System
   (b)   Globalisation of economy
   (c)   Consumer Protection Legislation in India
   (d)   Role of the marketing manager
   (e)   Characteristics of macro-environment

4

Social Role of Managers: Tasks, Responsibilities and Obligations

As discussed in chapter 2, the societal marketing concept postulates that management should take into consideration the social implications of their decisions and ensure that they contribute to the welfare of the society at large. This approach lays emphasis on showing equal concern for the interests of all the three parties, namely customers, organisation and the society. This concept requires a strong commitment on the part of the marketing manager to the matters of social concern and to the ideal of social responsibility. In this chapter, we will consider the social role of a manager and draw out a profile of marketing manager *viz.* nature of his job or his overall responsibilities and distinctive abilities and skills required to be an effective and efficient manager in the prevailing business environment, especially in Indian context.

4.1 SOCIAL ROLE OF A MANAGER

Management is the art of getting things done through people and hence directing the efforts of individuals towards a common object. It entails the coordination of human effort and material resources toward the achievement of objectives of the enterprise or the group goals. Thus management is a social process because the close cooperation of all the members of the group working in an enterprise is essential to achieve the organisational goals as well as the common goals. In other words, management has to seek the mutual cooperation of all the members of the group. The same view holds good for effective and successful functioning of a marketing department as well.

As discussed under the marketing concept in chapter 2, marketing management is an integrating process and the responsibility of implementing it to achieve organisational as well as group objectives falls on the marketing managers. Managers are, in fact, an activating element. It is they who convert disorganised resources of people, money and capital into integrated and coordinated business endeavour. And herein lies the need for securing professionalism in the management.

The need of professional managers in being felt not only in private business organisations but also in non-profit establishments like hospitals, educational institutions, training and development centres, space agencies, consultancy firms, media, print and electronic and enterprises providing other services. These organisations include:

  (i)   educational and research organisations,
 (ii)   cultural organisations like orchestra, museums, dramatic associations,
(iii)   healthcare institutions like hospitals, dispensaries etc.,
 (iv)   Professional and Trade Associations like Federation of Indian Chambers of Commerce & Industries, Institute of Chartered Accountants, labour unions etc.,
  (v)   religious and charitable organisations,
 (vi)   political organisations and parties,
(vii)   social and common cause bodies like environmental groups, consumer groups, and government organisations like Postal and Telecom Department, Fire Fighting Department.

Promoting a social idea, cause or practice in a target group is categorised as social marketing.

In today's world of increasing specialisation, complexity, size and multinational dependence, the task of achieving coordinated action is of paramount significance. Let us look into the role that managers, especially marketing managers, play in our society or are required to play to achieve organisational as well as societal objectives. The task of a manager in this context involves:

  (i)   integration of the business organisation with external as well as internal environment.
 (ii)   adaptation and innovation, and
(iii)   adjusting to shifting society's needs and national priorities.

4.1.1 Integration and Linkages

One key element of management is integration. No business organisation can be successful unless it is integrated with the environment surrounding it as well as with environment within the organisation. We have explained various facets of macroenvironment and microenvironment in chapter 3. Thus effective and successful management necessitates multiple integration or linkages.

(a) External Integration

Every organisation, be it a hospital, university or a factory, requires continuing exchange or give-and-take relationships with an array of contributors—customers, or consumers, suppliers, financiers and investors, advertisers, local governments and other institutions. These relationships continue for an indefinite period and benefits flow both ways from contributor to the organisation and *vice-versa*. In this context, a reference to *Barnard-Simon's Exchange Theory of Organisation* will be useful for our study. This study states that *the organisational equilibrium is attained when the utility of the inducements (i.e. of products, services, wages) is equal to the utility of the contributions (offering of value or service) i.e. value received must be equal to value paid.* The theory holds good for exchange relationships between the organisation and its customers as well as for relationships between the organisation and its employees. There are many facets of these complex relationships and it is important for the marketing managers to maintain these relationships on even keel. For example,

(i) **Customers** and consumers provide market and a firm owes a lot to them. Its survival and growth depend on them. A manager has to see that goods are produced and provided per real needs and demands of customers. It is to be seen that customers' minds are not debased through false advertising and market is not flooded with superfluous and superficial products, thus endangering the moral and cultural ethos of the society. He has to make it sure that goods produced are of a fairly good quality and are supplied unadulterated and at reasonable prices as and whenever they need them, if the firm has to earn their permanent goodwill.

In this context, there is no gainsaying the fact that a manager owes responsibility first to the company that employs him and he must make it sure that it earns reasonable profits so as to survive and grow. For this, he needs to take steps to reduce costs and increase yields but at the same time he has to safeguard the interests of the customers and ensure many things like providing quality and reliable product, safe product, with low fuel or energy consumption, fair price, fair advertising, attending to customers' grievances, mechanism for satisfactory after-sales service and secure their satisfaction. He has also to see that the firm does not indulge in hoarding, black marketing and unfair trade practices. This would involve questions of morals, values and ethics. We shall discuss the place of ethics in marketing management later in this chapter.

(ii) **Suppliers** provide materials, equipment and services. We have already noted the requirements of a marketing manager in this context in chapter 3 (under microenvironment). Now quality product can be produced with quality raw materials and reliable equipment only. Profitability will require that these are produced at minimum cost. On the other hand, it will require that suppliers are paid fairly suitable price well in time as per schedule and the firm does not claim an unfair discount.

The relationships of the organisation or manager with its contributors are dynamic. The varying consumer demands and desires would necessitate a change in the quality, specifications and delivery schedule of raw materials, products and other supplies and services. It will also result in change in cost, demand, and capacity of the supplier. Thus a manager will have to be in constant touch with his suppliers to maintain a mutually acceptable flow that may serve the interests of both the parties. They must be able to adapt and adjust to the requirements of each other and all the others. It requires cooperation of all the contributors involved in the process.

Backward Integration or Linkages

In order to get assured supplies of raw materials or services at competitive prices, some big organisations acquire a limited or complete control over firms or people supplying materials through contracts or by other means and establish linkage with them. This is known as *backward integration or linkage. Backward integration or linkage* is referred to as *"expansion of an enterprise achieved by acquiring control over firms or people supplying raw materials"*. We have given three case studies of yester years concerning establishment of backward integration or linkage in the appendix to this chapter to illustrate how these linkages are established.

(iii) **Marketing intermediaries** help the marketing department in promoting, selling, distributing the products to the customers or final consumers. As stated in chapter 3, they comprise distributors, wholesalers, retailers, agents, transportation firms, advertising agencies, financial institutions etc. Their cooperation and support on sustained basis are necessary to achieve marketing goals. A marketing manager is required to develop healthy and permanent relations with them.

(iv) **The government** including local governments provide functioning communities and many other facilities and services, infrastructural and otherwise, for the development of business. The government has the responsibility of socio-economic upliftment of the country. A business organisation can help the government in this task. A manager is to ensure that statutory obligations including tax obligations are honestly fulfilled in time. Besides he must see that his organisation contributes to government plans and programmes aimed at socio-economic upliftment and also in times of national calamity or disaster. Government is also a large purchaser of goods and services. A marketing manager has to see that he does not corrupt public servants or democratic process by using dishonest means.

(v) **Competitors** also influence the functioning of a business organisation. For smooth functioning of the organisation, it needs a congenial and healthy environment. A marketing manager should create healthy and cooperative inter-organisational relationships between different organisations. All his efforts should be directed at strengthening the competitive spirit in the market place. He should not resort to any unfair trade practices like price cutting, underselling, out-bidding or indulging in manipulations. Competition should not unnerve a manager, rather it should inspire him to serve the consumer even better.

Besides these contributors, **society** at large is also a very important contributing entity. Integration with societal needs has been discussed separately later in this section.

Integrating the organisation externally with its different contributors is a continuous and unending task and is concerned with much more than price negotiations, mediation or maintaining good relations. And as Newman and Warren suggest—dependability, timely agreements, good-will, numerous intangible benefits and inconveniences, adaptability to the needs of either party, honest dealing—all enter the picture. Most of these will come in the realm of values and ethics. Without them, cooperation among diverse groups will not be possible. Adjustments will have to be made to meet the changing situations and conditions.

(b)   Internal Integration

In order to effect external integration, the internal operations of business organisation need to be conducted efficiently. This would require setting targets and priorities in consonance with external needs and demands, proper and balanced allocation and deployment of resources, training and hiring people to suit manpower needs and synchronisation of operations and actions of various departments. A marketing manager's role in internal integration is pivotal.

4.1.2 Adaptation, Innovation and Invention

Besides integration, the task of a marketing manager also involves adaptation and innovation. Earlier a manager responded to change in factors and conditions as and when they occurred and actually happened. He adjusted and adapted to the conditions as they prevailed. For example, in case of reduction of demand, he would reduce production and retrench labour. He would not anticipate things and developments of new markets. In today's world where hopes and aspirations are mounting and the environmental conditions are rapidly changing, managers, especially marketing managers, are required to anticipate changes, set new goals and plan accordingly to achieve these goals. They have not only to adapt to the dynamic factors but also to initiate action to assure growth of the organisation. A marketing manager is expected to be a dynamic and innovating force. He must strive not only for increasing share in the market for existing products but also invent new products and create and make new markets by sponsoring research and development (R&D) programmes. The organisation must strive to command a substantial premium over competition by producing superior products at cheaper rates. It would necessitate finding new cheap substitutes for conventional raw materials through R&D programmes. He has to see that resources available at his disposal are utilised in such a way that the interests of all sections of the society are protected and promoted. Marketing, in this sense, is the process by which an organisation relates creatively, productively and profitably to the market place.

4.1.3 Society's Changing Needs and the Managers

With the advent of freedom in many countries and changing political scene in the world and with rapid advancement in the field of science and technology coupled with new innovations and inventions, not only the nature of production has changed considerably but the society's needs have also changed dramatically. For example, factories have started producing many hazardous goods which have harmful effects on health, ecology and environment. Health, education, welfare, pollution, maintenance of ecological balance, defence and security, poverty removal, social injustice, slums and other urban problems, human rights, exploitation etc. have become issues of society's concern. Some of the new priorities of modern management include:

 (i) Individual self-expression : People now want meaningful and satisfying jobs and demand share in high productivity.

 (ii) Pollution control (noise, water and air) and maintenance of ecological balance.

(iii) Racial and urban problems : These may include problems of social justice, enormous growth of slums, poor living standards, exodus to cities etc.

(iv) Health, education, welfare : An ever increasing share of resources is being directed into such things as medical care, education, public recreation, housing and other amenities,

 (v) Guns and butter : Our defence expenditures have increased enormously with the result that programmes concerning amelioration of the lot of the poor are being affected.

Manager's Responsibility

These new priorities affect the managerial task. Manager's services are now needed in non-profit enterprises and service sectors of the economy also. Managers owe special responsibility toward the society's changing needs. His responsibility towards society will include pollution free activity, beautification of land scape, financial aid or contribution to education, service, housing, care of backward classes and support to cultural, sport and recreational activities, creation of employment opportunities, producing products of social priority, support to charitable institutions etc.

Thus in modern business organisation, manager's job is not only to get things done through the people and with the people, his work goes much more beyond it. He owes responsibility towards various contributors and society at large. Thus alongwith his usual duties, he has a great social role to play and is required to contribute to the creation of an egalitarian society. The most important thing that is required of manager in this context is what is called 'social sensitivity' i.e. quality of reacting to managerial problems in terms of feelings of the people involved.

4.2 SOCIAL RESPONSIBILITY (SR) AND ITS FEATURES

In the previous section, we have discussed the social role of manager and the various facets of his social responsibility in a modern business organisation. The social aspect of business lies in the fact that very objective of business is to make available quality goods and services for the satisfaction of human wants. Thus business organisation has a social power and the avoidance of social responsibilities would result in gradual erosion of this power. Therefore social sensitivity on the part of a manager is of critical importance. From above study, the concept of social responsibility can be explained as follows.

The concept of social responsibility of an organisation implies that a business should accomplish its own objectives but at the same time it should also be sensitive to matters of social concern, should conform to social norms and values, and should contribute to solution of social problems.

Kenneth Andrews explains the concept as follows: "By social responsibility, we mean the intelligent and objective concern for the welfare of society that restrains individual and corporate behaviour from ultimately destructive activities, no matter how immediately profitable, and leads in the direction of positive contributions to human betterment".

H.R. Bowen has defined *"social responsibility as obligation (of the manager) to pursue those policies, to make those decisions, or to follow those lines of action which are desirable in terms of objectives and values of our society."*

The concept of social responsibility implies that management's responsibility is to organisation itself and to all interest groups and contributors. On the basis of the above study, we can enlist the following features of social responsibility (SR).

1. Social responsibility involves fulfilling obligations to various interest groups with which it interacts or which are concerned with the functioning of the organisation, directly or indirectly. It means external and internal integration.

2. A management must conform to moral and social norms as laid down by the society. These may vary from society to society. It is concerned with questions of values and ethics.

3. Management must be willing to fulfil public expectations.

4. Management must contribute toward efforts for solving problems faced by society at large.

5. Besides obedience to laws, it advocates cooperation with the government in its efforts to improve business standards through R&D programmes and to improve a lot of the people.

6. Social responsibility is the obligation of management as a group because a manager is not the complete master of his actions.

To make the concept of SR operational, it is not enough only that a manager should have social sensitivity, it requires much more than that because he is not the complete master of his activities. It demands commitment from the top, in respect of formulation of policies to that effect, institutionalisation of SR in decision-making process as part of process of integration, and evaluation of the performance in terms of social obligations.

Social Audit

Social auditing is concerned with assessment and evaluation of social activities performed by a business organisation or business enterprise. It is a device of appraising performance or results of social task undertaken by an organisation or a company's contributions to the society. *"Social audit is a commitment to systematic assessment of and reporting on some meaningful, definable domain of the company's activities that have social impact"*.

Of late, a number of industrial organisations in our country are increasingly showing interest in matters of social concern and are contributing towards solution of some social problems by undertaking community projects on voluntary and organised basis. Though reach of these programmes is very limited, social audit can take the following forms:

  (i) Drawing and describing a simple inventory of activities of social concern such as employment and training of backward classes, pollution control, support of minority enterprises, involvement in community projects by executives, sponsorship for higher studies, promotion of sports and the like.

 (ii) Compilation of socially relevant expenditures i.e. measurement of expenditures incurred on social activities by the company.

(iii) Formulation of specific management type programme of social audit, examining critically and objectively the organisation and management of specific programmes of social significance undertaken by the enterprise

(iv) Determination of social impact i.e. measurement of results and performance of social activities and their actual contribution to the society in terms of quantity or money values.

Social audits are presented in form of reports on the progress of social programmes undertaken by the organisation. It should contain objectives and goals to be attained, resources allotted and utilised, and achievement made in respect of each objective or goal etc.

4.3 ETHICS

The word "ethics" has origin in the Greek word "ethos" meaning character, standards, norms, morals and ideals prevailing in a group, community or society. Ethics has come to be known as a science of morals i.e. a study how standards of moral conduct are established and expressed. In plural, ethics are referred to as some standardised form of conduct behaviour of individuals understood and accepted in a particular field of activity. Ethics may be thought of in terms of a mass of moral principles or sets of values about what conduct ought to be. They give an idea what is right or wrong, true or false, fair or unfair, just or unjust, proper or improper. They can be understood with reference to value-oriented decisions and behaviour.

4.3.1 Business Ethics

Business ethics are concerned with the behaviour of a businessman in a business situation. Business ethics or ethical standards are the rules of business conduct by which the propriety of business activities may be judged. Ethical principles are dictated by the society and underlie broad social policies. These principles, when known, understood and accepted, determine generally the propriety or impropriety of business activities.

Ideas of right or wrong, fair or unfair, just or unjust are determined generally by custom and environment. Obviously the sources of ethical behaviour or conduct in any society are value-forming institutions i.e. family, school, state and religion, values and goals of the organisation, peers and colleagues, and professional codes.

Again the conduct or behaviour which is regarded and accepted as just or right by the society is called *ethical*, and behaviour which is considered as improper or against the standard norms is considered as *unethical*.

Some say that whatever is legal is ethical. They judge the business activities exclusively from the point of view of law. Now, this is an inadequate judgement because such a narrow interpretation may not receive acceptance of customers or people associated with the organisation or win the confidence of society at large including political parties. For example, though maximisation of profit is one of the primary goals of an organisation, but charging unfair prices or exploitation of customers are considered as unethical practices. Good behaviour is rewarded and bad behaviour is punished.

Whatever may be the compulsions, price fixing, piracy of products and copy rights, deceptive advertising, bribery, payment of kickbacks, corruption, employee theft, sexual harassment, discriminatory personnel policies, black marketing are regarded as unethical practices. They are examples of bad behaviour. Government has enacted a number of laws to check these practices and to bring in an element of ethics into business. What is of importance is that these laws are implemented in letter and spirit.

The social responsibility of business involves ethics and these must be reflected in the very philosophy of the business organisation. To be effective, sound ethics must be recognised by top management and reflected in policies of the firm. They must be such that they are understood and accepted by the members voluntarily. Right leadership, integrity, proficiency, and commitment to social values can change the expected behaviour of individuals. In this way a business organisation can best serve its own interests and those of society as well.

4.3.2 Codes of Conduct or Ethics for Managers and the Professional Groups

In order to regulate behaviour of its members and for the protection and continuity of the professional group as a whole, professional groups or associations in their codes of ethics specify behaviours required or expected of its members. The code gives a clue to what type of response is expected in a give situation. The code of conduct or ethics generally comprises norms and standards of discipline, honesty, integrity and professional ethics to be followed and enforceable by the professional association. A code of conduct for marketing managers is given in the appendix to this chapter. See case 4.1.

4.4   MANAGEMENT AND CHANGE

Management should not remain become content with its performance. It must search for new ways to help the organisation, its people and contributors. An important way to improve is by attention to the future, i.e. to what will and what may happen in future, say in the next twenty years of 21st century. In other words, management is concerned with the future and the change and challenges it is going to present. Change is inevitable and by anticipating new challenges, the management can be more proactive.

To be effective, a manager must understand future challenges and must be alert to the symptoms of change, may it be from any source whatsoever, because these challenges and change will directly or indirectly affect management theory and management practices and procedures. It may mean change in techniques, practices and procedures. It would also require acquisition of new knowledge and skills on the part of manager to meet new challenges.

In this context, managers must be agents of change. To be effective agents of change, it is very essential that they first overcome their own conservatism, introduce change at the work place, get people to accept it and manage the situation in such a way that people want the change. Thus what is needed in sequence is:

 (i) awareness or information in advance of challenges and pending changes,

 (ii) interest in introducing change.

(iii) evaluation i.e. proper discussion and consultation with people concerned for deciding response to the pending change.

(iv) Trial (a pilot project can be undertaken to this effect).

 (v) Action or adoption.

It is very important to note that change should not be introduced for change sake. It must lead to increased productivity and enhancement of the quality of life of the people and hence betterment of the society. We shall revert to the subject of handling change later in the chapter.

In order to meet future challenge, management and managers need to use the following approaches:

1. **Proactive approach** implies having a future orientation in order to anticipate challenges before they arise. It means that management should be alert and sensitive to emerging trends. Managers constantly must scan their professional and social environment for clues about the future. New developments may mean new challenges.

2. **Human relations** approach means that customers and other contributors should be treated fairly and with importance and dignity. As standards of fair treatment and dignity change over a time, the management should be sensitive to future developments.

3. **Systems approach** means that management takes place within a larger context, the organisation and its environment. Managers must view organisation objectives and contributors' needs as parts of the total system.

The emphasis should thus be on a proactive human-relation approach to management within a system frame work.

4.5 TASKS OF BUSINESS ORGANISATIONS AND MANAGEMENT

In the context of above study, business organisations will have to undertake many tasks, the first and foremost being that every business entity must be more than ready to adjust itself to the new era. It means not only the deepening of integration between national economies but also the spread of the process over ever larger parts of the world. At the micro level, management will have to strive for continued cooperation of employees at all levels and of other contributors such as financial institutions, banks, government authorities, shareholders, distributors, dealers and last but not the least the customers.

In order to meet these challenges, the business organisations will be required to have thrust on higher capacity utilisation, betterment of productive factors, cost reduction measures, quality upgradation and focus on niche markets.

Meeting future challenges and handling change will require change in management practices and the managements will have to undertake many new tasks. Future challenges require that management activities be considered as a system of connected activities each of which may be affected by the other and the external environment. A study of upcoming changes and challenges as discussed in previous pages helps us in enlisting some of the more important tasks that managements or managers will have to undertake to meet these challenges. These may include:

1. Balancing the organisational, personal, human and societal objectives.

2. Adoption of new information technology, collection, storage and processing of information, and using the new information to further the objectives of the organisation in a changing environment.

3. Making people aware of the challenges facing the management and of the pending changes so that they have time to discuss and adjust.

4. Training and developing workforce with a view to increasing their potential to perform and translating the potential into ability to meet new challenges.

5. Undertaking R&D programmes to uncover areas in need of improvement and to make improvements through new technology or better practices.

6. Recognising the organisation as an integral part of the socio-economic system and integrating the firm internally and externally.

7. Safeguarding interests of interest groups and contributors to ensure continuing growth of the organisation.

8. Encouraging innovations and creativity to meet organisational goals. Without a future orientation, organisations will not be successful.

9. Introducing and increasing professionalism in management so as to ensure competency and efficiency. It would require recruitment and selection of competent persons and their training and development.

10. Ensuring good and continuing two-way communication. Effective two-way communication is vital in any change situation.

4.6   PROFILE OF INDIAN MANAGER

From our study of nature and functions of management and from study of various approaches advocated by management pioneers and thinkers, we can draw out a profile of a manager *viz.* nature of his job or his overall responsibilities, working environment, knowledge required, and distinctive abilities and skills required to be an effective and efficient manager in the prevailing business environment, especially in Indian context.

A manager is appointed to achieve organisational goals through leadership. It is his primary task and the responsibility of its accomplishment remains constant. Besides other resources, his primary and key resource is the staff or workforce he has. As already stated, his primary responsibilities include examining organisational objectives, assessing available resources and their strengths and weaknesses, setting goals and results to be achieved, remedying deficiencies and deciding a strategy of action to achieve required results, which would require proficiency in planning, budgeting, organising, communication, leading and controlling. In addition, a manager has to train people continuously. At the same time, he must always be receptive to new ideas and is to strive for excellence.

For making best use of resources and for directing coordinated efforts, he must equip himself with technical ability relevant to the field of his business, proficiency in financial management and source utilisation and above all with human relations skills of implementing a plan of action. He needs to strike a balance of skills which demands perception, skill and judgement.

4.7   TASKS AND RESPONSIBILITIES OF THE MARKETING MANAGER

The overall task of marketing management is to plan, organise, implement and control marketing operations in such a way so as to achieve organisational and societal objectives effectively and efficiently. The process involves scanning of marketing environment, analysis of marketing opportunities, designing of marketing strategies and translating them into action to achieve above objectives. An effective marketing management aims at satisfying the needs of the customers or target markets by providing right type of products in right quality at right price and place at the times they are needed.

A marketing manager is required to perform several activities to fulfil this purpose. He needs to acquire a number of special abilities and skills to discharge his tasks and responsibilities efficiently and as a professional.

We have already referred to some key marketing activities which are required to be performed by a marketing manager in chapter 1. The tasks, responsibilities and obligations of a marketing manager can be enlisted as follows:

A. Tasks and Responsibilities

Marketing management is a process of planning, organising, directing and controlling the conception, pricing, promotion and distribution of ideas, goods and services in such a way that individual, organisational and societal objectives are achieved. Besides performance of managerial functions, this process involves following activities to be performed by a marketing manger.

(1) Devising marketing strategies and plans
(2) Analysing marketing information
(3) Conducting marketing research
(4) Designing and developing products in accordance with the demands of the target markets.
(5) Pricing the products and servicing fairly.
(6) Promoting the products through advertisements and other promotional measures and brand-building.
(7) Making goods and services available to the customers and consumers at right place and right times.
(8) Ensuring efficient post-sale service.

The list can be expanded and can be exhaustive one. We shall discuss all these tasks in details in later chapters.

B. Social Role and Obligations of the Marketing Manager

We have already discussed the social role of a manager in details in this chapter. However we can enlist the following social obligations of the marketing manager.

(1) Ensuring that the organisation complies with laws and regulations concerning consumers.
(2) Ensuring that the organisation and marketing personnel do not indulge in unfair trade practices.
(3) Participating in the design and conduct of periodic social audits.

In order to fulfil above tasks, responsibilities and obligations, a marketing manager requires necessary knowledge, and some distinctive abilities and skills which are briefly discussed here.

4.8   DISTINCTIVE ABILITIES AND SKILLS REQUIRED FOR A MARKETING MANAGER

While giving a profile of Indian manager in section 4.6 of this chapter, we have referred to distinctive abilities and skills required by a manager. The same holds good for our present purpose of study i.e. in case of marketing manager also.

Before enlisting abilities and skills needed by a marketing manager it will be pertinent to visualise what type of managers will be sought by business organisations. They will be looking for people:

1. Who can manage and lead a company of tomorrow.
2. Who will not merely anticipate future and manage it, but who will invent and engineer the future.
3. Who will create and make markets and not merely strive to increase market share.
4. Who are capable of discontinuous thinking with the courage and conviction to question the status quo and rewrite the rules afresh.
5. Who have the ability to manage ambiguity making sense out of confusion.
6. Who possess the quest to learn continuously.

Keeping these expectations in view, the following qualities and skills will be necessary to meet challenges of the time,

(a) (i) Intellectual ability and practical skill to keep abreast of relevant changes to redefine priorities when necessary

   (ii) Ability to analyse and evaluate information received and make decisions, if needed

   (iii) Ability to communicate effectively

   (iv) Skill in conceptual thinking of situations in all their details

   (v) A planning ability particularly in financial and marketing areas as envisaged above

   (vi) Ability to innovate and find new ways to meet the demands of customers and other contributors

   (vii) Managers will have to be creative, dedicated and hard-working.

(b) Human values like concern for people, respect for the dignity of individual, need for cooperation and ethical practices alongwith concept of social responsibility which enjoins on the manager to acquire human, social or interpersonal skills

These skills will increase his effectiveness in dealing with the people and help him in developing methods of encouraging cooperation for achievement of individual and organisational goals. These skills are briefly given here.

1. *Skill in understanding the basic human forces active in the organisation*: He must understand the fundamental concepts of motivation, leadership and behaviour that affect the activities of the people with whom he works. He must be aware of the factors that prevent or promote effective communication. In brief he must possess substantial knowledge about human behaviour in organisation.

2. *Skill in analysing complex human situations*: This will include ability to recognise and deal effectively not only with facts, but also with attitudes, opinions and sentiments including those that he himself brings to the situation.

3. *Skill in implementing a plan of action*: He must have skill in doing. This would include ability to gain greater cooperation from group of individuals by influencing informal leaders of the group.

All these skills must be brought together to accomplish organisational goals and to meet future challenges.

Change is normal, natural and inevitable in every organisation. It exists because the environment both inside and outside the organisation is dynamic and not static. Within the organisation many problems concerning production and introduction of new products, expansion and diversification, retirement of people and their replacement etc., arise that need lo be solved. Outside the organisation, there are thousands of environmental changes ranging from new products to new government regulations. Solving these problems will require an efficient planning and an effective business and marketing strategy. We shall discuss it in the next chapter.

APPENDIX TO CHAPTER 4*
*(For M.B.A. and advanced students)

CASE 4.1:
CODE OF ETHICS
(AMERICAN MARKETING ASSOCIATION)

Members of the American Marketing Association (AMA) are committed to ethical professional conduct. They have joined together in subscribing to this Code of Ethics embracing the following topics.

Responsibilities of the Marketer

Marketers must accept responsibility for the consequences of their activities and make every effort to ensure that their decisions, recommendations and actions function to identify, satisfy all relevant publics: consumers, organisations and society. Marketers' professional conduct must be guided by:

1. The basic rule of professional ethics: not knowingly to do harm;
2. The adherence to all applicable laws and regulations:
3. The accurate representation of their education, training and experience; and
4. The active support, practice and promotion of this Code of Ethics.

Honesty and Fairness

Marketers shall uphold and advance the integrity, honor, and dignity of the marketing profession by:

1. Being honest in serving consumers, clients, employees, suppliers, distributors and the public;
2. Not knowingly participating in conflict of interest without prior notice to all parties involved; and
3. Establishing equitable fee schedules including the payment or receipt of usual, customary and/or legal compensation for marketing ex-changes.

Rights and Duties of Parties

Participants in the marketing exchange process should be able to expect that:

1. Products and services offered are safe and fit for their intended uses;
2. Communications about offered products and services are not deceptive;
3. All parties intend to discharge their obligations, financial and otherwise, in good faith; and
4. Appropriate internal methods exist to requitable adjustment and/or redress of grievances concerning purchases.

It is understood that the above would include, *but is not limited to,* the following responsibilities of the marketer:

In the Area of Product Development Management

Disclosure of all substantial risks associated with product or service usage.

Identification of any product component substitution that might materially change the product or impact on the buyer's purchase decision

Identification of extra-cost added features.

In the area of promotions:
Avoidance of false and misleading advertising
Rejection of high pressure manipulations, or misleading sales tactics.
Avoidance of sales promotions that use deception or manipulation

In the Area of Distribution

Not manipulating the availability of a product for purpose of exploitation
Not using coercion in the marketing channel
Not exerting undue influence over the resellers' choice to handle a product

In the Area of Pricing

Not engaging in price fixing

Not practicing predatory pricing

Disclosing the full price associated with any purchase

In the Area of Marketing Research

Prohibiting selling or fund raising under the guise of conducting research

Maintaining research integrity by avoiding misrepresentation and omission of pertinent research data

Treating outside clients and suppliers fairly

Organizational Relationships

Marketers should be aware of how their behavior may influence or impact on the behavior of others in organizational relationships. They should not encourage or apply coercion to obtain unethical behavior in their relationships with others, such as employees, suppliers or customers.

1. Apply confidentiality and anonymity in professional relationships with regard to privileged information.

2. Meet their obligations and responsibilities in contracts and mutual agreements in a timely manner.

3. Avoid taking the work of others, in whole, or in part, and represent this work as their own or directly benefit from it without compensation or consent of the originator or owner.

4. Avoid manipulation to take advantage of situations to maximize personal welfare in a way that unfairly deprives or damages the organisation or others.

Any AMA members found to be in violation of any provision of this Code of Ethics may have his or her Association membership suspended or revoked.

Courtesy: The American Center Library, New Delhi

MANAGEMENT THOUGHT

I believe that nothing can be greater than a business, however small it may be, that is governed by conscience; and that nothing can be meaner or more petty than a business, however large, governed without honesty and without brotherhood.

The First Viscount Leverhulme

Founder of Unilever

CASE 4.2:
MARICO INDUSTRIES:
THE MARKETING FUNCTION
(TASKS AND RESPONSIBILITIES)

The **Vice President - Marketing** heads the Marketing Function. He is responsible for marketing, sales, new product development and acquisitions.
The VP-Marketing is supported by the following team of senior managers.

The General Sales Manager is responsible for achieving agreed upon sales targets, and strengthening the nationwide distribution network. He is supported by :
* Four Regional Managers heading the N.S.E. and W regions respectively.
* Sales Development Manager.
* Area Sales Managers/Executives assisting the Regional Sales Managers.

The Marketing Manager is responsible for continuously evaluating sales and profitability performance of existing brands, and designing marketing inputs to achieve predetermined profit goals. He is also responsible for generating new product concepts, carrying out market research, test marketing and product launches. He is assisted by Group Product Managers and Brand Managers.

The R & D Manager is responsible for developing formulations, products and packaging, based on product briefs and specification identified by the Marketing function. He is also responsible for transferring any labscale development into mass manufacture.

Source : The Marico Story (Marico Industries Limited).

CASE 4.3:
BACKWARD INTEGRATION
&
COMMUNITY DEVELOPMENT
(HINDUSTAN LEVER)

In 1964, a 30,000 TPA dairy plant was set up by Hindustan Lever at a cost of Rs. 2 crore in the backward Etah district of Uttar Pradesh, based on a survey indicating an abundance of cattle. The optimism that this investment would provide a fillip to the region's economy began to fade with poor capacity utilisation due to inadequate milk supplies. In spite of substantial investment in manpower, capital and technology for 10 years, the company's losses mounted. The task was to increase milk availability.

In 1976, the company launched an Integrated Rural Development Programme. For that, a relationship of trust based on mutual benefit had to be built with the farmers. Company personnel and management trainees went to live with the farmers to understand their problems and conceive simple solutions.

The low productivity of cattle was merely a symptom. The real issue was the farmer's beliefs, rooted in their traditions. Untrained in scientific animal husbandry, they could not give their milk animals balanced feeds. Small holdings and poor soil conditions made agriculture uneconomical. Awareness of balanced fertiliser use, animal health care and credit facilities was low. Above all, people could not accept the idea of selling milk, as it was against tradition.

To make relevant technology available, HLL set up a centre, the Krishi Pashu Vigyan Kendra (KPVK), to conduct trials and demonstrations of agriculture, animal husbandry, renewable energy resources, village level industries and so on. It also trained farmers in the better management of animals, bank loan procedures, the management of reclaimed land, and community development programmes. Farmers learnt intensive farming and remunerative crop rotation to grow more than one crop a year. Commercial crops like chicory and peas were also introduced.

The farmers now sell their milk directly to the company, bypassing middlemen. The village based milk collection centres have been further transformed into 'growth centres'. Through the villagers' involvement in managing their own programmes, the growth centres provide services like village-based payments, dispensing hybrid seeds, vaccinations and artificial insemination. Subsidies have gradually been reduced and farmers now pay for the services.

Today, the rewards of these efforts are visible. The company now sources 27,000 tonnes of milk per annum from Etah district compared to a mere 2,500 tonnes in 1976. With increasing use of better seeds and practices, the cropping intensity has gone up form 1.43 in 1976 to 2.22 in 1995. This has had a synergistic effect on improving milk productivity, which has gone up from 500 kg per animal per year to around 700 kg, and in some breeds, upto 1,400 kg. Community development programmes have resulted in visible improvement in the villages. The outbreak of cattle diseases has been checked with less than 1% of cattle dying of them.

The above experiences in Etah provide some lessons on how to increase raw material supply by providing a broad range of services. It entails shaping development in small steps and making simple ideas work for reviving a loss-making project. The experience presents a good example of establishment of backward linkages and social sensitivity.

Source: Linkages with agriculture (The Experience of the Food Processing Industry) (Brooke Bond Lipton India Ltd.).

CASE 4.4:
THE JAMNAGAR EXPERIENCE

A unique experience of Brooke Bond has been the introduction of an exotic crop like chicory in India. The use of chicory as a beverage was popular in France as 'French Coffee'. It was introduced in India to provide economy and richer taste. Commercial success was dependent on sourcing chicory locally instead of imports.

Suitable climatic conditions and low irrigation requirements led the company in 1956 to Jamnagar. In a drought-prone area where farmers have small holdings, the Company installed processing facilities for chicory, assured 100 per cent buy back, provided seeds free and implemented best agricultural practices. By 1960, it had 130 acres of land belonging to 50 farmers under chicory cultivation on a contract basis. Today, the area under chicory cultivation has grown to 4500 acres and the number of farmers to 1700. Their sincere efforts coupled with the use of appropriate technology have doubled the yield to 3 tonnes per acre.

One of the pillars of company's four-decade relationship with the farmers has been a willingness to contribute, albeit in a small way, to the efforts of the administration. This indicated the company's commitment to the farming community and the region's welfare. It is gratifying to note that other companies are now emulating pioneering efforts in Jamnagar to establish backward linkages.

CASE 4.5:
BACKWARD LINKAGES (PEPSI)

Pepsi installed a tomato paste manufacturing plant, alongwith the contract farming system, at Zahura in Punjab in 1990.

Though tomato was not a natural fit with the agro-climatic profile of Punjab, the state was chosen because of the progressive profile of its farmers, their large land holdings and water adequacy. Custom-made technology and its dissemination through farmers' linkages has ensured marketable surpluses.

Altogether 50 hybrid tomato seed varieties were tried on 53 acres for over two years from 1990 to test yields, disease resistance, climatic response and quality of fruit. Research focused on developing new agricultural implements. Consequently, tailor-made technologies and agricultural practices were developed to increase productivity and reduce the cost of cultivation.

With education and constant interaction, all contract farmers have adopted best practices. Today, the model encompasses the raising and supply of disease-free seedlings, tomato-specific agro-implements, training camps, crop health monitoring and post-transplantation care. Field offices train farmers on harvest prediction and quality control, and also provide critical services like payments directly to their accounts in the villages.

The results achieved illustrate the advantages of such backward linkages. The per acre yield of tomato has increased from a mere 8 tonnes to 20 tonnes. The peak available season has increased from 30 days to 60 days. The production of tomatoes in the state has increased from 30000 MT to 1.30,000 MT. Employment potential of almost 35000 jobs has been generated. Rapid progress has been made in creating marketable surpluses for processing.

The lesson is that without effective linkages, no agricultural system can be supported very long. However, there are problems. In times of glut, some factories cannot accept the entire produce, and farmers are put to great hardship. Similarly, companies have little recourse when contract farmers sell their produce in the fresh market at relatively higher prices. Some kind of quasi-legislation which protects the interests of both the farmer and the processor will provide a major fillip to this activity. Introduction of more crops like peas for processing will help forge stronger relationships.

Source: Linkages with agriculture : The Experience of the Food Processing Industry (Brooke Bond Lipton India Ltd.)

EXERCISES

1. Explain the social role of a manager in modern marketing.
2. What is meant by social responsibility? Enlist its important features.
3. What is meant by terms "ethic" and "code of ethics"? Draw a code of ethics for marketers in your organisation.
4. Explain the tasks and responsibilities of a marketing manager and the skills needed to accomplish them.
5. Write notes on the following giving examples
   (a) External integration
   (b) Backward linkages
   (c) Business ethics
   (d) Marketing Functions, Tasks and Responsibilities.

5

Strategic Planning

5.1 INTRODUCTION

In modern times, the number of goods and services to be produced is very large and varied. The recent advancements in the field of science and technology, transport and communication coupled with innovations and inventions, and entry of multinationals as a result of liberalisation and globalisation of economies have widened the scope of business activities. The factory system, mechanisation of production processes and techniques based on division of labour and the specialisation, and union-management frameworks have made the task of business enterprises and organisations more varied and more complex. For example,

CASE 5.1: COMPLEXITY OF BUSINESS

(Hindustan Unilever Limited)

In last 50 years, Hindustan Lever now Hindustan Unilever has witnessed a phenomenal growth. It has diversified its activities from its original base of vanaspati, soaps and personal products through the addition of synthetic detergents, animal feeds, dairy products, ice creams speciality and bulk chemicals, fertilisers, hybrid seeds and several export lines. During recent years, it has got amalgamated in itself Brooke Bond Lipton India, which itself was India's biggest foods and beverages company with its business based on agricultural raw materials, namely tea, coffee, chicory, fruits and vegetables, edible oils, milk, cereals, animal feeds, sugar. Earlier the erstwhile Tata Oil Mills Company Limited (TOMCO) was amalgamated with HLL, now HUL.

(Contd.)

Even at present HUL has six subsidiary companies, namely Indexport Limited, Lever India Exports Limited (formerly Shekhar Engineering Industries Limited), Hind Lever Chemicals Ltd. (Formerly Stepan Chemicals Ltd.), Industrial Perfumes Limited, International Fisheries Ltd., Nepal Lever Limited. The company HUL had a net turnover of about Rs. 13718 crorcs during 2007. Earlier HUL had formed with Lakme Limited a joint venture named Lakme Lever Limited which is manufacturing and distributing colour cosmetics and other personal care products.

The above example amply suggests that a large corporate organisation or company may be controlling or having a holding stake in several business units, running different businesses and producing different types of products and services in widely spread plants, factories or units. An effective and efficient management implies perfect coordination and organisation of human efforts and material resources to achieve the objectives of the organisation and for the future growth of the company. It requires formation of some strategic plans at organisational level, whatever be the structure of organisation. Obviously the planning strategies of the business units will be shaped and influenced by the planning strategy of the parent organisation or company or in other words by the corporate strategic planning.

Considering at micro-level, a business unit or enterprise undertakes several activities such as technical activities, financial activities, commercial or marketing acclivities, security activities, accounting activities, managerial activities etc. in order to achieve its objectives. Marketing, no doubt, is the key business activity, yet it should not be forgotten that marketing is but one important activity out of several. Therefore any marketing strategy must fit in the overall strategy or basic plan of the enterprise. It should not be formed in isolation, rather it is to be integrated into the overall strategy of the firm or enterprise so that the enterprise has what we call a *market-oriented strategy*.

It is clear from the above study that strategic plans are needed to be developed at various levels of structure of a corporation or an organisation. These strategic plans will take different forms keeping in view the tasks and responsibilities to be performed by authorities at various levels. For example, corporate strategic plan will specify the corporate mission, analyse environmental conditions, identify strategic business units, and allocate resources to them and make decisions and policies to achieve the specific mission and guide the entire enterprise to ensure growth and profitability. Similarly each business unit will develop its own strategic plan and determine its own objectives and course of action to accomplish them. Integrated into this overall strategy will be a marketing strategic plan at marketing department or product level.

In this chapter we will study briefly the features of corporate strategic planning and business strategic planning. In the next chapter, we will discuss marketing planning and marketing strategy as developed in the process of marketing management. But before doing that, we must understand the concepts of planning strategy and strategic planning.

5.2 STRATEGIC PLANNING: THE CONCEPT AND DEFINITION

Planning is a process through which an enterprise tries to make best possible use of its resources towards achieving its goals. The objective of planning function is to predetermine a course of action to be pursued to accomplish organisational goals. Therefore the primary tasks of planning are developing a strategy to carry out the long term objectives of the organisation and decision making at every level or stage. Thus formulation of a strategy is central to systematic planning.

Strategy is the overall basic plan or process of determination of company's objectives and goals keeping in view the environmental changes and determination of courses of action and allocation of resources to achieve those objectives. It is a specific action plan of a firm to achieve specific goals over a period of time. It aims at mobilising, deploying and utilising resources in an integrated and optimal manner. We can enlist here a few important features of a strategy such as

1. Strategy is an overall plan of determination of long-term objectives, formulation of necessary guidelines and policies and adoption of selected courses of action in pursuance of the objectives.
2. It is aimed at determining enterprise's capability to cope with environmental factors and conditions especially problems of competition, survival and growth.
3. Strategy is a comprehensive concept. It is a long range programming of company mission and consists of policies and decisions.
4. Strategy is future-oriented and forward-looking. It is the vision of what enterprise would be in the future.
5. Strategy is a matching action between company's strengths and weaknesses with external opportunities and competition.

In the light of the above discussion, we can define strategic planning as follows:

Strategic planning is a process of defining objectives for the entire organisation and determining appropriate strategies for achieving these objectives. It is the process of determining objectives of the organisation, charting out plan of action to achieve them, allocating resources to different businesses activities to be undertaken (i.e. determining investment or business portofolio) toward this end, and developing a set of guidelines and policies concerning them keeping in view the environmental conditions and opportunities. Strategic planning, thus, is a managerial process of effectively matching goals and resources of a company with target market opportunities.

Survival, profitability, sustained growth and excellence have been the cherished goals in that order of every business organisation and strategic planning is the process that prepares the ground work to achieve these goals.

We shall examine the process of strategic planning as it can be undertaken at corporate level and business level in the following pages.

5.3   CORPORATE STRATEGIC PLANNING

As stated above, corporate strategic planning is a process of determining mission or objectives of the corporation (i.e. company) keeping in view the environmental situation, allocating resources to different businesses or business units and developing guidelines and policies to achieve the corporate mission or goals and hence satisfactory profits, sustained growth and excellence. The importance of corporate planning lies in the fact that only it can give shape to the corporate management's vision about the future direction of the organisation.

Obviously, this exercise is undertaken by the top management at the corporate headquarters.

Corporate strategic planning process involves the following steps:
1. Assessing and analysing the situation
2. Determining corporate mission
3. Determining composition of business and selecting strategic business units
4. Evaluating current funding and business portfolios
5. Identifying business development alternatives or new business areas.

Assessing and Analysing the Situation: This includes an examination and analysis of environmental factors and conditions, company capabilities in terms of resources, available and likely to be available, company's own strengths and weaknesses, its position in the industry, growth prospects of the industry, competition and other key factors required for success, growth potentials and profit prospects etc. It is on the basis of this situation assessment that the company would be able to determine its mission and objectives and develop a suitable and effective strategy.

Developing and Determining the Corporate Mission: An organisation is set up to accomplish some purpose or fulfil some specific objectives. A statement of mission is a statement of corporate collective direction and desire. It should help to build synergistic relationships with the external stake holders. Defining the corporate mission in clear, specific and unambitguous terms is the foremost responsibility of the corporate management. Strategic planning involves establishing clear and specific objectives for the company as a whole. An organisation cannot grow or prosper in orderly or progressive manner unless clear, well-defined and specific objectives are set to guide its activities. Besides, the mission statement should express explicitly its philosophy, its policies and the vision of what it would be in future, say over the next ten to twenty years. See the following example.

CASE 5.2: COLGATE-PALMOLIVE'S VISION AND MISSION (1990)

Our Vision
   Be the Company of First Choice in Oral and Personal Hygiene by Continuously Caring for Consumers and Partners.

Our Mission
   We aspire to be the Company of First Choice through building together a Successful Organization driven by Competence, Shared Values and *personal accountability* - a proactive learning organization which will generate *profitable growth.*
   *By focusing on our consumers, both urban and rural, anticipating and responding to their changing needs.
   *By ensuring undisputed leadership in oral care categories
   *By enhancing oral health in the country and promoting daily usage of hygienic, safe and affordable oral care product for 500 million people by the year 2000.
   *By establishing sustainable positions in personal hygiene categories.

Determining Composition of Business and Selecting Strategic Business Units: After analysis of situation and determination of the corporate mission, understanding and determining of composition of business is essential for developing corporate strategy. It is necessary to understand varying types of businesses and business activities a corporate is running, the different products it or its subsidaries are producing, and markets, single or multiple, it is serving. This will help the company in grouping similar business areas in strategic business units (SBUs). The corporation needs to identify and define business or businesses in terms of the customer groups, customer needs and the technology to be used and group similar business areas into strategic business units to be managed as viable, independent and separate units of the company. It is dividing the entire business into strategic units, each to be managed separately by a manager who is proficient in his managerial work.

Strategic business unit (SBU) is an organisational unit within a corporation that can perform all the basic business functions in meeting the needs of specific market with a product, a product line or a mix of related lines and can be planned and managed separately as an independent business profitably in competition with its rivals. The formation of strategic business units helps the corporation in allocating appropriate resources to them to achieve their respective objectives. Again, strategic business units can also be formed by combining two units producing similar products or serving the same target market or customer group in order to reap management, market, operating or the other advantages. Allocation of resources and determination of business portfolios will depend on such formulation.

Evaluating Current Funding and Business Portfolios: Strategic planning necessitates evaluating current funding and business portfolio which implies evaluation of existing strategic business units. This needs to be done in order to decide which of the SBUs needs heavy investments for expansion and which is a burden on the company and its resources, and needs to be closed or disposed of. Evaluation of a business unit involves assessing its past performance, examining its present situation and forecasting its future potential. It involves assessing unit's strengths, weaknesses, anticipated risks and opportunities and its relative position in the market or industry, profitability etc. Several approaches have been developed to analyse business units and hence the business portfolio. Three most widely used analytical models or approaches are — the Boston Consulting Group's Growth Share Matrix, The General Electric's Market attractiveness—competitive position, Business Strength Matrix and Profit Impact of Marketing Strategy (PIMS) model.

1.   Boston Consulting Group's Growth Share Matrix

A popular approach to assess and analyse business units and hence business portfolios was developed by Boston Consulting Group (BCG), a consulting firm and this model is known as Growth Share Matrix. As the name suggests, this approach evaluates the SBU by rating it in terms of its market growth rate and its market share relative to the biggest (largest) competitor. In fact, in this approach, a single factor, namely average *annual growth rate* measures the market attractiveness and a single factor namely *market share* of the unit relative to its largest competitor measures the business strength of the unit.

BCG's growth-share matrix is represented diagrammatically by a grid in which growth rate is shown along vertical axis and relative market share along horizontal axis. A market growth rate of more than 10% or so in sales or turnover can be considered high growth rate while growth rate of less than 10% is regarded as low growth rate. The business strength of SBU is indicated by the market share relative to its biggest competitor.

A relative market share of 0.2 means that the turnover of the unit is 20% of that of its largest competitor. An index of 3.0 means that the turnover of the unit is 3 times that of the turnover of the largest competitor. An index of 1.0 indicates equality of business strength and is the dividing line between high and low relative market share. The growth share grid is shown in figure 5.1

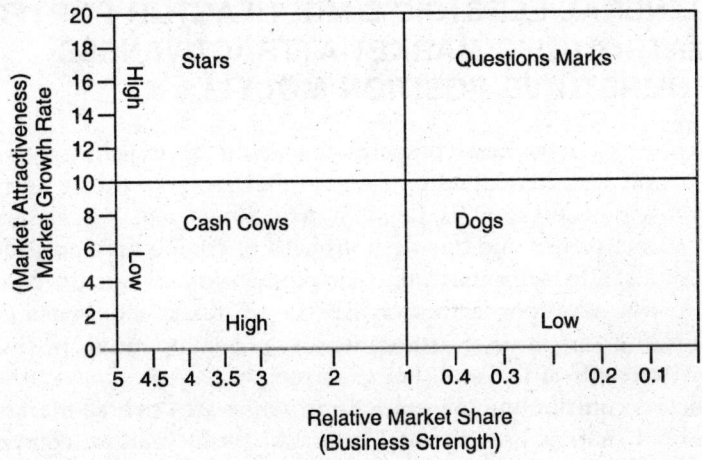

Fig. 5.1 Boston Consulting Group's Growth Share Classification.

As illustrated in the above figure, BCG's matrix classifies business units into the following four categories.

(1) Stars: These are business units having high growth rates and high relative market share. They need huge investments to keep up the tempo. They may not earn immediate profits but have great future potential for the company.

(2) Cash Cows: These business units operate in low growth market but have high relative market share and are leaders. They do not need large funds for their expansion but earn substantial profit due to economies of scale and bring enough cash for the company.

(3) Question Marks: These business units operate in high growth market but have low relative market share. They need heavy investments for expansion and use lot of money but have an uncertain future, though bringing cash flow. The company needs to have a hard look at a question mark unit and has to decide rationally whether to invest more funds or not in the unit. It may be possible to turn a "question mark" unit into a star by devising suitable strategy of funding and promotion.

(4) Dogs: The units that operate in low growth market and have low market share relative to their competitors are called "dogs". They yield very low profits or produce losses. They are a drag on the company and need to be phased down, sold out or liquidated. The company must have some good reasons if they are to be allowed to continue.

A balanced portfolio would mean many cash cows and stars and too few question marks and dogs.

2. GENERAL ELECTRIC'S MULTIFACTOR PORTFOLIO MATRIX (GE'S MARKET-ATTRACTIVENESS-COMPETITIVE POSITION MODEL).

GE's market attractiveness-competition position approach to evaluate a business unit was developed by General Electric. Like BCG's approach, here also a business unit is rated in two dimensions i.e. in terms of market attractiveness and business strength or competitive position. But instead of a single factor indicating each dimension, as was the case with BCG's model, multiple factors or aspects are taken into consideration while finding out market attractiveness and competitive position or business strength of the unit. For example;

(a) Factors contributing to *market attractiveness* are - overall market size, annual market growth rate, historical profit margin, competitive intensity, technological requirements, energy requirements, inflationary influences, environmental impact etc. A weighted index pertaining to measures of these factors is computed in each case to indicate the market attractiveness of the business unit.

(b) Similarly, the factors underlying *competitive position* include market share, growth, product quality, brand reputation, distribution network, promotional effectiveness, productivity, capacity, productive efficiency, unit costs, material supplies, R&D performance, managerial personnel etc. Again a weighted index of all these aspects is commuted in each case to indicate the competitive position or business strength of the business unit.

In GE's model, market attractiveness is categorised as low, medium and high. Similarly competitive position is also categorised as low (weak), medium and high (strong). EG's market attractiveness competitive position grid is shown in figure 5.2 which comprises 9 cells.

As illustrated in the above figure,

A. Units in cells 1, 2 and 4 are strong and the company should invest in them and grow.

B. Units in cells 3, 5 and 7 are medium. The company should pursue a policy of selectivity/earnings. They need to be protected and built on selective basis. Investments may be concentrated in segments where

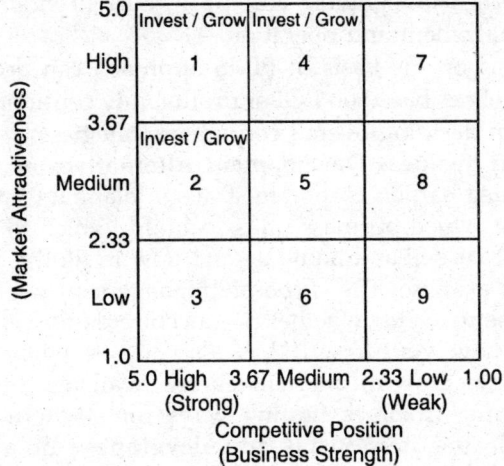

Fig. 5.2 GE's Market Attractiveness - Competitive Position.

profitability is good. What is needed is selective use of resources and focus on earnings.

C. Business units that fall in cells 6, 8 and 9 and are weak in overall attractiveness and the company should have a hard look at them. They fall under the category Harvest/Divest. They need to be disposed of or liquidated unless the management can find ways to bring them out of the weak position into a favourable one.

3. Profit Impact of Market Strategy (PIMS)

Another approach to analyse business unit was developed by strategic planning institute (SPI), Massachusetts, under Sydney Schoeffler, Robert Buzzell and Donald Heany and called PIMS. This model was built as a result of study of extensive data i.e. market pertaining to some most important variables and factors that affected profitability (i.e. return on investment and cash flows) and gathered from thousands of business units. They identified a number of variables that had an impact on profitability and studied the relationship. They found that the *increase in relative market share* of its served market i.e. market share of the SBU as percentage of the combined market shares of its largest 3 competitors has positive impact on the profitability. The identified variables affecting profitability (ROI) include investment intensity (turnover or income per unit of money invested), productivity (value added per employee), market position or share, growth of served market, quality of product or service, innovation, vertical integration, cost push, current strategic effort and quality of management and operation.

The analysis on the basis of PIMS approach can provide resource allocation guidelines based on ROI or profitability consideration and help the company in deciding overall corporate strategies.

Identifying Business Development Alternatives or New Business Areas: As stated earlier, corporate strategic planning involves making decisions about which business units to build, which to maintain, and which to divest i.e. sell and liquidate and how to fill the gaps arising as a result of this exercise. The corporate management while planning its strategy has also to decide whether to concentrate on its core business or to enter into new ventures. It has also to see what expansion or development (i.e. growth) alternatives are available. In other words, strategic planning involves finding ways for strengthening existing business, exploring new vistas and developing or acquiring new businesses and hence making decisions regarding its growth. A company has three options or alternatives for its development and growth, namely intensive growth options, integrative growth options and diversification growth options.

 (i) **Intensive Growth Options:** These options imply strengthening and improving the performance of existing businesses. They may take the form of making efforts to increase market share of current products in existing markets i.e. *market penetration* (ii) designing developing and marketing new products for existing markets i.e. *product development* and /or, (iii) finding out and developing new markets for existing products i.e. *market development.*

(ii) **Integrative Growth Options:** Integrative growth implies integrating and acquiring additional units and discovering additional profitable sources. We have already referred to backward integration in chapter 4.

 Backward integration means acquiring control over suppliers to earn more profit or to obtain supplies at lower costs while forward integration implies acquiring profitable distribution channels. Horizontal integration involves taking over or acquiring one or more competitors.

 About more than a decade back Hindustan Lever, now Hindustan Unilever, acquired Tata Oil Mills Limited and got amalgamated into it. Vertical integration implies making or manufacturing the goods rather than buying them. It includes backward and / or forward integration. Vertical integration means single ownership of supplier, producer and /or wholesalers and retailers with a view to improve performance and reap higher profits.

(iii) **Diversification Growth Options:** These alternatives involve seeking new products and new markets for them. It involves entry into new businesses and markets through internal development or acquisition. By adopting this strategy, the company can spread business risks over a number of business areas. This movement may be towards related product areas (*horizontal diversification*) or unrelated product market areas (*conglomerate diversification*).

In the previous pages, we have examined the process of corporate strategic planning, and matters involved in devising an effective corporate strategy. As we have seen, the essence of corporate strategy is securing an advantage over competition. It is through an effective strategic planning that a company can improve its performance and maintain its momentum for growth.

5.4 BUSINESS STRATEGIC PLANNING

Business strategic planning is an exercise which is done at Business Unit level. It is the process of determining the mission and specific objectives of the enterprise keeping in view the environmental condition's, evaluating various courses of action and devising policies and programmes to achieve the enterprise objectives. In simple words, business strategic planning is the process of devising effective business strategy to achieve the unit's goals and objectives. These strategies are to be prepared under the guidelines set by the corporate strategic plan.

We have already defined the term strategy in section 5.2. Alfred Chandler defines a business strategy as *"the determination of basic long-term goals and objectives of an enterprise and the adoption of courses of action and the allocation of resources necessary to carry out these goals"*. Again, every business or organisation must have a mission. It will have to compete for relevant resources with other organisations to accomplish its mission. It might have to combine its efforts with other enterprises to achieve its mission and while doing so it will need to consider environmental factors, i.e. elements of change, growth and adaptation. Strategy is the unit's plan to deal with these factors. Thus business strategic planning process involves devising a master strategy i.e. *a basic overall or grand design of determination of long-term goals and objectives of an enterprise, the framing of necessary policies and the adoption of selected courses of action for translating objectives into performance results.*

The business strategic planning is concerned with giving overall direction in which human and physical resources are to be deployed and applied in order to maximise the chance of achieving the enterprise goals and objectives. Strategy development is the core of business strategic planning process.

As is clear from the above study, the business strategic planning process consists of the following steps;

1. Defining the business mission
2. Determining objectives and goals of the enterprise
3. Developing an effective strategy keeping in view environmental factors.
4. Implementing the strategy and feedback

Defining the Business Mission

Within the confines of broader corporate mission, each business unit or enterprise needs a central mission or purpose expressed in terms of services it will render to the society. It must specify its responsibilities towards stake holders, the customer needs and wants to be served by the unit's products and services, scope of its activities and extent of specialisation within each product-market area etc.

Determining Objectives and Goals

Setting objectives is the foremost responsibility of a manager. *The objectives are the goals i.e. the end results to be achieved* and need to be expressed in clear, specific and unambiguous terms. In fact it is these goals that guide the efforts of the enterprise and its components. Effective management is always management by objectives (MBO). Traditionally, profit maximisation has been the sole objective of a firm or enterprise. Other objectives can be sales maximisation, output maximisation, satisfaction maximisation or achieving a certain minimum or target level of profit or return on investment.

Thus a specific goal or objective of a business unit can be to earn 15 per cent return on investment (ROI) or to achieve 20 per cent market share or leading position amongst its competitors or increase its turnover by 10 per cent or development of new products or brand building or it may be a mix of these objectives. The crux of the problem is that as far as possible, these goals should be expressed quantitatively and should be realistic and achievable.

The management at business unit level is also required to set the specific objectives and goals to be achieved by various functional departments in consultation with them so that they fit in the overall objectives and targets set by the business unit for itself.

Developing a Strategy (Strategy Making)

Development of a strategy involves the following steps:

1. Identifying specific roles or niches suitable for the company in view of society's needs and company resources.

2. Integrating various roles with other company efforts to obtain synergistic effects.
3. Expressing the plans in terms of targets.
4. Setting up sequences and timing of changes

Step I. Identifying Suitable Roles

Identifying suitable roles for business unit, i.e. selecting what products or services will be provided by the enterprise (product mix), which markets or customers are to be served by it (target markets), what new products and services will be developed, and identifying new markets in case of changed conditions, is not an easy task. It will require a study and analysis of company's own strengths and weaknesses, its position in the industry, growth prospects of the industry, competition and other key factors required for success i.e. environmental factors.

Finally those roles or fields will be identified where strengths correspond with success factors. In other words, what is required is to analyse the factors and conditions (i.e. determinants) that may vitally affect a company's strategy.

Determinants of strategy: The factors and considerations that affect a strategy are as follows:

 (i) **Demand for firm's goods and services:** An enterprise's strategy will depend on the demand for its goods and services and estimate of its growth over a time. Again desire for a service, possibility of substitutes, ability to pay, business cycles, structure of markets are the factors that will affect the demand. All these factors require analysis. All these factors are likely to undergo change over a period of time. Therefore an effective strategy demands analysis of all these factors and also what change can be in the offing.

 (ii) **Supply of services:** Another factor determining a strategy is the supply of services related to the demand i.e. what would be probable supply of services and conditions under which they will be offered. This would require appraisal of capacity, costs and technology available with the enterprise or likely to be available with the firm.

(iii) **Competitive conditions in the industry:** Another factor that affects the strategy is the nature of competition faced by the enterprise or it is likely to face. The size, strength and attitude of other companies, presence of trade associations and help rendered by them, and government regulations regarding subsidies, patent laws influence the nature of competition. All these factors need to be analysed before devising a strategy.

(iv) **Key success factors:** Every enterprise must identify those key factors that will be crucial for its future success. The key success factors will differ form one enterprise to another, from one segment to another. These key success factors could include leadership R&D, low costs, adaptability to local needs or creative advertisement.

(v) **Growth potential and profit prospects:** Analysis of above factors would provide a forecast of the growth potentials and the profit prospects. An appraisal of all these factors is vital for framing the firm's strategy as this appraisal will give the strengths and weaknesses of the enterprise. The task before the strategist for identifying suitable roles is how to integrate the strengths with firm's key success factors.

(vi) **Market strengths of a firm:** Market strength of an enterprise includes its market position i.e. share in market sales - areawise, product-wise and customer-wise, relative standing of firm's products and their quality and its marketing system.

(vii) **Supply strength:** A firm's ability to supply goods needed by customers will depend on its *resources*, financial, material and human, production facilities available with it, and its *research and development programmes*.

(viii) **Character of management:** The financial capacity of the management, its quality i.e. vitality of key managers to take risk, their enthusiasm to seek for growth and excellence, and their desire to ensure stable employment of workers can be helpful in developing suitable and effective strategy.

An analysis and appraisal of above factors and conditions will help in identifying desirable roles.

Step 2.   Combining Efforts for Synergistic Effects

An enterprise makes efforts in a number of directions to render services and serve its identified roles. They will have far greater effect if these efforts are made together in cooperation rather than separately. While developing a strategy, efforts need to be combined so as to obtain what is called *"synergistic effects"* i.e. the combined effect of efforts. This means that the proposed strategy should be built on integration and coordination of various efforts being made by different functional departments.

Step 3.   Expressing Potential Results of Strategic Plans in Terms of Strategic Targets or Selected Criteria

It must be ensured that the strategy formulated achieves results or targets in consonance with the organisation objectives. The strategy might have to be modified if results do not come upto expectations.

Thus the strategy sets forth: what services are to be provided and to whom, how will these services be created or obtained and what criteria will be used to get the desired results.

Step 4. Setting up Sequences and Timing of Changes

This step consists of deciding what to do first and what should be the sequence of efforts and activities, and how fast to move and what to postpone. This would require detailed scheduling of activities and appropriate time frame so as to translate decisions into action and desired results. Strategy provides targets and major moves alongwith general guides for sequence and timing. The long-range programme spells out fully the steps taken out to execute that strategy. Developing a strategy and translating it into managerial action requires decision-making at each stage.

Implementing the Strategy and Follow up

All the above steps upto now are preparatory. The real thing needed is to translate the strategy or decisions into action. Implementing the strategy is most important. The question is how and when the strategy is to be implemented. Can the targets and target dates be marked? Who is to do it, who's to be contacted and when he is to be contacted? Can the responsibility of taking action on a decision be delegated? Whether those who have to put the strategy into effect possess requisite abilities and skills and so on?

It must be kept in view that various functional managers have to implement the strategy within the framework of policies of the organisation. Implementation of the strategy will involve developing programmes, setting schedules, specifying procedures and determining budgets. One has to work within the limits of a budget to keep down cost.

A follow up is also needed to know what has really happened as a result of implementation of the formulated strategy. How far the strategy has achieved the goals? The implementation needs to be monitored. If the results do not come upto expectations or goals are not achieved as planned or environmental conditions change in the meantime, then the strategy may need modifications or even goals and objectives may have to be revised.

Business strategy lays groundwork and sets guidelines for various functional departments including marketing department to develop their own strategies and plans within the ambit of broad business strategy. We shall study marketing planning in the next chapter.

EXERCISES

1. What is corporate strategic planning? Describe the steps involved in this process.

2. Define business strategic planning? Explain the process.

3. Describe various approaches to evaluate business portfolios.

4. Explain the following models

    (a) Boston Consulting Group's Share Matrix

    (b) GE's Multifactor Portfolio Matrix

    (c) PIMS Approach

5. Write notes on

    (i) Strategic business unit (SBU)

    (ii) Stars and cash cows

    (iii) Strategy making

    (iv) Master strategy

    (v) Deteminants of a strategy

    (vi) Corporate mission

    (vii) Strategic planning

6. Colgate-Palmolive produces and markets a broadline of consumer goods like tooth paste, soap, cosmetics etc. In the wake of your study of colgate's corporate mission in this chapter, discuss its influence on the marketing management function of the company.

7. "Strategic planning draws from all the functional areas of an organisation." Discuss the contribution of marketing management in this context.

8. Define the competitive domain for Colgate-Palmolive by discussing its version of mission as given in this chapter.

6

Marketing Planning

6.1 MARKETING (MANAGEMENT) PROCESS AND MARKETING MIX

In chapter 1, we have defined marketing management as "the process of planning, organising, directing and controlling the conception, pricing, promotion and distribution of ideas, goods and services to create exchanges that satisfy individual and organisational objectives". According to this definition, in addition to the performance of management functions *viz.* planning, organising, directing, controlling, a marketing manager is also required to execute or perform some operative functions or marketing activities, namely conception, pricing, promotion and distribution of products in order to satisfy the needs of customers. Thus marketing management is the process of planning, organising, directing (implementing) and controlling these operative functions, i.e. marketing activities in order to fulfil individual, organisational and societal objectives.

It is clear from the above definition that the first and primary task of a marketing manager is the planning of marketing activities, that is planning the conception, pricing, promotion and distribution of ideas, goods and services that may serve the needs of various customer groups. Planning is essentially a decision making process and under it a marketing manager is required to find out the needs of the target market and make decisions about the design and quality of product or service to be offered by the firm to meet the customer needs (*conception*), the price to be charged for the product (*pricing*), the advertising, publicity and sales promotion programmes to be undertaken (*promotion* of *products*), and the distribution channels to be used to make the products available to the customers (*physical distribution* or placing of the products). Thus planning involves making decisions about programmes or strategies pertaining to various marketing activities that constitute the marketing process. All these four factors or 4 Ps i.e. product (the design, quality, packaging etc.) the price, the promotion and the distribution (i.e. the place) influence the demand of the products and are regarded as components of what is termed as *marketing mix*. It is the package of all these four Ps or four components of the marketing mix or the marketing programme that would have synergistic effect on achieving the end results. An appropriate package or combination of these components of marketing mix will yield successful results.

Marketing mix is a combination of a number of marketing-decision variables or factors (or inputs) employed by an organisation to achieve its marketing objectives in a marketplace.

A number of classifications of the marketing decision variables have been suggested by thinkers but the most popular classification has been given by E.J. McCarthy who has categorised these variables into four factors known as 4 Ps of marketing, namely product, price, promotion and the place (distribution), so much so that *marketing mix is referred to as the combination of four basic elements or inputs, namely product, price, promotion and the place.* They form the core of a firm's marketing efforts. These inputs or factors are inter-related and the decision to change one is bound to affect other in one way or the other thus effecting a change in the marketing mix.

The Four Ps of The Marketing Mix: The marketing variables or mix ingredients categorised under each 'P' are given here.

1. Product: Features, quality, design, brand, trademark, size packaging, guarantees/ warrantees, services (presale and after-sales) R&D, new products.

2. Price: List price, discounts, allowances, credit terms, payment period, flexibility.

3. Promotion: Advertising, publicity, personal selling, sales force (selection, training, motivation), sales promotion.

4. Place: Distribution channels, intermediaries and middlemen, locations, storage, warehousing, inventory levels, transportation.

As we have seen, marketing management is the process of planning, organising, executing and controlling of the ingredients of the marketing mix. These ingredients are critical inputs in the marketing process. Management of these ingredients of the marketing mix, thus, forms the core of our study in this book. We shall discuss these elements in details in later chapters.

6.2 MARKETING PLANNING

Broadly speaking, planning is a process in which a firm tries to make best possible use of its resources towards achieving its goals. Specifically, the process of planning involves analysing current situations, anticipating and predicting future conditions, establishing planning premises, determining enterprise objectives, formulating and developing plan of action or strategy to achieve the objectives, budgeting i.e. allocating resources correctly. Seen in broader context, marketing planning is the process of making a coordinated set of decisions about a marketing strategy for one or more target markets.

Marketing planning is a process that consists of analysing current situation and information about marketing opportunities, forecasting and establishing planning premises, selecting target markets, determining marketing objectives, designing and developing marketing strategy or course of action to achieve these objectives and allocating resources to the ingredients of marketing effort i.e. marketing mix and developing procedures and policies.

The above definition is a comprehensive one and enlists the steps involved in the process of marketing planning. A marketing plan is a central instrument of the marketing department for directing and coordinating marketing efforts. The major steps in developing the marketing plan are examined here in brief.

1. Assessing and Analysing Current Situation and Marketing Opportunities

This step includes examination and analysis i.e. scanning of environmental factors and conditions, such as growth, government policies, competitors' policies, company capabilities in terms of resources available and likely to be available, firm's position in the industry, structure of market and nature of competition, customer sensing i.e. customer reactions to firm's business or product, firm's own strengths and weaknesses and possible marketing opportunities that are available to the firm. It requires gathering and analysing a vast amount of information and data especially pertaining to all aspects of product-market i.e. product situation, size and growth of the market, competitors' position, distribution situation etc.

One of the main purposes of scanning environmental conditions is to identify marketing opportunities and evaluating company's strengths and weaknesses.

A firm's marketing opportunity is an attractive arena or product-market in which firm would enjoy an advantage over its competitors, provided that it takes appropriate marketing action in that direction.

Analysis of marketing opportunities involves defining, describing and estimating the size and nature of each product-market of interest to the business unit as well as the nature and extent of competition in that product-market. It also implies identification of strengths and weaknesses of each major competitor. Knowledge of markets and competitors is essential in designing and developing an effective plan of action or marketing strategy.

2.   Forecasting and Establishing Planning Premises

Forecasting means estimating and predicting future conditions and events. The second step involves predicting the trend, event or situation that is most likely to occur and should be based on the accurate and reliable data and information on the market, product, competition, distribution etc. Forecasting is the assessment of future conditions such as future demand for products, sales, costs, organisational requirements. An appraisal of future prospects is an essential element of planning and provides planning premises on which a manager founds his planning. Planning premises are basic planning assumptions about production's sales, profits, costs, firm's position and reputation in industry.

The job of planning becomes easy if the planning premises are established clearly and accurately. Some thinkers regard forecasting and establishing planning premises as the first activity or step in planning.

3.   Selecting the Target Market

Deciding and selecting what customers to serve is very important in chalking out a marketing plan. Each customer group offers a distinct marketing opportunity. The marketing manager has to identify and select that group or market segment which offers best sales potential and profit potential vis-à-vis resources available with the firm. It is, in fact, a question of matching the product with the needs of the customers. In other words, a marketing planner has to identify that set of customers who have interest, income and access to a particular market offer. It is selecting that segment of qualified available market that a company decides to pursue. The target group may be upper middle, middle, lower middle or ordinary class. It is only after this decision that a manager can prepare an effective marketing plan and set the marketing goals. We shall study about market targeting in a later chapter.

4.   Determining Marketing Objectives

Developing objectives is the foremost responsibility of a manager. The objectives are the goals i.e. the end results to be achieved. They are the goals towards which all activities are directed. It is the objectives or goals that guide the efforts of the enterprise and its components. *Only after identifying and determining the objectives to be achieved can a manager devise the plans, methods and strategies to attain them.* Therefore they need to be expressed in clear, specific and unambigious terms. Marketing planning involves setting clear well-defined and specific objectives for the business unit as a whole and for all product-markets and all levels of management. They need to be defined down to the smallest organisational unit. In addition to the broad objectives for the whole organisation, goals with specific details should be determined for departments and divisions of the organisation to guide the subordinate units. Every body in the firm must know the end result he or she is expected to achieve. Objectives need to be expressed in money terms as well as in units or weights of the product to be marketed. For illustration, take the following case of a large company.

CASE 6.1:
MARKETING OBJECTIVE (EXAMPLE)

An Indian Corporate company manufacturing vanaspati, hydrogenated oils, soaps, synthetic detergents and personal products like toothpastes, skin cream, hair oils has the following financial and marketing objectives for the next year.

Total Capital Employed   **= Rs. 600 crore**

(Shareholders' funds, share premium, suspense account and loan funds)

Total sales revenue achieved during last year   **= Rs. 3600 crore**

Financial Objectives

- To earn a net rate of return on investment of 28 percent after taxes.
- To generate profits before taxes @ 40 percent of capital employed i.e. gross profits to the tune of **Rs. 240 crore.**
- To generate profits after taxes @ 28 percent i.e. net profits of **Rs. 168 crore,** assuming 30 percent of gross profits as corporate tax, wealth tax etc.

Marketing Objectives

The above financial objectives need to be converted into marketing objectives. These goals are estimated in terms of volume of sales and gross sales revenue. In above case the market objectives are as follows:
- Increasing gross turnover by 10 percent i.e. achieving sales revenue of **Rs. 3960 crore.** Details are given hereunder.
- Achieving sales target of 72000 tonnes of vanaspati hydrogenated oils

at average realised price of Rs. 39000 per ton achieving sales revenue of **Rs. 280.8 crore**

* Achieving sales target of 3.5 lakh tonnes of soaps at the average realised price of *Rs. 20000* per ton i.e.
* Achieving sales revenue of **Rs. 700 crore**
* Achieving sales target of 4 lakh tonnes of synthetic detergents at the average realised price of *Rs. 60000* per ton i.e. achieving total sales revenue of **Rs. 2400 crore.**
* Achieving sales target of 60 crore units of personal products at the average realised price of Rs. 10 per unit i.e.
* Achieving total sales revenue of **Rs. 600 crore.**

Thus the total market goal is to achieve total sales revenue to the tune of **Rs. 3980.8 crore**.

The above illustration refers to the identification and determination of marketing objectives of a large company with an annual turnover of about **Rs. 3600 crore** and that also in the area of consumer goods of daily use. Smaller units and firms can set the objectives in the same way.

Again the marketing department of the unit can split the overall marketing objectives into specific goals according to market structure i.e. marketwise or areaswise depending upon the potential of the target market. It must be noted that there is a hierarchy of objectives and they are generally plural. An enterprise can not grow or prosper in an orderly or progressive manner unless clear, well-defined and specific goals are set to guide the activities.

5. Designing and Developing an Appropriate Marketing Strategy or Plan of Action

After having set the marketing objectives, the next step in planning is to identify and discover the most appropriate course of action that may achieve the objectives. There may be a number of alternate courses of action or strategies to achieve the marketing objectives. For example, a firm may increase its sales-revenue by increasing price, or by increasing market share or by dealer network or number of sales personnel. The planner has to select that strategy that may be most promising and ensure maximum customer satisfaction while achieving the set marketing goals.

We have already explained how to develop a strategy in chapter 5 in section 5.4. We have also referred to 4 Ps i.e. the ingredients that form a marketing mix so crucial for framing a marketing strategy.

A marketing strategy which is the basic plan of action to achieve marketing objectives, in fact, consists of basic decisions about the ingredients of marketing mix, namely product, price, promotion and place (distribution) or takes the form of programme positioning strategy.

Programme positioning strategy is an integrated mix of strategies aiming at combining the components of marketing mix (i.e. product, price, promotion and distribution strategies) to achieve the marketing objectives in each target market and involves formulation of a number of derivative and supportive plans.

6. Formulation of Derivative and Supportive Plans

After deciding the final course of action i.e. the basic plan, the next step in planning is to draw up detailed and specific plans to implement selected course of action to achieve the objectives. These plans are derived from the basic plan and support it to achieve the specific objectives. These plans may pertain to different components of the marketing mix. These plans are in the form of programmes, schedules, budgets, procedures and policies. Thus other activities of planning function include programming, scheduling, budgeting, developing procedures and policies.

(i) Programming means determining action steps necessary to achieve the objectives.

(ii) Scheduling involves determining time required to perform the action steps.

(iii) Budgeting means allocating resources to implement the action steps according to the schedule and hence to achieve the end result.

(iv) Developing procedures means specifying the way in which work is to be accomplished. It is suggesting the standard way of performing the work.

(v) Developing policies involves framing principles and rules for action i.e. standing rules to be followed during performance of work to ensure accomplishment of objectives.

7. Allocation of Resources and Budgets (Budgeting)

A budget is a plan specifying allocation of resources necessary to carry out the steps (or the performance) within the time limits and so to achieve the desired result. Budgeting implies estimating man power, money, materials, equipment, tools and facilities required to implement a programme effectively according to the schedule. In this sense, a budget is numerical statement of men, money, materials and equipment required to carry out a programme over a period of time. A fresh budget is needed when the period is over. In context of marketing planning, budgeting would require:

(i) allocation of funds and resources i.e. determining budget for marketing plan or programme as a whole.

(ii) allocation of funds and resources for determining budgets for various components of the marketing mix (product, price, promotion, distribution channels and personal selling) and

(iii) allocation of funds and resources for determining budgets for various constituents of each component of the marketing mix i.e. allocation of funds to various programmes within the main programme concerning each component of the marketing mix.

Market expenditures as reflected in the above budgets are generally expressed in percentages of the total sales revenue or percentages of turnover. Taking the above case, the marketing budgets can be as follows:

CASE 6.2: MARKETING BUDGET

Planned target of Total sales revenue = **Rs. 3960 crore**

Let us assume that the budget estimated for the entire marketing plan is 15 per cent of the turnover,

Then Marketing Budget = **Rs. 594 crore**

The marketing budget can be split as

Advertising and Sales Promotion (3% of turnover) =
Rs. 118.8 crore

Distribution and General (10% of turnover) = **Rs. 396.00 crore**

Marketing Research R&D (1% of turnover) = **Rs. 39.6 crore**

Brand building (1% of turnover) = **Rs. 39.6 crore**

Total = **Rs. 594.0 crore**

Note : These estimates can be further split on monthly or on quarterly basis.

In the above exercise, we have calculated the overall allocation of funds for the marketing programme and split budgets for various components of the marketing mix. These separate budgets can be further split into budgets for different sub-plans and programmes to be udertaken within each element of the marketing mix.

It would be very necessary for the successful implementation of the plan to secure willing cooperation and participation of the marketing personnel in its framing as well as in its execution. It is important that the details of the plan are communicated to them properly. Sound planning is a cooperative effort. Moreover, action plans need to be re-evaluated, reviewed and modified, if need be, from time to time. Planning is a continuous process and needs to be revised in light of changed situation so as to achieve desired results.

In the above pages, we have discussed how to develop a marketing plan for a business unit. Within the broad guidelines of the main marketing plan, marketing department needs to develop separate plans and programmes for various components of the marketing mix to attain the end results. As stated earlier, marketing management is, in fact, an integrated process of planning, organising, directing and controlling the action programmes concerning various components of the marketing mix *viz.* product, price, promotion and place (physical distribution). We shall study these aspects of marketing management in further details in later chapters.

EXERCISES

1. What is meant by marketing mix? Describe its basic elements.
2. Define the term "marketing planning". Describe the major steps involved in it.
3. Illustrate with an example how financial and marketing objectives are formulated in an organisation?
4. Explain the four P's of marketing mix.
5. Write notes on:

   (a) A firm's marketing opportunity

   (b) Programme positioning strategy

   (c) Budgeting for marketing planning
6. Discuss the factors which have a bearing on marketing mix decision.
7. You have been asked to develop specific objectives for the marketing function of your company. What information would you need as a market manager to develop such objectives.
8. Using your study in this chapter, draw a brief marketing plan for a new brand of Cola soft drink.

7

Marketing Information System and Marketing Research

In this era of globalisation, management of information and knowledge has assumed added singnificance. Information today is an inseparable entity for any business organisation trying to devise its strategy, policies and procedures to achieve its objectives. Information, in fact, is the primary asset for the company. Besides other resources in the form of men and materials, information is a fundamental resource with the quality of adding value to the activity and needs to be managed effectively. The success of a business activity would no doubt depend upon the quality and selectivity of information and the ability of the decision maker to evaluate its usefulness from vast amount of data. Proper decision making is not possible in the absence of adequate accurate information.

In chapter 3, we have defined the concept of a marketing information system and have explained briefly the need for building a sound marketing information system in the organisation to achieve the organisational objectives. In fact, building a sound marketing information system that may provide reliable and adequate relevant information to the decision makers at the time it is needed is a great challenge to modern marketing management. In this chapter, we shall study the concept of marketing information system and its components in further details and learn how such a system can be built.

7.1 MARKETING INFORMATION SYSTEM—THE CONCEPT AND COMPONENTS

Marketing information system (MIS) is defined as "a continuing and integrating structure of men, equipments and procedures to collect, analyse, evaluate, interpret and distribute pertinent, timely and accurate information about the marketing environment to those who make market decisions".

Thus a marketing information system is an integrated combination of information, information processing and analysis, equipment and tools (i.e. software and hardware) and information specialists who analyse and interpret the collected information and provide it to decision makers to serve their analysis, planning and control needs.

The information needed by the decision makers or marketing managers usually pertain to the marketing environment i.e. the target markets, distribution channels, competitors, industry, politics and microenvironment factors. It is usually obtained by marketing managers on regular basis from internal and external sources. Such information may include customer brand loyalty, effectiveness of advertising, studies about product's trials or information about some specific developments in the markets. Such information may include sales and cost analyses, usage rates, customer analyses, market share.

They may also need information about some special problems or situations which may involve undertaking studies or research. Marketing managers also need information about changes in marketing conditions, on changes in competitors' strategies and customer preferences. Such information comes under the head of marketing intelligence. The needed information may be obtained through marketing research or upon request or by purchase form suppliers of standardised research services or from those information suppliers who render customised services. A manager needs the required information in an analysed form so that he may be able to make better decision. This would require the help of a set of statistical tools and models with supporting hardware and software to investigate relationships between various marketing factors. *This support system is also a part of the integrated information system.*

It is evident from the above study that a marketing information system (MIS) consists of four subsystems, namely *internal information system, marketing intelligence system, marketing research and marketing decision support system.* A brief description of each subsystem of the organisation's MIS is given here.

1. Internal Information System

The internal information system or internal records system is the backbone of any marketing information system. This information may consist primarily of sales and cost reports and provide data on other aspects like orders, inventories, cash flows, receivables and payables. A large number of companies are developing highly sophisticated internal information systems for providing reliable, timely and adequate information to improve their strategic marketing planning, implementation and control.

2.   Marketing Intelligence System

Effective marketing management necessitates acquiring knowledge and information about what is happening in the product-market. A market manager needs to have information about developments in marketing conditions, changes in customers' requirements and preferences, change in competitors strategies, if any and emerging marketing opportunities. In other words, marketing managers must keep themselves informed about changing conditions in the macroenvironment and the task environment. The management needs to develop an intelligence system within its MIS. The management can obtain such information from its sales force, network of dealers or distributors, suppliers, news papers, trade journals, reports and advertisements issued by competitors. They can obtain such information from external sources such as syndicated research services and organisations and host of other sources. Again, the firms can set up a special cell or unit to collect marketing intelligence.

A marketing intelligence system consists of procedures and sources to obtain and use strategic information about relevant happenings and developments in the marketing conditions and macroenvironment.

The need for having a reliable and efficient system for gathering intelligence and providing it to the marketing manager for use at proper time cannot be overemphasised. Timely provision of intelligence is an important element in the formation of an effective marketing strategy.

3.   Marketing Research

Marketing research is the systematic investigation of the facts and specific problems relevant to various aspects of marketing. American marketing association has defined marketing research as follows:

Marketing research is the systemic gathering, recording and ana-lysing of data about problems relating to the marketing of goods and services.

In other words, it is a step by step process of acquiring, analysing and interpreting information relevant to marketing decision making body.

Marketing research is conducted by the marketing research departments of the organisations or by market research organisations and firms hired to do so. Marketing research involves investigation and is usually conducted to solve some specific marketing problems faced by the company and therefore it needs careful planning.

The Scope of Marketing Research

Marketing research involves specific study of problem and opportunity areas and may consist of market surveys, product preference tests, sales forecasting, advertising and promotion effectiveness studies, studies of business trends and marketing potential. The need for marketing research arises when the needed information is not obtainable from company's internal records, marketing intelligence, trade and business publications, journals or standardised sources. Market research or study will give information specifically needed by a marketing manager to make a marketing decision. Therefore market research or study may cover any aspect of the various marketing functions (namely the product, pricing, promotion and distribution), environmental situation or social responsibility area.

According to a survey of marketing research activities of 599 American Companies conducted in 1980's, it was found that most of the research activities were undertaken in areas of sales and marketing, advertising, products, business and pricing, corporate responsibility etc. They included studies about market potentials, market share and sales analyses, distribution channels, store audits, promotional measures, advertising effectiveness, competitive advertising, consumers preferences, consumption patterns, product acceptance, designing and packaging, business trends, pricing, plant and warehousing locations, consumer's rights to information and knowledge, ecological impact, legal restrictions; social values and policies. The scope of marketing research is so wide and comprehensive that many commercial research organisations have been **set** up to undertake specific research activities to serve the needs of their clients.

Again, a marketing research study may be exploratory in nature or it may be fullscale research study. The study can be undertaken by firm's marketing research division or it can be purchased from an outside research orgainsation, or services of external organisation can be hired to conduct the required research study.

The Marketing Research Process

The major steps in planning and conducting an inquiry or marketing research are:
  (i) Formulating the problem and defining research objectives
 (ii) Designing and developing a research plan
(iii) Collecting the data or information
 (iv) Analysing and interpreting the information
  (v) Reporting results or findings.
  (i) **Formulating the problem and defining research objectives:** This step consists of two activities— (a) defining the problem clearly and

correctly, and (b) determining the purpose for which research is being taken i.e. indicating the information needs or in other words identifying research objectives. The effectiveness and usefulness of a marketing research study depends mainly on a careful formulation of the problem and identification of the specific research objectives. In case of faulty formulation of the problem, the research findings may mislead rather than help the marketing manager in making decisions and may result in unnecessary wastage of money, time and efforts.

Only by focusing on the real problem and research objectives, we can understand the scope and nature of the research study and the type of the inquiry to be conducted. This step comes under the ambit of planning an inquiry or a research study and is essential for arriving at correct findings or results.

(ii) **Designing and developing a research plan:** The second step is to design or develop a plan of action to be implemented for collecting, analysing and interpreting data and information. It implies preparing and providing a blue print of the research work. The designed research plan indicates the sources of information, the methods of sampling and collection and the tools and instruments of research.

(iii) **Collection of data/information:** After determining the problem and the research objectives, and designing a research plan, the next all important task is to collect needed information and data. The data to be collected are generally of two types *viz.* primary data and secondary data but it must be ensured that collected information is relevant, accurate, reliable and adequate. It is to be seen that no errors or inconsistencies creep into the data. Defective schedules should be rejected or sent back for necessary correction.

(iv) **Analysing and interpreting the data:** The fourth step in conducting a research study is to analyse and interpret the collected data. Analysis of the information or data involves presenting them in summarised form (usually in the form of tables, graphs, averages, percentages etc), comparing the differences and examining relationships between the variables. In other words analysis of data will involve placing them in relation to other facts. This also involves use of advanced statistical techniques like measures of dispersion, correlation, regression analysis, sampling tests, and other models to arrive at the findings. Use of computers can be made in computing some important statistical parameters.

Interpretation of data is the task of drawing correct conclusions from the collected and analysed data. It must be noted that data are not an end. As referred to here, interpretation of data is the task of drawing valid conclusions and inferences in a field of study and usually involves formulation of predictions.

(v) **Reporting results or findings:**   The final step is to come out with results or findings in clear terms so as to help the marketing managers make better decisions. Main function of the marketing researcher is to draw proper conclusions and inferences and communicate them in intelligible terms to the marketing managers.

Marketing Research Suppliers

Generally, large organisations have their own marketing research departments or units. Smaller companies hire market researchers to undertake specific studies or get it conducted through marketing research firms. Several organisations, big or small, purchase research services from external sources. A number of marketing research firms provide syndicated and/or *customised research services* or specialised services in some aspects of the research such as data collection, analysis etc. Syndicated services are subscribed to by the organisations who need such services and in this way the cost of such research/studies is shared by them.

Marketing research is an expensive exercise and utmost care is needed in conducting the same. Moreover, the findings and conclusions drawn and supplied by research service firms should be thoroughly evaluated and scrutinised before using for making decisions. It has also to be seen that the marketing research service firm has the requisite capability, competence and experience in handling research projects. It has also to be kept in view that the research services are obtainable at reasonable cost. Studies show that large companies spend about 1 to 2 percent for their sales on marketing research.

4.   Marketing Decision Support System (MDSS)

Marketing information system is more than a data retrieving record keeping system. As the name suggests, MDSS is a system that is devised to help the manager with accurate analysed information for making quick effective marketing decisions. The analysis of retrieved data will involve use of models and statistical tools to investigate relationships between various marketing variables with the support of information processing capabilities i.e. computerised system, both hardware and software. These capabilities will include capabilities for data retrieval, classification and comparison, statistical analysis and computer modelling and simulation. We can define the marketing decision support system as follows:

A marketing decision support system (MDSS) is a set of statistical tools, and techniques models with hardware and software support that helps the manager in analysing the information properly and using it in making better and effective marketing decisions.

MDSS is a sophisticated integrated system which emphasises on analysis of data with the help of statistical techniques and methods, use of models, and use of computers and a set of relevant software programmes. The models may represent the relationships between various marketing variables such as sales-advertising relationship, price elasticity, media mix, marketing mix-budgeting etc.

EXERCISES

1. What do you mean by marketing information system? Enlist its main components.
2. Describe the marketing research process.
3. Write notes on:
   (a) Scope of marketing system
   (b) Marketing intelligence system
   (c) Marketing decision support system (MDSS)
4. Compare and contrast the use of standardised information services and special research studies for evaluating the performance of a new product, say a fast food item.
5. What could be some research tasks for the following?
   (a) Product decisions
   (b) Distribution decisions
   (c) Advertising decision

8

Market Segmentation and Market Targeting

8.1 INTRODUCTION

We have stated that a market consists of all the potential customers sharing a particular need or want who might be willing to engage in exchange to satisfy that need or want. Thus it is the presence of a group of people with a need or want, with the purchasing power and the willingness to buy a product or service to satisfy that need or want or it is a group of people with demand for a product that forms the market. On the other hand, offer or availability of product or services to satisfy the needs or wants of customers is an essential condition of engaging in the above exchange. In other words there exists a product-market relationship in this exchange.

Product-market denotes a combination of a product or service and groups of customers having demand for that product or service to satisfy their need.

As observed earlier, different groups of consumers or customers have different needs and wants depending upon their socio-economic conditions and a firm or organisation cannot satisfy the needs and wants of all types of customers in a market simultaneously. Nor all the customers would need the same type of a product. Again no product or service could have such universal appeal as all customers in a market may be equally attractive to buy the product. Besides, there can be other competitors in the market offering the same types of products or substitutes to satisfy a particular need of those customers. However, a firm or organisation can certainly meet the demands of a particular group of customers who want a product of a specific quality. In other words, it can at least satisfy a particular need of a target market, i.e. a group of potential customers of a product and it is towards this particular group that a firm can direct all its marketing efforts and devise its marketing strategy to attain its objectives. Determination of target market is thus an essential element in devising an effective marketing strategy. This would necessitate dividing all potential customers into such distinct groups that will respond similarly to a particular marketing mix or who might require separate products or qualities of a product and herein lies the need for market segmentation.

8.2 VARIOUS MARKETING APPROACHES

We have already explained some major marketing concepts and approaches in chapter 2. Broadly speaking, the organisations or sellers follow three approaches, namely mass marketing, product-variety or product differentiation marketing and target marketing. Each approach necessities a separate marketing strategy.

(1) Mass Marketing

In this approach, the producer or seller regards all potential customers or the entire market in a product-market as a single group. He produces a simple product on mass scale and goes in for mass promotion and mass distribution. Obviously he aims at supplying the product at a low price and attracting the attention of all customers to his product using the same marketing mix or strategy. The producer designs and develops his product in accordance with the needs or requirements of what we can call *a typical or average customer.* For example, most of the large undertakings in the country are following mass marketing approach for each of their brands of ordinary soaps and detergents and their marketing mix is same for all potential customers of a particular product or brand, irrespective of the size of the market.

> *Mass market strategy consists of attracting the entire market i.e. all the potential buyers in a product-market by using the same marketing mix for the product to appeal to all of them.*

No attempt is made to cater to any differences in tastes or likings among the potential customers. A typical or average customer is regarded as a representative of the entire market in a product-market. For example, Pepsi-Cola, Coca-Cola Maggi-noodles etc. are now goods of mass consumption.

(2) Product-variety or Product-differentiation Marketing

In this approach an attempt is made to concentrate on differences among potential customers in a product-market by introducing variety and differentiation in the product to appeal to different tastes and likings of the customers in a particular product-market segment. Herein, a producer or a seller produces and offers several products or a variety of products differing in features, size, colour, quality, brand name, style etc. to cater to different tastes of the buyers in the product-market. The appeal here is to the potential customers in the same product-market (market-segment) but there is variety in the offers to the buyers. Usually, the products offered are in the same price-range. This is also done to distinguish products of the seller from those of the competitors.

CASE 8.1: PRODUCT-VARIETY

Hindustan Unilever, Colgate-Palmolive, Procter and Gamble and many other large companies in various fields follow product-variety marketing approach. For example, Hindustan Unilever offers many brands of soaps in personal wash categories such as Lux, Lifebuoy, Liril, Rexona, Hamam, Jai to the customers. Colgate-Pamolive offers a number of brands in the tooth paste category, leading among them being Colgate Dental Cream, Colgate Gel, Cibaca Top, Cibaca Fluoride, Cibaca Gel. Bombay Oil Industries offer Saffola, Sweekar in different packings in the category of refined edible oils. The examples can be multiplied.

The seller will determine separate marketing mix and strategy for each variety or brand of the product in question.

(3) Target Marketing

In this approach, the emphasis is on designing and developing the products to fulfil the specific needs of one or more particular groups or segments of the customers. Target marketing involves division of the potential customers in a product-market into different groups and design and develop products to fulfil specific needs of one or more of these segments called target markets and employ corresponding marketing mixes and strategies to achieve organisational goals. It is toward this target market i.e. specific group of potential customers within a particular product-market that an organisation directs the marketing efforts.

EXAMPLES:

(i) This particular book is designed specifically to fulfil the needs of that particular group of scholars in the field of management who want to acquire proficiency and excellence in marketing management. The group of scholars referred to as above thus forms the target market for this book.

(ii) In order to meet the needs of upper strata, Hindustan Unilever has launched the beauty bar Dove with moisturising cream in the personal wash category.

Target market strategy consists of dividing all the potential customers into small groups, identifying one or more segments to enter i.e. selecting a target market that will respond similarly to a particular marketing mix i.e. to the combination of product, price, advertising and distribution network, and using specific marketing mix and strategy for that target market.

In short, target marketing involves three steps — *market segmentation, market targeting* and *product positioning.* In this chapter we are concerned with the study of the first two i.e. market segmentation and market targeting.

8.3 MARKET SEGMENTATION

Market segmentation is a process of dividing the entire heterogeneous market for a product into several submarkets or segments, each of which tends to be homogeneous wholly i.e. each of which has common or similar properties, characteristics or needs.

It is sub-dividing a market into homogeneous subgroups of customers, each subgroup being selected as a target market to be reached with a distinct marketing mix. The segmentation is meaningful only when each segment is served by a separate marketing mix. If customers belonging to two segments or separate groups respond similarly to a particular marketing mix, then we can not say that there has been any segmentation. Marketing segmentation is based on the assumption that different groups of customers in a product-market respond differently to various marketing mix offers and each segment or target market needs to be served by a separate distinct marketing mix. This brings us to cultivating still smaller segments with specific needs of their own or what we call market niches.

Market Niches

Some firms select for themselves a small distinct segment or group of customers to serve. This small segment is called a market niche.

A market niche is a small special market segment which a firm cultivates for itself with unique products or services and serves through a unique or special marketing mix.

The difference between a market segment and market niche is that whereas the former may represent a substantial part of the product market, a market niche represents only a small market having special needs and requirements to be served with distinctive products and services. Again, whereas a market segment may attract many competitors, a market niche is too small to attract many competitors.

For example, in the case of passenger-cars, the high income and affluent group of customers needing a super luxury car, may it cost Rs. 25 to 30 lakh, forms a market niche. Mercedez-Benz is cultivating at present this market niche with its specially manufactured car E 220 Sedan with its diesel version E 250 Sedan. It has rather a unique marketing mix to offer to this segment or market niche with little or no competition in the market.

Again in baby-food category, infants falling in 0-6 months age group may form a market niche. They would need special types of feeds in addition to mother's milk. There are few companies in the market that are serving this segment or market niche.

The process of segmentation would require surveying the market, analysing it and identifying segments on basis of certain criteria and profiling each segment. Obviously it will involve gathering comprehensive information and data on various facets of the market in question.

Each segment should be homogeneous internally but different externally.

8.4 BASES OF CONSUMER MARKET SEGMENTATION

In the examples given above, we have seen that a product-market can be divided into segments on the basis of income, social status, age, education or personal characteristics of the consumers. Market segmentation is essentially a classification task done on the basis of some variables i.e. differences in the preferences, characteristics and/or responses to marketing mix offer. There are many methods or ways to group customers in a product-market into different market segments. Some major bases for classification are explained here.

(i) **Geographic Segmentation:** It refers to the division of the product-market into market segments on the basis of region, state, density of population, size of the country or city, climate or season, or other geographical variables e.g. North, South, East and West regions, rural, urban, A class, B Class, C class Cities and so on.

(ii) **Demographic Segmentation:** It refers to formation of market segments on the basis of age, sex, family size, income, social level, occupation, education, religion, community, language, race, nationality or other demographic variables e.g. infants, children, adults, old, male and female, lower, middle, high income, professionals, white collar, blue collar, school, college undergraduates, postgraduates, Hindus, Muslims, Sikhs, Jats, Japanese, Indian. American, French. The classification may also be made on the basis of more than one demographic characteristics e.g. *low-income adults segment*

(iii) **Psychographic Segmentation:** In this segmentation, customers are divided on the basis of social class and status, life style, value system (interests, attitudes, beliefs) and personality characteristics e.g. lower, middle, upper strata, simple, fastidious, playboys, conservatives, liberals, preferences, fashionable, health conscious, status seekers, self-confident.

(iv) **Behavioural Segmentation:** In behavioural segmentation, consumers are divided into segments on the basis of behavioural variables, namely their attitude, buying behaviour and response to a product. These also include occasions of need, quality and benefits expected of a product, brand loyalty, usage rate, buyer-readiness stage etc. Segmentation on the basis of degree of brand loyalty is also called *marketing factor segmentation.*

(v) **Volume Segmentation:** It refers to division of product-market into various segments on the basis of the quantity of purchase e.g. bulk buyers, small scale buyers, regular buyers. They would need different marketing mix offers and have to be treated differently.

However the process of segmentation will not be complete unless a detailed profile is prepared for each segment.

8.5 ESSENTIAL REQUIREMENTS FOR MEANINGFUL AND EFFECTIVE SEGMENTATION

The main purpose of market segmentation is that the firm may direct its marketing efforts towards specific segments effectively. Segmentation will serve no purpose if this object is not fulfilled. Segmentation can be meaningful if

(i) there exist customer differences in preferences for seller's brand,

(ii) if identified differences would match with specified segment,

(iii) if the selected segment has enough demand potential so that effective marketing programmes could be framed,

(iv) selected segment is stable enough to allow time for designing and executing marketing strategy.

Market segmentation is effective if it meets the following requirements

(1) The segments are relevant to the marketing requirements of the firm

(2) The size, purchasing power, sales potential and profit potential of selected segments can be measured

(3) The segments are substantial i.e. they are large enough as to respond favourably to the firms' marketing efforts and prove profitable.

(4) The segments are accessible i.e. they can be reached and served by firm's marketing mix offer.

Thus relevance, measurability, sizeability, profitability and accessibility are the main characteristics of an effective segmentation.

8.6 BENEFITS OF SEGMENTATION

Product-market is segmented to serve the consumers better and design and develop appropriate marketing strategies to meet the needs of the selected segment. Marketing segmentation is, therefore, beneficial both to the marketer as well as the customers. It has the following benefits:

1. It helps to distinguish one customer group from another in the product-market and guides the seller to select that segment as target market that has a substantial potential.

2. It enables the marketer to know the specific needs of the customers in target market and devise attractive marketing mix offers to meet those needs.

3. It results in concentration of marketing efforts that are directed to the target group and helps in better utilisation of marketing resources.

4. It helps him in devising separate marketing strategies for separate segments.

5. It helps in avoiding wastage of marketing efforts and resources in unprofitable segments.

6. It helps the firm in assessing the strengths and weaknesses of the competitors and framing its marketing policies and programmes accordingly.

7. It helps the firm in the acquisition of a greater competitive capability and distinctive excellence.

8.7   BASES FOR SEGMENTING INDUSTRIAL OR PRODUCER MARKET

As stated earlier, industrial market consists of all the individuals and organisations that acquire goods and services for use in producing other products and services for sale or supply to others. Broadly, the marketing segmentation of industrial market is also done on the basis of many variables used for segmenting consumer market such as geographic, demographic, behavioural factors. They are summarised as:

1. **Demographic variables:**   Nature, size and location
2. **Operating variables:**   Technology, user status (quantity of purchase), customer capabilities
3. **Purchasing approaches:**   Degree of centralisation, power structure, nature of existing relationships, purchasing policies of customers, purchasing criteria of the customer - quality, service or price.
4. **Situational factors:**   Nature of urgency, specific application, size of the order.
5. **Personal characteristics:**   Similarity in value system, attitude toward risk, nature of loyalty to their suppliers

Usually, an industrial market is segmented using more than one of the above criteria.

8.8   DEVELOPING CUSTOMER PROFILES

After segments are formed, the next logical step is to draw and develop a profile of each customer segment. Such a profile consists of detailed description of each customer segment in terms of various factors, demographic, geographic, psychographic, behavioural attitudes, loyalty etc. Each segment profile needs to be described in terms of size, age-group, different benefits desired or expected, usage rate or frequency and nature of use, income level, brand loyalty, general purchase policy, attitudes, value system.

8.9   EVALUATING CUSTOMER SEGMENTATION

After segmenting the product market and profiling each market segment, the next step is to assess and evaluate the demand potential and expected profit from each segment.

Estimation of segment potential involves forecasting or predicting demand of the segment for the product. It is influenced by the size and growth of the segment and firm's competitive position. In evaluating a segment, the marketer has to assess the size, growth and profit potential of each segment. He has also to analyse the positions of the competitors in each sector alongwith his company resources. While evaluating segments, it has also to be seen how far the segment to be selected would contribute to the accomplishment of the company objectives and goals.

In evaluating each segment, the company should also take into consideration various threats a segment might confront from within or outside of the segment. These threats include segment rivalry, new entry of competitors, introduction of new products, increasing purchasing power of customers or growing bargaining power of suppliers that may result in increase in cost, or reduction in quality or demand for the product.

While assessing the potential of a segment, the company must also gauge the threat the company may face from its competitors. For example Maruti had retained its dominate position in this segment of medium car till 1995. With the entry of Hindustan Motors, Telco, Pal Peugeo with their versions of passenger cars for the medium sector, the competition has become tough. So each segment needs to be evaluated in context of the above threats as well.

In evaluating a segment, managers have also to gauge how far their marketing programmes will create attitudes and brand loyalty for its product consistent with the company's objectives and image.

It is after evaluation of all the segments formed as a result of segmentation that the management can be in a position to select the most favourable or attractive or lucrative segment/segments as its target market/markets and devise its marketing mix offer for the selected segment.

8.10 MARKET TARGETING

Market targeting or target market selection is the practice of selecting one or more segments worth serving and entering into.

After all the market segments in the product market are evaluated, the most crucial step on the part of the company is to decide about the segment or segments that offer the best marketing opportunities for the company's product and to devise specific marketing mix or mixes accordingly. Thus market targeting is prelude to the development of its marketing strategy to serve the target market or selected segments. Market targeting is selecting segment or segments in which a firm could achieve a strong market position.

In nutshell, market segmentation strategy is dividing the product-market i.e. all potential customers of a product into segments that will respond similarly to a particular mix and selecting segment or segments attractive to the product and offering best marketing opportunity for the company's product and then devising a specific marketing mix for the target markets.

As stated above, the market targeting may cover a single segment or two or more segments. In single segment coverage, the company uses a simple specific marketing mix for the target market and directs all its marketing efforts to this single target market. There is single-segment concentration of marketing efforts.

Multiple-segmentation strategy or multi segment coverage strategy is the targeting of two or more market segments in the product-market and devising different marketing mix offers for each segment.

8.11 SOME ALTERNATIVE MARKETING STRATEGIES

(i) **Market Specialisation:** It is the strategy wherein a company satisfies a large number of needs of a particular market segment e.g. college libraries need all sorts of books i.e. books on different subjects for different courses. There are book distributing houses that supply all types of books and have market specialisation. Again there are suppliers of all types of equipment, tools, instruments and wares for laboratories, hospitals, sports organisations etc. who have acquired market specialisation.

(ii) **Full Market Coverage:** It implies that a firm tries to satisfy almost all the needs of all customer groups in product market. A firm needs extensive resources to cover the entire product market and so this strategy can be adopted by large organisations only.

Hindustan Unilever (soaps and personal care products), Pepsi Cola (Drink market), Telco (vehicles). Philips (electronics) are a few illustrations. These organisations cover the entire market adopting two strategies *viz.* undifferentiated marketing and differentiated marketing.

(a) **Undifferentiated Marketing:** This marketing strategy serves the entire product-market i.e. all groups of customers or consumers with a single marketing mix offer. Undifferentiated marketing does not give any weightage to the differences among various groups of customers and assumes that the firm's marketing offer would appeal to almost all groups of customers. These firms undertake extensive advertising and build a large distribution network.

(b) **Differentiated Marketing:** In this marketing strategy, a firm tries to serve different segments in the product-market with a different marketing mix offer or programme for each segment.

Maruti Udyog Limited, General Motors, Pal Peugeo (all manufacturing cars), Videocon, Philips, BPL, Panasonic (all electronics - TVs, Music Systems etc.) design and develop different marketing mix offers and programmes for different segments in their respective product markets. Colleges, universities and professional institutes offer different marketing mixes to different groups of students. This strategy is costly as it involves incurring a number of costs like production costs, promotion costs, administrative costs, inventory costs etc.

(iii) **Customised Marketing:** In customary marketing, goods were made to order. Every individual would get the goods like shoes, shirts, suits, prepared or manufactured in accordance with his needs. But now goods are produced on mass scale. We have already explained mass marketing approach in an earlier section. In order to meet needs of typical or average customers, the producers produce goods of standard sizes to meet individual needs of customers. Many large companies are designing and developing their products and services on large scale to cater for each individual's need. Bata shoes, Liberty shoes, VIP hosiery products, readymade garments are a few examples to illustrate.

The ultimate form of target marketing strategy is called *mass customisation* in which goods are produced on mass basis but are tailored to meet the individual needs of each customer.

Mass customisation is the ability to prepare on a mass basis individually designed products to meet each customer's requirements. This strategy helps the customer to get the product prepared exactly in accordance with his want or need. Here are a few illustrations:

EXAMPLES:

Big hotels and restaurants prepare base materials on a mass scale and serve vegetarian and non-vegetarian dishes per needs and tastes of each individual customer.

Asian Paints Ltd., the paint manufacturers, have come out with Insta Colour paints (base material paste produced on a mass scale) and at their centres, a customer can get prepared a paint paste in any one out of 625 colour shades per his individual liking within 10 minutes with the help of a computer terminal.

Commercial marketing research suppliers provide customised research service e.g. Marg and other organisations of the type conduct all types of surveys including poll surveys for various customer-organisations including political parties.

EXERCISES

1. Explain various marketing approaches.

2. What is meant by market segmentation? Explain the bases of consumer market segmentation.

3. What is the importance of market segmentation? What are the essential requirements for segmentation to be meaningful and effective?

4. What do you mean by market targeting? What is multiple segmentation strategy?

5. Write notes on

   (a) Market niche

   (b) Bases for segmenting industrial market

   (c) Customer profile

   (d) Differentiated marketing

   (e) Customised marketing

   (f) Mass customisation

   (g) Product market

   (h) Target marketing

6. "Why do companies change the market target strategies over a time?" Explain.

7. Explain the conditions under which segmentation may not be feasible.

9

Analysing Markets and Buyer Behaviour

9.1 BUYER BEHAVIOUR—THE CONCEPT

Goods and services produced by producers are bought by individuals and organisations for consumption or use in markets. In chapter 1, we have referred to various types of markets, namely consumer market, industrial market, reseller market, government market etc. In order to devise effective marketing plans or strategies, marketers need to study and understand the characteristics of each of these markets (buyers, target segments) and how they make purchasing decisions or in other words how they generally behave and react towards a marketing offer. This study and understanding would also help in knowing the needs and wants of a particular type of market and market opportunities available. Thus effectiveness or soundness of marketing depends on proper understanding of buying behaviour of the customers or target market. We can broadly define "buyer behaviour" as follows:

Buyer behaviour refers to the people's or organisations' conduct, activities and actions together with the impact of various influences on them towards making decisions on purchase of products and services in a market.

This definition suggests that:

(1) Buyer behaviour concerns study of activities and actions of people and organisations that purchase goods and services for use. It involves *gathering and analysing information* on these activities.

(2) There are a number of factors that influence the buyer behaviour or in other words buyer behaviour is influenced by the *buyer's environment*.

(3) Buyer behaviour is essentially and typically a *decision making process*.

Such an information on buyers i.e. concerning environment and how they make decisions will help in predicting their behaviour and in designing a sound marketing strategy.

It is obvious that different groups of customers or markets have different needs, wants, or requirements depending upon the socio-economic conditions or their respective environments and so behave differently. The buying behaviour of the consumers is different from the industrial buying behaviour. We shall, therefore, study and understand them separately in this chapter.

9.2   CONSUMER MARKET AND BUYER BEHAVIOUR

The consumer market comprises all individuals and households who buy and use products and services in order to satisfy their personal needs and wants. For different types of consumers, there can be different consumer markets or submarkets depending upon their age, income, educational levels, ethnic characteristics and tastes. Though there is no cut and dried formula for success in marketing but a lot depends upon correct reading of the target market and consumers constituting that segment. Understanding consumer behaviour, thus, lies at the heart of marketing. In the light of this study, we can define consumer behaviour as follows:

> *Specifically, consumer behaviour is the conduct or behaviour that consumers exhibit before they purchase a product, during the purchase period, while using that product, and even after the product has been used (evaluating product performance).*

Obviously, study and understanding of consumer behaviour would involve reading of the psyche of the consumer — a complete insight into the subtle influences that mark and determine the behaviour of the consumer.

In the previous section we have referred to *three component model* of studying and understanding buyer behaviour, *viz.* buyer's environment, buyer's use of information and purchase decision process. We shall use the same model for studying the consumer behaviour. We examine here each component.

A. Factors Affecting Consumer Behaviour

There are a number of factors that influence the consumer behaviour, make him buy what he buys, where he buys and when he buys. These are–cultural factors, social factors, personal factors, psychological factors.

(i) Cultural Factors

Broadly speaking, culture refers to the traditions, laws, rules, conventions, beliefs, morals, customs and other capabilities and habits acquired by a human being as a member of society. The basic elements of culture are values, norms, symbols (e.g. language), folklore, religions, ideas.

(a) **Subcultures:** Specifically, *culture is referred to as "shared ideas, beliefs, values and ways of behaving among large groups of people living in a defined geographic area."*

Within a culture, there may be several subcultures with their specific beliefs, values, norms, folklore, traditions and conventions being shared by their members. For example, in a multi-religious multi-racial society, there may be religious groups, racial groups, nationality groups. Again, the traditions, conventions and cultural values and customs may vary from region to region within the same country. All these cultural factors and values influence the behaviour of consumers and a marketer must plan his marketing mix in conformation to the culture and subculture of the region. For example, red colour is considered auspicious and therefore brides in India usually wear red-coloured costumes. However, India is a big country with one culture but several diverse subcultures. Take the case of a Bengal- bride, she would be happy to wear an off-white silk saree with red border whereas a bride in Kerala would wear an off-white saree with a golden border. A marketer must know all these things if he has to achieve success.

(b) **Social Stratifications / Social Classes:** Within a culture, there exist social classes i.e. groups with different buying habits. Now-a-days, social stratification is being done on the basis of level of education, occupation, income, wealth. Even a few decades back, the basis of social stratification in the country was caste.

Social classes refer to large groups of people relatively homogeneous in a culture or society who share similar values, interests, life styles and behaviours. These classes enjoy different status and esteem in the society. They generally form homogeneous groups and have distinct life styles and product preferences. These social divisions have a powerful influence on people's behaviour. In India, people belong to upper class, middle class, lower middle class, lower class, working class etc. Again there are backward classes. Even among them there is a creamy layer. Marketers need to take into consideration the differences in the buying behaviour of various classes. After all, the product preferences, leisure-needs, fashions, clothing of various social classes cannot be alike and marketers have to devise their marketing mix accordingly.

(ii) Social Factors

Social actors or factors such as reference groups, family, person's diverse social roles and statuses also have profound influence on the buying behaviour of the consumer.

(a) **Reference Groups:** *A reference group is a small social group to which a consumer belongs or aspires to belong and whose beliefs, values, norms, and behaviour are accepted by him* such as family, relatives, friends, colleagues etc. Obviously such groups have powerful influence on the attitude and buying behaviour of their members. For example, a lady may like to purchase a sari or a pair of sandals which her colleague in the office had bought.

Then there are *aspirational groups* to which a person aspires to belong. They also influence person's buying behaviour. For example, recent popularity of MTV styles among the Indian teenagers proves how an aspirational group influences the buying behaviour of consumers.

(b) **Family:** Family is defined as two or more people living in the same household and related by blood, marriage, or adoption. This is rather the most important reference group, where, not individual but a number of family members may be involved in making purchase decisions. Depending upon the type of products to be purchased, different members of family may assume dominate role. For example, while buying a motorbike, scooter, the husband would have a dominant role whereas traditionally the wife makes purchases for the kitchen. Some of the purchases are made jointly e.g., furniture for the house, carpet or buying gifts for children. A study about use of *Janam Ghuthi* and gripewater in rural areas found that "it is the rural women who made the decisions and it's men who buy".

Again, in case of working women, the wife's employment has great influence on household's consumption patterns. As expected, wife's employment influences family expenditures on time saving devices like washing machines, dishwashers, microwave ovens. Family life cycle also has great influence on the buying behaviour. For example, a newly married couple is more likely to go in for more frequent purchases rather than elderly people.

Conflict and disagreements in family over purchase: When a purchase is to be an outcome of a joint decision, a conflict or disagreement may arise in respect of purchase priorities, attractiveness of purchase alternatives, timing of the purchase etc. Family members may differ on what product to buy with the available money. The resolution of the disagreements would involve influencing each other or entering into some sort of accommodation. This would have a bearing on who makes the final decision.

(c) **Roles and Statuses:** The role one plays as a member of a particular social group or the status or position one holds in a group also influences the buying behaviour. Most often, the same individual has to perform different roles in different positions. For example, a person who is husband in the family may be working as a manager in his office and as a secretary of a club. He will be exhibiting different types of buying behaviour and making different types of purchase decisions while working in respective positions as a husband, manager or secretary. A consumer makes purchase decision that he thinks reflects his role and status.

(iii) Personal Factors

Personal factors such as age, occupation, economic circumstances, life styles or living patterns and personality affect the buying behaviour of a person. The type of clothes one wears, the kind of music one listens, the kind of books one reads— all reflect an individual's personal considerations or factors. For example, a sixty year old lady in our country, generally, does not wear short dresses and the colour of her dress will not be jazzy. A marketer has to keep things in view while designing a marketing plan.

Again, a person's self-concept also influences his buying behaviour. A person's self-concept or what does he think about himself i.e. his own perception about himself is a very powerful factor influencing his behaviour. An ideal self-concept is what a person thinks he should be like. For example, the self concept of a person may be that he is very sober. Obviously, he would buy simple and sober clothes and would never like to wear outlandish dress. This means that consumers tend to buy what they think would suit them.

(iv) Psychological Factors

Among the psychological factors, motivation, perception, learning, beliefs and attitudes are the most influential factors.

(a) **Motivation:** The word "motivation" has been derived from the word "motives". Motive is that which moves the will to act in a particular way. These motives may be one's needs, wishes, purposes, objectives, and capital goals. They act as the driving forces which impel a person to act towards fulfilling them.

We have already given *Maslow's hierachy of needs* in chapter 1. These needs are — physiological needs, safety or security needs, social needs, esteem needs and self-actualisation needs. The basic premises of Maslow's theory are :

(i) The behaviour of any person is dominated and determined by the most basic groups of needs which are unfulfilled.

(ii) The individual will systematically satisfy his needs, starting with the most basic and moving up the hierarchy.

(iii) More basic needs groups are said to be proponent in the sense that they will take precedence over all those higher in the hierarchy. What would a hungry man buy first—food or an expensive shirt?

Herzberg's theory of motivation tells us about motivation factors (factors causing satisfaction) and maintenance factors and has implication for the marketers to try and analyse and identify satisfiers and dissatisfiers and devise marketing mix accordingly.

(b) Perception: Perception means the way a person or an individual discerns or interprets a situation, information or message and forms an opinion. Different persons may have different perceptions of the same information or message. There are three perceptional processes and an understanding of these can help in predicting consumer behaviour.

(i) *Selective attention or noticeability:*   It means that people notice only what interests them or is related to their needs. So marketers shall strive to transmit those messages that are captivating enough to catch the attention of prospective consumers.

(ii) *Selective distortion:*   It is a process in which the person manipulates or twists the information or message in a way he likes it, draws a personal meaning and interprets it according to his preconceived notions. It is very important that messages are well-meaning.

(iii) *Selective retention:*   Human mind has a limited and short memory and people remember only those messages which support their attitudes and beliefs. Astute marketers send those messages which reinforce the beliefs and attitudes of the target market and get favourable and positive responses.

(c) **Beliefs and attitudes:** Attitudes are preformed dispositions that are hard to change and belief is a well-set thinking existing in a mindset. Attitudes are influenced by personal experience and information gained from various personal and impersonal sources. A marketer has to take these attitudes and beliefs into consideration while devising a marketing strategy.

It is evident from the above study that a person is subjected to a complex set of factors which ultimately lead him to make a purchase.

B. Buyer's use of Information (Information Processing by Consumers)

In this context, it is also necessary to know how a buyer receives, interprets and uses information or message. For information and message to be effective, exposure, attention, comprehension, acceptance and retention are important steps to make a use of it.

The consumer needs to he exposed to the information repeatedly. It must catch his attention. A consumer can understand the information by *encoding* it i.e. by interpreting the meaning of information by matching it with previous knowledge lodged in his memory. Again the message needs to he accepted and retained before it yields positive results. The marketer must keep in view the consumer information process before devising a marketing plan.

Relevance, repetition, reinforcement and reasoning must form the core of information and message if the marketer needs to achieve good results.

C. Purchase Decision Process

Consumer buying decision process is a process in which a consumer decides whether a product or service will meet a need or want well enough to warrant purchasing and using it, when, where and how to make purchase, and determines satisfaction obtained from the purchase.

Researches have proposed a five stage decision process. However this process is not applicable in case of purchase of items of every day use and impulsive buying. These stages are explained here.

Stage 1. Need Recognition / Arousal

The first stage in the buying process is that of the awareness of the need i.e. when a person feels deprived of something, recognises need of it and wants to buy it. The marketer needs to identify the stimuli that would trigger the need arousal. He should develop such information or message that may stimulate consumer interest.

Stage 2.   Search for Information

After a consumer feels a need, he would need information which products or brands would satisfy that need. Need arousal is followed by search for ir.formation about the products. It costs both time and money to find and use reliable information. The time spent on the search for the information varies, from one product category to another. High involvement shopping (i.e. shopping of expensive and technical products) requires more search for the information.

The sources for such information about products, their benefits and uses may be — advertising, in-store displays, sales packages, sales persons or personal sources like opinions of friends, relatives, colleagues and others. It is in the interest of the marketers to identify the sources of information that could influence the buying behaviour of the consumer.

Stage 3.   Weighing of the Known Alternatives (Evaluating the Alternatives)

While gathering information about a particular product, a person may come across a number of brands available in the market. A customer would buy only that brand which he thinks offers a better value for his money. This process is complex as consumers evaluate different brands on many levels. It could he done on the basis of product attributes and benefits, price, distribution, utility function etc. A market is made up of many buyers and marketers need to identify major buying styles. For example, the question arises whether the target market is more sensitive to price or it is more quality conscious?

Stage 4.   Making the Final Decision (Purchase Decision)

It must be kept in mind that there are so many factors that go on to make an impact on the buyer. Making a decision is not an easy task. On the stake is the hard earned money of the customer. A person will always try to minimise the risk associated with the purchase of a product by making a choice of a brand about which he has received favourable information and about which he feels that it would provide him full satisfaction. Products of every day use do not require much thinking but high value products do involve a five stage process.

In final decision making, attitude of others and situational factors make their presence felt. Favourable attitude of the people close to the buyer has a major influence on decision making.

Stage 5. Postpurchase Behaviours (Use Out-comes)

By far, the most important of all these stages is the last stage of post-purchase behaviour. The main aim of marketing is to satisfy the need of the customer and in this context the last stage of buying process assumes greater significance. A marketer should be interested not only in selling the product but also show that the "company cares" and cares enough in providing absolute satisfaction and delight to the customer. For it, customer is the king and he must be satisfied.

How to achieve this objective? It is done by studying the post—purchase behaviour of the customer and accordingly providing him prompt after-sales services. This is of prime importance to the marketers because researches show that a dissatisfied customer tells about the products to about 11 people and a satisfied customers tells about the products to only 3 people. The marketer has to ensure that his products do not get bad publicity.

In the above study, we have explained various factors that influence consumer behaviour and the stages through which a consumer moves while making purchase decision. This under-standing of consumer behaviour will help the marketer to devise and develop an effective and efficient marketing strategy for the target market.

9.3 INDUSTRIAL MARKET (BUSINESS MARKET)

Industrial market comprises all individuals and organisations that purchase products and services for end use i.e. for use in the production of other products and services that are sold or supplied to others. Industrial market is also called business market or producer market. The industries include agriculture, forestry, fisheries, mining and quarrying, manufacturing, electricity, gas and water supply, construction, transportation, stores and communication, trade, hotels and restaurants, banking services, public administration and community services.

Products that are generally purchased by industrial buyers or in other words industrial goods are categorised as *installations* i.e. expensive equipment, capital items, buildings, *accessory equipment* i.e tools and equipment, *raw materials* i.e. basic commodities, *component parts and materials* (intermediary goods), *supplies* for maintenance, repair and operational products and services. An industrial market has following features that distinguishes it from consumer markets.

(1) An industrial market has fewer purchasers.

(2) It consists of larger buyers.

(3) Demand for industrial goods is a derived demand and depends upon the demand for its products or services.

(4) There is scope for reciprocal demand.

(5) The total demand is, by and large, insensitive to price.

(6) There is a close customer-supplier relationship

(7) Purchasing is usually done by professionally trained persons.

Industrial markets and buyers can be grouped in terms of sales volume, number of employees, geographic location, size of order, repeat orders and products produced and sold. Industrial markets can be classified by product (pharmaceutical companies, food companies, chemical companies) or by size (large, medium and small).

9.4 INDUSTRIAL BUYING DECISIONS

(a) Industrial Buying Behaviours: Definition

Industrial buying refers to the buying activities of organisations, including governmental and non-profit organisations that purchase products and services for end use i.e. to maintain and expand their operations. Manufacturers, wholesalers, retailers, governments and other organisations serve as markets for the industrial sellers. It is these organisations that have to make buying decisions.

Specifically, industrial buying behaviour refers to the conduct and actions of purchase department, purchase personnel or agents of an organisation towards making decisions on selection of products and of the supplie'rs, purchase of products and post purchase evaluation including communication.

Before analysing the industrial buyer's behaviour, we should know — who influences and makes the industrial buying decisions and what types of these decisions are?

(b) Who Participates in Buying Decision?

It is not only the employees in materials or purchase department of an organisation who participate in or make buying decision. A number of other persons are also involved in influencing and making purchase decision. The decision making units of an organisation is called "Buying Centre".

According to Webster and Wind, *"a buying centre comprises all those individuals and groups who participate in the purchasing decision process, who share some common goals and the risks arising from the decisions."* In other words, Buying centre comprises all persons, inside or outside an organisation, who are involved in making a purchase decision, including the persons who influence that decision in some way. In most companies the purchase job is entrusted to the employees who act as purchasing agents. However as seen above, the decision making often goes beyond the purchase department.

The different roles played by a buying centre can be described as below:

- **Initiators:** Those who initiate the buying proposal and determine the specifications of the product.
- **Influencers:** Those who influence the purchase decision e.g. technical persons.
- **Deciders:** Those who decide on the requirements of the product and/or on the suppliers or vendors the company will deal with.
- **Users:** Those who use the purchased products.
- **Buyers/Purchasers:** Those who negotiate, decide terms of purchase and complete the purchase of the products according to the specifications.
- **Gate Keepers:** Those who alanyse the company's needs and recommend likely matches with potential vendors. They have power of preventing the vendors or information reaching the members of buying centre.

There is another role, that of *approvers* i.e. the persons who authorise the proposed actions of deciders or purchasers. It is for the marketer to find out where the power lies in the buying centre so as to clinch the deal.

(c) Purchase Decisions by Industrial Buyers

The study regarding buying decisions will necessitate knowing industrial buying situations and types of buying decisions that need to be taken.

(i) Industrial Buying Situations: Resellers have identified three types of buying situations. These are also called buy-classes. They are — new task, modified rebuy and straight rebuy.

(1) **New Task:** The purchase decision concerns a product or service which is being bought for the first time and of which buying centre has practically no purchasing experience. It has to seek such information and make extensive efforts to evaluate alternatives. It poses a great challenge e.g. buying a delivery van for a manufacturer.

(2) **Modified Rebuy:** The situation concerns buying a product or service that would replace the one that is being presently used or the situation may involve modification or change of terms of purchase such as price, delivery etc. e.g.: replacement of a computer with a superior model. This would require gathering of information on several alternatives.

(3) Straight Rebuy: The purchase decision concerns a product or service that is already in use or has been purchased frequently and needs only a reorder from the same vendor by the buying centre as a matter of routine. The centre already possesses the required information and purchase experience.

(ii) Types of Industrial Buying Decisions: Industrial buyers or buying centres have to make several decisions. Three most important purchase decisions are:

(1) Authorisation to purchase the product or service

(2) Determining product specifications

(3) Selecting vendors/suppliers.

We shall explain them when we discuss the buying decision process in the next section.

9.5 ANALYSING INDUSTRIAL BUYER BEHAVIOUR

In section 9.2 of this chapter, we referred to a three component model of studying and understanding the buyer behaviour *viz.* buyer's environment, buyer's use of information and purchase decision process. We shall use the same model in studying and understanding the industrial buyer behaviour with a difference that we shall combine the last two components in one. The components of our study here are — factors influencing industrial or organisational buying behaviour and industrial purchase decision process.

A.   Factors Influencing Industrial Organisational Buying Behaviour

There are several influences on industrial buyers besides economic influences. Researchers have divided these influences into four categories — environmental, organisational, interpersonal and individual. These factors are briefly described below.

 (i) Environment Factors: There are several factors in the environment of the purchasing company that influence its purchasing. These factors include governmental regulations, changes in level of demand for buyers' products, economic conditions, technological changes and innovations, cost of funds, shortage of supplies and competitive challenges. The industrial marketers and suppliers need to study these influences and how they would affect their buyers' purchasing decisions. They will have to adjust their offers accordingly.

 (ii) Organisational Factors: Organisational purchasing decisions are also influenced by organisational goals and objectives, its established policies and procedures for purchasing, organisational structure and materials systems. These factors and characteristics show how a particular company is different from other companies. The objectives and goals throw light on what type of products are needed by the buyer and what would be the criteria for evaluating products. Organisational structure helps in understanding who wields the power or responsibility of making purchase decision. The supplier needs to take these organisational characteristics in view. Similarly established procedures of various companies for making purchase decisions also need careful consideration.

(iii) **Individual Characteristics:** It is, in fact, the individuals that make purchasing decisions and form what is called buying centre. Their characteristics such as personal goals, positions, traits and capabilities, education, experience, opinions and attitudes, values and life styles also affect the purchase decisions. A supplier needs to learn about and understand these characteristics to make his offer suiting the nature of decision makers. He needs to keep a record of the same.

(iv) **Interpersonal factors or buying centre interaction:** Buying centre or purchase decision making unit consists of many members, each having different authority, status, degree of empathy or persuasiveness. Each interacts with the other and there is possibility of conflict arising out of such interaction. These interpersonal relationships and factors also affect the purchase decision. Suppliers need to find out the personalities involved in the process and interpersonal relationships to understand the buyer behaviour.

Industrial marketers should take all these factors – environmental, organisational, interpersonal and individual, that affect purchase decisions, into consideration and accordingly adjust their marketing plans in such a way as may yield best results.

B. Industrial Purchase Decision Process

A typical industrial purchase decision process consists of following steps or phases:

(i) Recognising need or problem

(ii) Identifying or searching suppliers

(iii) Determining product specifications

(iv) Evaluating suppliers and selecting the most suitable ones

(v) Negotiating and placing a purchase order

(vi) Evaluating product supplier performance

Step 1. Recognising Need/Problem

The industrial purchase process starts with recognition of a need or problem that may arise out of any situation concerning the operation of business e.g. company may need new machinery to produce a new product or there may be a need to replace old machinery or there might have been the problem of break down or a company may be going in for a technological change. These needs and problems are to be met and solved by acquiring products and services. This would also involve determining the general characteristics and quantity of the product to be acquired.

Step 2. Identifying Suppliers

After recognising a need for a product the industrial purchaser has to decide whether to make the product or purchase it from a supplier. In case of decision to purchase from outside supplier, the purchaser has to find alternative sources of supply i.e. he has to search for most suitable and qualified suppliers who may be capable of supplying the requisite technical quantity of product of the required quality. Such information can be collected from directories, financial reports, brochures, product samples, salespersons, suppliers themselves. The purchasers have also to find out suppliers' facilities. Search for qualified suppliers with adequate capabilities is of crucial importance.

Step 3. Determining Product Specifications

Before asking for any additional information and soliciting proposals from identified supplier, the buying centre has to determine technical specifications of the product, its performance, characteristics, quantity required, delivery and installation schedule, price range. In this step, the purchaser sets the standards which should be fully met by the products' suppliers. The supplies not measuring upto these specifications are liable to be rejected.

Step 4. Evaluating Various Suppliers and Selecting the Most Suitable Ones

This step involves asking additional information about qualified product suppliers to evaluate them. Specific detailed proposals for supply of products per specifications will be invited from them. Before it, a purchaser may think to shortlist a few suppliers in the light of information gathered earlier. The proposals submitted by the vendors are reviewed. It would require a vendor-analysis i.e. analysing each vendor on the basis of its competence and capability to make supplies according to the specifications of the product determined above.

In the same way, the products to be supplied are also evaluated. The tool used to evaluate the products is called value-analysis.

Value analysis is a procedure used by industrial purchasers to evaluate the worth of a product to be purchased. It is based on assessing the functions of the product in a use-situation and the value of those functions.

The criteria to analyse a vendor orgnaisation can be — its technical and production capabilities, financial strength, pricing, product reliability, delivery reliability, service capability, image in market as a supplier etc. After receiving all alternative proposals, the purchaser will select the supplier or suppliers that are found to be most appropriate.

Step 5. Negotiation and Placing Purchase Order

Negotiation is the stage of bargaining. During bargaining, the purchasers and sellers enter into negotiation. The purpose of negotiation is to conclude a purchase agreement or develop a contract specifying the terms. The process involves presenting proposals and counterproposals and get the best terms and concessions from other party. Generally, two approaches are adopted by purchasers and sellers while negotiating a deal, namely competitive bargaining and coordinative approach. *Competitive bargaining* implies use of threats, persuasive arguments, promises and other means to win concessions. *Coordinative approach* strives to solve the problem and depends upon cooperation and mutual trust. Seller's approach also influences the buyer behaviour.

After the terms are settled as a result of negotiation between the two parties, the buyer writes and places the final order/orders with the selected supplier or suppliers specifying technical specifications of the product, quantity required, price, cash-credit terms, delivery schedule, return procedures, warranties and so on.

Governmental institutions approve rate contracts for various products which suppliers have to supply to government organisations throughout a specified period say a budget year. The contracts specify specifications of the products and the approved price at which they are to be supplied by the suppliers.

Step 6. Evaluating Product/Supplier Performance

Evaluation of product and supplier performance is carried out after the products are supplied. As demand for products by manufacturers or organisations is a derived demand, the best way to evaluate product performance will be to collect information about product performance from the end users or consumers who use products made out of these supplies. Again, the evaluation of supplier performance involves assessing how far they have complied with the terms of purchase agreement and the quality of service in sticking to the delivery schedule etc., rendered by them and how much the buying centre is satisfied with a supplier. In case of poor performance, a supplier may be dropped.

To conclude, an industrial marketer must carefully analyse the industrial buyer behaviour. He must learn the personal characteristics and traits of individuals in the buying centre and how they select product and suppliers. Only then they can design an effective marketing plan. They need to upgrade negotiating skills and marketing capabilities. The large organisations in reseller markets (wholesalers, retailers) go through a buying process similar to one described here for industrial and organisational purchasers.

EXERCISES

1. What is meant by consumer behaviour? Explain the factors which affect the consumer behaviour.
2. Explain different stages of consumer purchase decision process.
3. Define industrial buying behaviour. Describe the factors which influence industrial buying behaviour.
4. What is meant by a buying centre? What are the different roles which a buying centre plays.
5. Write notes on
   (a) Social stratification
   (b) Competitive bargaining
   (c) Value analysis
   (d) Buyer behaviour
6. Explain various steps followed in industrial purchase decision process.
7. Buying is a decision process with many steps. What is the significance of this view for planning and controlling marketing strategy? Discuss.
8. Using the concept of consumer behaviour, describe the probable similarities and differences among consumer purchasing of Car, VCR, tooth paste and ice cream.

9. Compare consumer and industrial buying behaviour.

10. You find that a number of members of a customer's buying centre disagree over the purchase decision. What information will you seek to obtain and how this information will help you in your marketing strategy aimed at this customer.

11. Apply various stages of decision process to your decision of admission to your present institute.

10

Analysing Competition

In chapter 3, while discussing microenvironment, we have studied that one of major tasks of a marketing manager is to identify the company's competitors, know their strengths and weaknesses and find out the nature of competition faced by the enterprise or it is likely to face. This involves collection of full and reliable information about the prevailing competitive challenges. Analysis of this information is necessary to develop an effective marketing plan and decide on the marketing mix. Examination of these aspects forms the subject of our study in this chapter.

10.1 COMPETITION AND ITS TYPES

Competition faced by a firm refers to the efforts and offers made by other firms or individuals to obtain market share in a market consisting of customers who have the same or similar needs or wants.

Competitors thus refer to those firms or individuals who are seeking to satisfy the same customers and customer needs and offer similar products or services, close substitutes or brands to them.

Any similar product or service or all such products or services produced by other firms or individuals that meet the same customer needs pose competition and should be considered a threat. Without considering the nature and extent of this threat, it is not possible to devise a meaningful and effective marketing strategy. There is no gainsaying the fact that competition is often a driving force behind the planning of company's marketing strategies. Careful and imaginative analysis of competition is instrumental in building a better understanding of the competitive threats and devising the marketing programme accordingly.

Companies generally face two types of competition — brand competition and product-type competition.

Brand Competition: Brand competition is that market condition (situation) in which various brands of the same product type strive for market share (or compete) in a market consisting of all potential customers sharing a particular need or want. For example, Lux, Palmolive, Cinthol, Ganga are the brands of same product-type soap that vie for market-share in the same market. Similarly tooth paste brands Colgate dental cream, Close-up and Forhan's manufactured by Colgate-Palmolive, Hindustan Unilever and Geoffrey Manners respectively compete with each other though they serve the same markets. Recently, a major battle was waged by Procter & Gamble and Hindustan Lever through brand competition. In the struggle for the larger market-share, while P&G used Vimal Seal (Textile) to advertise and promote its detergent New Ariel, Hindustan Lever followed suit by taking help of Bombay Dyeing to promote its detergent Surf Excel. But managers are concerned not only with brand competition but also with much broader product-type competition.

Product-Type Competition: Product-type competition is that market environment (situation/condition) in which brands of different product-types strive for market share in the same market i.e. in a market consisting of all potential customers sharing a particular need or want. In other words, it reflects a competition among different product-types or dissimilar brands that are capable of satisfying the same needs for wants.

The product-type competition may comprise competition among:

(i) different product forms like Ice Creams — vanilla, pistachio, choclate, pine apple etc.,

(ii) close substitutes like soaps and detergents; tea and coffee,

(iii) broad generic product types like Idli, Dosa, Potato Chips, Biryani, Hamburger to meet your hunger or soft drinks (aerated), syrups, squashes, butter milk (lassi); or cars, jeeps, motorcycles, scooters, bicycles etc.

Competition among Needs and Wants

A consumer or customer has limited resources in terms of money and cannot satisfy all his needs and wants. Customers have thus to decide on the priorities of needs and wants to be fulfilled first and need to postpone the fulfilment of needs or wants thought to be less important to some later period. In this situation, different needs and wants vie with each other for priority. The individuals or organisations will buy those products or services that satisfy those needs which secure their highest priority i.e. on which they set buying priorities. For example, a family might decide to do major home repairs than to buy a car or it may spend money on purchase of some other durables. *This is a broad form of competition in which companies compete for the same consumer resources.*

10.2　MARKETS AND COMPETITION

We have studied that a market consists of all potential customers, individuals and organisations that share similar needs and wants and respond similarly to marketing offer. We have seen above that there are a number of competitors who vie for market share in a market. Again, there is brand competition among brands but equally important is to understand the product type competition or competition from dissimilar brands or different product types. The nature and extent of competition will depend on how an organisation defines its market. If a market consists of customers wanting a particualr product to satisfy a particualr need, the competition will be limited to brand competition but if it is a larger generic-class market, then the competition will be broader (product type competition), as all companies manufacturing various products or dissimilar brands serving the same need will vie with each other to sell their products or services.

Therefore, identification of competition or competitors requires to define the market first. Market in this sense is the arena in which companies compete for sales and market-share. The nature and extent of competition will depend on how an organisation defines market and surveys the competition that could be confronted in that arena. This defined arena would tell the organisation which companies are or will be its competitors or in other words the companies that are affecting its marketing opportunities.

10.3　ANALYSING COMPETITION

We have briefly explained above the nature of competition that is generally faced by the firms and the product. The degree and severity of competition varies from market to market. In the largest context, companies and products vie against each other indirectly for the limited resources available with customers and in a narrow context there is a brand competition. However, a company is generally confronted with brand as well as product type competition. In order to formulate an effective competition marketing strategy, an overall study and analysis of competitive environment as it obtains is of critical importance.

Two approaches can be followed in analysing competition i.e. competitive environment. First, analysing competition from the entire industries i.e. analysing the whole industries that offer products or services to satisfy the needs and wants of a given market. This would provide a general understanding of the competition coming from similar brands or other product types. The second approach is to analyse key competitors whose offerings and decisions influence the firm's market opportunity. The procedure for analysing competition from the entire industries is almost the same as for analysing key competitors, though it is the findings of industry analysis that help management identify firm's key competitors. We shall briefly explain here the two approaches.

10.4 ANALYSING COMPETITION FROM INDUSTRIES (ANALYSING AN INDUSTRY)

Analysis of competition from an industry involves the following steps:

(1) Defining an industry

(2) Describing important marketing aspects of the industry, namely its size, structure, marketing strategies.

(3) Evaluating the industry i.e. its market coverage, its strengths and weaknesses.

(4) Predicting changes in its marketing strategies and analysis

(5) Results and findings of analysis.

(1) Defining an Industry

An industry is defined as a group of organisations offering competing brands of the same product type or products that are close substitutes of each other (product types) e.g. electronic industry, auto-industry, steel industry, pharmaceutical industry. For the task of analysis, an industry needs to be defined in terms of the product/products/product line manufactured or marketed, level of channel (whether manufacturers, distributors, retailers), territory or geopgraphical area of marketing presence, location of the industry, organisational structure (company divisions).

Let us take the case of **Audio-System Industry.** We can define it in terms of its characteristics in the following way

Product — Audio Systems

Level — Manufacturing/Assembling/Importers

Marketing Area — India

Location — India, Japan, UK

Product Line — Radios, Cassette players, CD players, Two-in-One, Three-in-One, Walkman, Music System etc.

(2) Describing Important Marketing Aspects of the Industry

After having defined the industry, the next step is to gather and analyse information about marketing aspects of the industry in question, namely the size and growth of the industry, structure i.e. nature and degree of competition among firms as result of their number and size and product differentiation, and the marketing strategies and tactics being adopted by them. A brief explanation of each aspect is given here.

(a) **Size and Growth of Industry:** The information to be gathered and analysed will pertain to industry sales, profits, number of firms, number of employees and their growth in recent years. Alongwith it the size of market also needs to be described.

(b) **Industry Structure:** Another important marketing aspect of the industry to be described is its structure.

The structure of an industry means the nature and degree of competition among the firms comprising the industry, and degree of homogeneity or defferentiation of their products. The information to be gathered is whether there are one, two, few or many sellers or in other words, what is the competitive structure of the industry. Industry can be categorised on the basis of absence or presence of competition such as monopoly, Oligopoly, duopoly, monopolistic competition or perfect competition. A brief explanation of these types of industry structure is given here.

Monopoly: Monopoly or absolute monopoly is a situation in which there is only one producer or seller of a product or service and there is no substitute of the product. Monopoly industry is one firm industry and in this situation the firm has a complete control over its supply and supply price. There is absence of competition. The aim of a monopolist is to earn maximum profit.

Oligopoly: Oligopoly is that type of industry structure in which there are a few sellers and the price policy of a firm is affected by price policies of other firms. Oligopoly is of two types–

(i) Perfect Oligopoly in which products are homogeneous and

(ii) Imperfect Oligopoly in which various firms differentiate their products. There is cut throat competition and firms incur heavy selling costs on advertising. There is difference in size of firms and there is lack of uniformity in price structure.

Duopoly: In such industry structure, there are only two producers of a homogeneous product. There can be cut-throat competition if no agreement exists between them. However to avoid competition, sometimes they divide the market on regional basis and agree not to compete with each other in each other's market area.

Monopolistic Competition: This is a type of industry in which there is a large number of producers and their products are so differentiated that the products produced by one firm cannot be regarded as close substitutes of the products produced by others. Product differentiation is done on the basis of texture, design, size, brand, trade mark, advertising and publicity. There is freedom of entry or exit of firms.

Perfect Competition: This is a situation in which—

(i) there is a large number of sellers and buyers so that each firm produces only a small proportion of the product and no single firm can affect the market' price,

(ii) the product can be sold at a fixed price,

(iii) the product is homogeneous,

(iv) the demand for product is perfectly elastic,

(v) there is absence of artificial restrictions or interference,

(vi) there is freedom of entry or exit of firms,

(vii) there is perfect mobility of products and factors,

(viii) the sellers have complete knowledge of market conditions,

(ix) there is absence of selling or advertising costs.

Pure Competition: It is a type in which the degree of competition is not as high as in the case of perfect competition. It is a condition a bit lower than perfect competition. In this type, first six conditions of perfect competition are fulfilled.

(c) **Marketing Strategies:**   Another important aspect of industry is the typical marketing strategies adopted by the industry. Information on marketing strategies pertain to target markets served by the industry, industry's marketing objectives and industry's marketing mix i.e. information about industry's products and brands, general pricing policies and behaviour, promotional approaches, advertising, research, innovations, distribution channels and services rendered. In case of expansion of marketing arena and international trade, the influences of government economic policy on the industry's marketing strategies should also be recorded and analysed.

(3)   Evaluating the Industry

The information collected above can be helpful in assessing the industry's market coverage, strengths and weaknesses. This information will tell how far and upto what extent the market demand is met by the industry as a whole and whether or not there are any untapped marketing opportunities. The lesson to be learnt is how can be the firm's offering superior to the competitors' products to meet the consumers' demand, keeping in view the requirements of target segment.

Again, *industry strengths* mean those elements of marketing offers or products that meet the customer's perceptions and satisfy the market requirements, whereas *industry weaknesses* mean those elements of marketing offerings that do not come to the customer perceptions and do not satisfy market requirements.

Thus knowledge and information about the capacity and efficiency of the industry to meet market requirements can be helpful in planning marketing strategies.

(4) Predicting Changes in Industry Marketing Strategies

In order to predict changes that are likely to be effected in future in industry marketing strategies, it will be necessary to analyse past trends and how the industry generally reacts to market conditions. Besides, it would necessitate monitoring marketing strategies of those who have substantial market share. This also involves anticipating rationally the possible industry reaction to varying marketing environment. The information gathered above can be helpful in knowing how the industry will react to changing marketing conditions.

(5) Results and Findings of an Industry Analysis

The industry analyses help marketing managers in knowing the complete competitive environment besides providing information on industry's performance, progress and growth, efficiency and capability to serve the market. They provide information on market requirements, the extent to which these are being met by the industry as a whole and untapped demand, if there is any.

As stated earlier, analysis of industry as a whole is not enough. In order to devise a truly competitve marketing plan or strategy, a careful analysis of the key competitors is also necessary.

10.5 ANALYSING KEY COMPETITORS

We have discussed so far how analysis of an industry is conducted. Alongwith the analysis of industry, a firm also needs to study and analyse the nature and extent of competition coming from some individual companies. It may not be necessary to analyse every individual competitor. However, those competitors whose marketing strategies are going to affect the firm's marketing strategy must be analysed. These rivals may be termed as *key competitors*. As referred to earlier in this chapter, the process of analysing the key competitors is almost the same as in case of the entire industry. We shall briefly explain these steps here.

(1) Identifying Key Competitors

The first step to analyse market competition is to find out the identity of key competitors. This would involve identifying the competitors that are addressing their offerings to the same market i.e. the companies that are serving the same target market or group of customers. For example, in case of tooth pastes, Colgate-Palmolive have key competitors in Hindustan Unilever (Close-up, Pepsodent) and Geoffrey Manners (Forhan's). Another criterion to identify key competitors is to name out the competitors that have achieved success in meeting the demand of the market. Success is usually measured in terms of size and growth in sales, market share, profits and public image.

(2) Describing Important Marketing Aspects and Capabilities of Key Competitors

After having identified the key competitors, the next step is to gather and analyse information about their capabilities and other important marketing information. This information will pertain to their financial position and growth, *i.e. financial profile*, technical and operating capability, marketing objectives and marketing strategies and allocation to marketing mix. Much of this information is available from their annual reports and profits and loss statements and from other market sources. Again, competitors' capabilities are assessed in terms of marketing objectives, expressed in terms of mission of business, profitability, market-share growth, cash flow, capacity utilisation, new investments, technological leadership, quality leadership, service leadership.

Vertical integration is another criterion to assess a competitor's capability. Hindustan Unilever Limited, which has achieved phenomenal growth in India, offers a classic example of vertical integration. We have already referred to this fact in case studies in chapters 4 and 5. Hindustan Unilever has almost financial control over Brooke Bond Lipton, Pond's India Limited and has started a joint-venture with Lakme and is in a very advantageous position.

(3) Evaluating Key Competitors

The information collected above can help in assessing the competitor's coverage i.e. competitor's ability to meet the market requirements, its strengths and weaknesses. The information regarding the extent of market requirements being met by the competitor will throw light on the strengths and weaknesses of the competitor and the quality of competitor's marketing programme. These profiles have a bearing on their capability, strengths or weaknesses as well.

(4) Predicting Changes in Competitors Marketing Strategies (Competitors' Response to Company's Marketing Strategy)

The next step is to anticipate changes in the marketing strategy of the key competitors. This would need study and analysis of past trends and how each of the key competitors generally reacts to the changing marketing conditions or competitive environment.

Different competitors respond to changing marketing conditions differently depending upon their business philosophy, attitude and mentality. These are four types of competitors.

(a) **The laid-back competitor:** Such a competitor is slow in reacting to competitor's change in strategy. His reaction is neither quick nor strong to competitor's moves.

(b) **The selective competitor:** Such a competitor responds to only certain types of threats while he does not react to other types of moves. For example, a competitor may react to price cuts and not promotional campaigns.

(c) **The tiger competitor:** Such a competitor is very sensitive and reacts quickly and strongly to any competitive attack on its arena. Such a competitor reacts sometimes even to matters which may look even trivial. It is always advisable not to attack a tiger. The scene becomes sometimes quite interesting when two tigers try to outstrip each other.

(d) **The stochastic competitor:** It is a competitor whose reaction is usually not predictable. He may or may not respond to a competitor's move.

The information about how the competitors would react to the company's moves will help it in planning and developing its marketing strategy.

It is evident from the above study that marketing managers need a lot of information and data about their key competitors which need to be analysed and interpreted correctly so as to be useful for purposes of devising a meaningful and effective marketing strategy. Thus decision makers need to design a competitve intelligence system. We have already described the marketing intelligence system as a component of the company's overall marketing information system (MIS) in chapter 7. This would help companies to formulate their competitive strategies and decide which competitors to attack and which competitors to avoid.

10.6 COMPETITVE CHALLENGES AND MARKETING PLANNING

In the above pages, we have explained the need for analysing competition and finding out the strengths and weaknesses of the key competitors so as to devise an effective marketing strategy. A successful and effective marketing strategy must meet two conditions - First, it must satisfy customer needs and wants i.e. it must fulfil customers' marketing requirements Secondly, it must strive to make its offering match or better than the competitors' offerings, and this superiority should be perceived by the customers as such. In other words, it must effectively position the company and its offer against the competitors in the target segment. Thus one important aspect of marketing planning is positioning company and its products in the target market. We can define positioning as follows:

> *Positioning refers to the use of marketing strategy to design the company's image and value offer in such a way as may give customers a reason to buy it from the company rather than a competitor.*

Positioning means making customers understand and appreciate what the company stands for in relation to its competitors. For example, a company may be known for its low-price position, high quality, high service or advance technology position.

The most effective way to tackle competition for a company is to devise a marketing strategy that may aim at matching the strengths of competitors and exploiting their weaknesses in the target market or target segment. This would require careful selection of the target market or target segment. Of course, the best way to tackle competition is to make your value-offers/products have differential and distinct advantages over competitors' offerings. In other words, company's marketing mix must be perceived by the customers to be offering superior value than those of the competitors.

The crux of the matter is that whereas a marketing strategy to be effective and successful should aim at matching the strengths of the competitors and exploiting their weaknesses or '*holes*', in positioning it in the target market, it should also cover well the target market i.e. it should also meet the customers market requirements. Thus the marketing strategy should be both competitor-centred as well as customer-centred. In other words marketing strategy should have a good balance of both competition and customer dimensions and thus be truly market-oriented.

EXERCISES

1. What is meant by the term 'competition'? Describe the types of competition faced by a firm.

2. Explain the steps involved in analysing competition from industries.

3. How can a marketing manager analsyse the key competitors of his firm?

4. What are the types of competitors in respect of their reactions to competitors' moves?

5. Write notes on the following:

   (a) Brand competition

   (b) Positioning

   (c) An Industry's strengths and weaknesses

   (d) Product type competition

6. Compare the competitive impact that would be experienced by a firm which decreases its price if the company were in a monopoly or a pure competitive industry. Discuss your answer.

7. Illustrate with example how with the help of advertisements is a company positioning its products in markets?

8. A company wants to launch a new non-cola soft-drink in markets in India. What type of competition it would face and from which quarters? Explain your answer.

11

Measuring and Forecasting Market Demand

Estimating and forecasting market demand are essential tasks for analysing market opportunities, devising and developing an effective marketing programme, and monitoring and controlling the marketing effort. Formulation of various budgets to be allocated to different marketing activities also depends upon the demand forecasts. Obviously estimating the demand in markets is an important responsibility of marketing managers.

11.1 MARKET DEMAND—A FEW CONCEPTS

A market consists of all the actual and potential customers who share particular need or want and are willing and able to buy a product to satisfy that need or want.

The *demand* for a product at a given price is the amount of it which will be bought per unit of time at that price. Demand implies three things:

(i) desire to possess a thing (or interest of the buyer),

(ii) ability to pay price or means of purchasing it (income), and

(iii) willingness to use these means for purchasing it (access).

Demand function is the functional relationship between demand for a product and the factors that influence this demand. These factors are — the price of the product, prices of substitutes and complementary goods, level of income, tastes and preferences, habits, size and composition of population, distribution of income, promotion and advertising etc.

Market Demand

Market demand for a product is thus the total volume of the product which will be bought at a price by a market i.e. all actual customers in the market per unit of time.

Market refers to a set of customers. Customers sharing different needs and wants make up different types of markets. Therefore, there would be different market demands of various specific groups of customers. Estimates of market demand may be computed in terms of sales volume of a single product, product line, an industry or a company. It may be computed areawise i.e. on the basis of district, zone, region, state, country or at world level. Again, these estimates may be for a short term or may be in the form of medium range or long range forecasts. Thus all these levels and types of demand need to be defined.

All demand estimates are dependent on the prevailing or future market conditions as the case may be. Demand also depends on the nature or content of marketing programme or offer.

The above study gives us an apt definition of market demand.

Market demand for a product is total volume that would be bought by a defined customer group in a specified geographical area in a given time period in a defined marke-ting environment under a specified marketing programme.

The *"specifics"* referred to in the above definition describe and answer the question which market is to be measured. This definition also suggests that different levels of marketing effort (industry marketing expenditure or marketing programmes) in a given marketing environment also affect the market demand or in other words the estimates of market demand vary with the varying levels of marketing expenditure or marketing effort.

Market Forecast/Sales Forecast

Market forecast indicates how much of a product is likely to be bought by customers during a specified future period in a specified market at specified prices. It is expected market demand. In context of sales, it is called sales forecast. Sales forecast indicates how much of a product is likely to be sold during a specified future period in a specified market at specific prices. Market forecast is also defined as estimated market demand corresponding to a given level of industry marketing expenditure in a defined marketing environment.

Market Potential

It is a quantitative estimate of the total possible sales by all the firms selling the product in a specified market. It indicates the maximum demand or the ultimate potential for that product assuming that industry is making very high level of effort (i.e. marketing expenditure).

Symbolically, market potential can be estimated as follows

$$Q = n \, q \, p$$

where Q is total market potential; n, the number of buyers in a specified product-market under given market conditions; q, the quantity purchased by an average buyer and p, the average price per unit.

Company Demand

Company demand is defined as a company's portion or share of market demand and indicates estimated company sales at various levels of company marketing effort or company spending on marketing.

$$Q_A = S_A \times Q$$

where Q is the total marketing demand, S_A is market share of the company A and Q_A is the company A's demand.

Company Sales Forecast

The company sales forecast is the expected volume it would sell during a future period with its specified marketing plan and tactics in a given marketing environment.

Corresponding to its marketing effort, the company sales forecast indicates the estimated sales expected during a future period at a particular level of company marketing effort (i.e. spending on marketing).

Related to the company forecast are two other concepts, namely sales quota and sales budget.

(a) **Sales Quota:** A sales quota is the volume of sales set to be achieved as a goal in case of a product or product line or by a company sales department or salespersons. This is a forecast which is set a bit higher than the expected sales so as to stimulate sales effort.

(b) **Sales Budget:** *Sales budget is a forecast of total volume of sales with its break up productwise, areawise and periodwise, i.e. quarterly or monthly, as converted into money.* It is prepared on the basis of past experience, anticipated market trends and conditions, economic climate or nature of competition, price trends, extent of advertisement etc. Sales budget shows the monetary value of anticipated sale units of the product or products. It is used for making decisions on current purchases, production and cash flows. It also serves as the basis for developing other budgets such as production budget and cash budget. A sales budget is generally set a bit lower than the company forecast.

Company Potential

It is the quantitative estimate of the total possible sales of product/ products that is the maximum amount of volume that a company can sell in a specified market with marketing effort relatively more than competitors. It is a portion of market potential in accordance with company market share. The company becomes almost a monopolist when it captures the entire market.

11.2   SALES FORECASTING/DEMAND ESTIMATION TECHNIQUES

There are two types of techniques/methods that companies use to predict future sales. These are—informed judgement techniques and formal mathematical methods.

A.   Informed Judgement Techniques

As the name suggests, informed judgement techniques are forecasting methods that rely on the opinions or judgments of knowledgeable and expert people to predict future sales. These techniques use what is called '*soft data' and* do not employ mathematical models. There are three types of informed judgement techniques, namely jury of executive opinion technique, sales force composite technique, buyer intention surveys.

(1)   Jury of Executive Opinion Method

In this method, opinions of managers of functional departments working in the organisation (such as sales, production, finance, materials, marketing managers), outside experts, consultants, dealers, distributors, advertising agencies are solicited, pooled and weighed to provide a sales forecast. These experts usually have sufficient working experience about company products and markets and can provide reliable opinion. The persons selected for soliciting opinion form a jury of experts. Though the estimates provided by these experts are likely to vary from person to person, yet these can be averaged to find a single estimate.

Sometimes, the demand estimates are arrived at by mutual decision among a group of such experts. However such a course has its own limitations because of clash of opinion or group dynamics; However, some organisations solicit opinions from experts without giving them opportunity to have face to face interaction. This is called *Delphi Technique.*

Delphi Technique is a method in which individual experts acting separately give their judgement in a systematic and independent fashion. These opinions are pooled, reviewed and revised to arrive at the final estimate.

(2) Salesforce Composite Techniques

Salespersons are in close contact with the customers. They know their needs, preferences and requirements and so are the best judges of their purchasing decisions. In this method, each salesperson estimates future sales in his or her sales territory. Such opinions about possible sales are pooled. Their estimates are scrutinised and reviewed and adjusted. The total of all these adjusted estimates form a composite salesforecast. It must be borne in mind that though salespersons may provide accurate estimates per their knowledge and experience but they might not be fully aware of the conditions that might affect customers' purchasing decision.

(3) Buyer Intention Surveys

As the name suggests, in this technique, a company may conduct survey to find buyers' intentions or buying plans for the given future period. Buyer intention surveys are marketing research studies conducted to elicit buyers' opinions about their planned purchases of a particular product or brand for the given future period. Some companies take the help of some outside organisations to conduct surveys to elicit buyers' intentions and their purchase plans. This is particular in case of consumer durables and industrial buying. The surveys are based on filled-in questionnaires asking their degree of likehood or probability to buy the product in the said forecast period. The survey is done on sample basis and then estimates are computed for the entire target market. This method is based on the assumption that buyers are able to anticipate their needs and are willing to take the surveyor into confidence about their intentions and plans.

B. Formal Mathematical or Modelling Techniques

Formal modelling techniques are methods which rely on mathematical models and equations and use *"Hard data"* (i.e. data which are collected from company's actual sales accounts and other records and are tangible in nature) to predict future sales. Again, there are three types of mathematical modelling techniques, namely analysis of time series or fitting trends, customer aggregation techniques and descriptive models, the first being the most important method to give projections of the future estimates.

(1) Analysis of Time Series Technique

A time series is a set of observations made at different times. It gives the measurements of a phenomenon, say sales, over a period of time. The importance of time series becomes obvious once it is realised that a major problem in modern business is that of making correct estimates for the future which cannot be done unless data representing changes over a period of time are systematically and scientifically analysed.

Broadly speaking, there are two types of well-defined movements operating in a time series.

(1) long-term or secular trend

(2) short-term movements or fluctuations

To measure long term trend, the short-term variations need to be removed and irregularities should be smoothed out and to study short time changes, long time variations must be removed.

This technique involves predicting sales forecast using past data of the company over a period of time. Historical data or time series show periodic changes in the variables over a period. Thus we can have past sales, data of a company over a number of years say from 2002-03 to 2008-09 or from 2000 to 2008. These variations are not regular. The sales may change from period to period (from year to year) due to varying marketing conditions. Then there are seasonal fluctuations, cyclical oscillations, and erratic events that affect sales from period to period and are responsible for introducing irregular element in the time series. *This technique of forecasting is based on the assumption that historical data or time series of past sales contain a long-term secular trend which is discernible and can be found out.* This trend forms the basis of estimation of future sales growth.

In order to find long-term trend, what is required is to eliminate short-term seasonal, cyclical and irregular (erratic) fluctuations from the series. Statisticians use a number of methods to find the trend and remove short-term fluctuations such as method of moving averages, extrapolation etc.

A technique most often used in statistics is that of fitting a curve or a line. This technique is based on regression analysis and is also known as *method of least squares.* In this technique, a line or curve is fitted to the data in an historical series. This is technically a line of the best fit. We know that a line is represented by a first degree algebraic equation. In this method, we find an equation which will give estimate of sales corresponding to future years.

Let us suppose the equation is $y = a + bx$ (1)

where y is the estimate sales, x the corresponding year and a and b are unknown constants to be found out.

To find out the value of a and b, the method suggests formation of *normal equations* which are

$$\Sigma y = na + b \ \Sigma x \tag{2}$$

$$\Sigma xy = a\Sigma x + b\Sigma x^2 \tag{3}$$

By solving equations (2) and (3), we can find the values of a and b and insert the values in equation (1) to find the trend. The following examples illustrate the method.

EXAMPLE 11.1: Find trend Sales Forecast for the year 2009-2010 and 2010-2011

1	2	3	4	5
year	sales (y) (Rs. in crores)	x (2005-06 = 0)	x^2	xy
2002–03	81	–3	9	–243
2003–04	106	–2	4	–212
2004–05	160	–1	1	–160
2005–06	210	0	0	0
2006–07	237	1	1	237
2007–08	284	2	4	568
2008–09	350	3	9	1050
Total	1428	0	28	1240

The data pertain to 7 years

Let line to be fitted with $\qquad y = a + bx \qquad$ (1)

Normal equations $\qquad \Sigma y = na + b\Sigma x \qquad$ (2)

$\qquad \Sigma xy = a\Sigma x + b\Sigma x^2 \qquad$ (3)

Inserting values in (2) and (3)

$1428 = 7a + b \times 0; \quad a = 1428/7 = 204$

$1240 = a \times 0 + b \times 28; \quad b = 1240/28 = 44.29$

the equation is

$y = 204 + 44.29 \ x$

trend sale forecast for 2009-2010, $x = 4$

$y = 204 + 44.29 \times 4 =$ Rs. 381.16 crores

trend salesforecast for 2010-11, $x = 5$,

$$y = 204 + 44.29 \times 5 = \text{Rs. } 425.45 \text{ crores}$$

These sales forecasts can be modified keeping in view changes in marketing conditions likely to arise in the forecast year and combining anticipated short term fluctuations into them.

EXAMPLE 11.2: *2008 report of a company* shows the following 9-years record of its turnover (gross sales). Find trend

year (x)	gross sales (y) (Rs. in crores)	x (1991 = 100)	x^2	xy	
2000	953	−4	16	−3812	
2001	1021	−3	9	−3063	
2002	1216	−2	4	−2432	
2003	1460	−1	1	−1460	−10767
2004	1776	0	0	0	
2005	2087	1	1	2087	
2006	2436	2	4	4872	
2007	3240	3	9	9720	31779
2008	3775	4	16	15100	
$n = 9$	17964	0	60	21012	

Let the trend equation is

$$y = a + bx \tag{1}$$

Normal = ns

$$\Sigma y = na + b\Sigma x \tag{2}$$

$$17964 = 9a + b \times 0$$

$$a = 17964 \div 9 = 1996$$

$$\Sigma xy = a\Sigma x + b\Sigma x^2 \tag{3}$$

$$21012 = a \times 0 + 60b$$

$$b = 21012/60 = 350.2$$

required equation is $y = 1996 + 350.2\ x$

For 2009-2010,　　$x = 6$

$$y = 1996 + 350.2 \times 6$$

$$y = 1996 + 2101.2 \ = 4097.2$$

For 2010-2011,　　$x = 7$

$$y = 1996 + 350.2 \times 7$$

$$= 1996 + 2451.4 = 4447.4$$

The above trend sales forecasts tor year 2009-2010 and 2010-2011 are Rs. 4097.2 crores, and Rs. 4447.4 crores respectively.

A second degree curve $y = a + bx + cx^2$ can also be fitted to data. It would give more refined measure.

These forecasts are revised and modified by taking marketing conditions likely to prevail in the forecast period into consideration.

In order to find out effect of each component of time series i.e. of cyclifical oscillations (except of irregular or erratic movements), the effects of other components are eliminated or removed. In this way, forecast of each component *viz.* trend (T), cyclical movements (C) and seasonal fluctuations (S) are computed and combined together to find the estimates of total sales.

$$\text{Sales forecast} = T \times C \times S$$

The study of procedures of analysing each of the above components is beyond the scope of our study in this chapter. However you can study these procedures in any good book on statistics.

(2) Customer Aggregation Models

This method depends on forecasting of each component of sales in the target market and then aggregating these component forecasts. The estimated future product sales is arrived at by aggregating components forecasts. We know in a simple sales model.

Product Sales = Number of buyers × Average quantity purchased

In this technique, sales are broken down into their major components *viz.* number of buyers and average quantity purchased. Generally, companies gather data on the characteristics of the target market i.e. of the typical buyers, the product use, the competitive challenge in the form of availability of substitutes, life of product and other allied matters. These data are used to estimate the future product demand. In this technique demand is estimated by forecasting each component. For example, in a simple basic model, the components of demand are:

(i) the number of people or organisations in the target market (N)

(ii) all people in the target market will not buy the company product as there are a number of substitutes or competitive products. Thus the second component of the product demand is the probability or likelihood among the target market people who would actually buy the product in the forecast period (L)

(iii) the other component of demand is the average quantity of the product purchased by a typical buyer (R)

To predict future demand, it will be necessary to forecast each of the above components.

If S is the estimated future product demand. Then the estimated demand is

$$S = N \times L \times R$$

To get estimates of sales in terms of money, the estimate demand is multiplied by average price per unit of the product.

Sales volume (in money) = $N \times L \times R \times P$

This method is very useful in situation where sales data of past years are not available.

EXAMPLE 11.3:   Demand Forecast (Customer Aggregation Technique)

Suppose the number of people in a target market,

$N = 5000$

There is likehood of 25 per cent of the target market who would buy the company product i.e. $L = 0.25$

the average quantity purchased = 24 gross per annum (R)

estimated demand $= N \times L \times R$

product-demand $= 5000 \times 0.25 \times 24 = 30000$ gross p.a.

Let us suppose that the average price is Rs. 1200 per gross

estimated sales forecast = $30000 \times 1200 =$ Rs. 36,000,000 p.a.

The company's product sales forecast for the next year will be Rs. 3.6 Crores

$$S = N \times L \times R \times P$$

There are several other causal description models and typical statistical techniques also to estimate and predict demand that are based on multiple correlation and regression analysis and discovery of effects of important factors such as price, income, population and promotion etc. We have already referred to these factors while defining demand function in the early part of this chapter. However, the study of these techniques is beyond the scope of this chapter.

11.3   SELECTING A FORECASTING TECHNIQUE

It is not easy to recommend which technique out of those mentioned above would be most appropriate to predict demand estimate. The method would vary with varying situations. Large companies use formal mathematical or modelling techniques for the purpose, though sometimes they also use one of the informing judgement techniques to complement formal modelling techniques. So a company can use more than one techniques if situation warrants so.

However, a few decisions need to be made and criteria determined before selecting a forecasting technique e.g., required level of accuracy of estimates, availability of reliable data, cost of the technique, time to be allotted for the purpose, past experience of application of the technique of the forecast.

Evaluation of the technique would involve assessing the accuracy of the predicted or forecast sales with the sales that actually occur. Evaluation would also involve knowing the forecasting error and diagnosing the reasons for the same and taking corrective steps to improve the level of accuracy and avoid forecasting error.

EXERCISES

1. Explain the terms—market demand, market potential, company demand, company potential, company sales.
2. Describe various forecast techniques used to estimate future sales of a company.
3. Write notes on
   (a) Jury of Executive opinion method
   (b) Analysis of Time Series technique
   (c) Customer Aggregation technique of Demand forecast
   (d) Delphi technique
   (e) Informed Judgement Techniques
   (f) Buyer Intention Surveys
4. As a district marketing manager, you have been asked to supply sales forecast information within two months. What procedure will you adopt to procure accurate forecasts?

12

Product: Concept and Decisions

12.1 PRODUCT: DEFINITION AND CONCEPT

Marketing is defined as the conception, pricing, promotion and distribution of ideas, goods and services to satisfy the wants and needs of various customer groups. The conception of an idea, good or service is the first and foremost activity in the marketing process. We have already stated that *product* is one of the four components of the marketing mix, the other three being price, promotion and place. We shall discuss this most important ingredient of the marketing mix offer in this chapter.

A marketing manager is required to design and develop such products as may satisfy the needs and wants of various customers. It implies that what is offered as a product must match with the needs and wants of the customers or in other words, it must fit in the customer's overall perception in terms of benefits or satisfactions it would provide. It is in this context that we need to define product.

A product is anything that can be offered to a target market and has the potential of satisfying their need or want.

Taking customer perception in view, we can define product as follows:

A product is anything that is potentially valued by group of customers for the benefits and satisfactions it provides. In other words, anything which a customer perceives as having value-in-use is called a product.

Products may, thus, include objects (physical goods), services, places (locations), institutions and even ideas or concepts. For example, tooth paste, car, hotel facilities and services, arranging pilgrimages, marketing research services, life insurance are all products from a marketing viewpoint.

It must be kept in view that it is not only the product that a customer is buying but it is, in fact, *the benefits and satisfactions provided by it that a customer is purchasing* and basically a product is to be seen in this context.

It is clear from the above study that the first and foremost task of a marketing manager is to match the conception of a product with the customer's overall perception of it. Different persons may have different perceptions of a product. For example, Maruti Esteem may be a symbol of status for a person, while for another it may be a means of comfortable and reliable transportation.

Levels of a Product/Customer Product Perception

As stated earlier, it is the benefits and satisfactions that a customer is purchasing. A product passes through an hierarchy of levels as it provides more and more additional benefits and satisfactions through its development.

The first level or stage of the product comprises the basic benefits or services that a customer purchases. It can be called *fundamental benefits or satisfactions* that a product delivers. It is the *Core benefit*. For example, cleanliness of teeth, transportation, sleep at night etc. can be designated as *Core benefit level* of the product. These needs can be satisfied in different ways or by using different products.

The second level or stage of the product is to have a product/products that may deliver the core benefit or service. In other words, the core benefit and satisfaction need to be transformed into basic versions of the product i.e. *Generic Product*. Generic product represents a tangible offer by the marketer to deliver the core benefit or service expected by a customer. For example, tooth paste, tooth powder for cleanliness of teeth, bicycle, scooter, car or bus for transportation, a lodge, hotel, guest house or *dharamshala* for night rest. All these are generic products.

The third level of a product is reached when the product acquires some tangible attributes and gets an identity of its own through a name (brand) or other attributes. This level is termed as *expected product* i.e. a product having attributes normally expected by the purchaser. These attributes include quality, style, brand, packaging and other features e.g. Colgate Dental Cream, Close-up, Cibaca, Vico-Vajradanti, Maruti 800, Vespa, Ashok Leyland Bus, Moti Mahal Hotel, Adarsh Lodge etc.

The fourth level product marks the addition of more benefits or services i.e. of elements beyond the specific product itself. These additional elements may be reflected in additional features or in additional benefits. This level is called *augmented or extended product*. These additional features distinguish the company's offer from competitors' offers and provide additional benefits and services to the customers. An augmented or extended product may offer additional benefits such as warranty service, status oriented promotion, distribution convenience like delivery arrangements, financing etc.

It needs to be noted that additional features would mean additional costs and these will be realised from the customer in the form of enhanced prices. Moreover, the *augmented* product may lose many customers who may like to buy only the expected product.

The final or fifth level is the *potential product* i.e. the ultimate form of the product that may be evolved in future. It is the tomorrow's product with all its augmentations, transformations and improvements and unexpected features. We have already referred to Hindustan Unilever's end objective which is to delight the consumers and customers by satisfying their needs and aspirations.

Most often, competitive market offers core benefits, expected products and augmented products and it is the task of marketing manager to blend these three levels in a consistent manner to satisfy the needs of the target market. Thus three levels *viz.* benefits and satisfactions, expected product and augmented product together create the customer's overall perception of the product.

A large number of multinational companies like Hindustan Unilever, Procter & Gamble, Maruti, Hindustan Motors, Taj and Oberoi Groups of Hotels, electronic companies, are offering what can be called *globalised products.* Some companies consistently and continuously go on adding additional elements and benefits to their products.

Take for example the case of tooth pastes in dental oral care. Colgate is offered in many brands, namely **Colgate Total,** claimed to be the all family tooth paste, **Colgate Herbal** the first fluoridate tooth paste in the country, and **Colgate Gel** with clenzones, an antibacterial cleaner—a

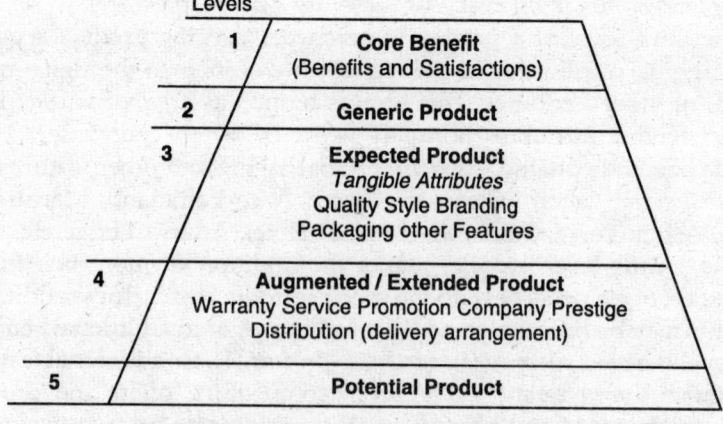

Fig. 12.1 Customer Product Perception.

combination of teeth cleanser and mouthwash. Producers also attempt to augment their products by changing the packaging, colour, perfume, fragrance etc.

Thus product is something more than a physical commodity or service. It has a distinct identity and personality. The benefits and satisfactions it delivers, the brand name, the package, the prestige and image of the manufacturer and the marketer i.e. a combination of the tangible and intangible, all together constitute what is called the product offer.

12.2 PRODUCT CLASSIFICATION

We have already given the classification of goods and services (i.e. of products) in details in chapter 1. The marketing manager has to devise different marketing strategies for different types of products. The marketing strategy for marketing soaps and detergents will differ from that of marketing cars or computers.

12.3 DEVELOPING A PRODUCT POLICY

In this context, the management has to make a number of decisions i.e. what products should a company produce? How are they to be differentiated to obtain competitive advantage? How should the product be positioned in the market? A product policy sets guidelines for making decisions about the above dimensions.

Broadly speaking, an overall product policy of a company involves marketing decisions about various products to be offered (i.e. about product mix and product lines), product differentiation, branding, packaging, positioning the product in the market and development of new products. We shall study all these aspects in further details in the following pages.

12.4 PRODUCT MIX

Product mix is the entire collection of product lines and items that a particular company offers for sale to customers or buyers.

A product mix denotes the range and variety of the products of a multi-product organisation and is also called *"product Assortment"*. It is thus a set of all product lines and items offered by an organisation. Table 12.2 illustrates the product mix of Ashok Leyland.

TABLE 12.2:   Ashok Leyland's Product Mix (Range)

1. **BUSES:** CHEETAH (57 SEATS), VIKING-222 (48 SEATS), **VIKING** (66 SEATS), TITAN (DOUBLE DECKER), (78 SEATS), INTEGRAL (52 SEATS), LEOPARD LUXURY A/C, ARTICULATED (82 SEATS), CHITAL MINI BUS

2. **TRUCKS :** COMET MINIR, COMET HAULAGE, COMET TIPPER, CZ 1611, TUSKER SUPER, TUSKER TURBO, COMET NORMAL, TAURUS, CARGO 709, CARGO 900, CARGO 1312

3. **HEAVY VEHICLES :** GAUR 2516, HIPPO TRACTOR, BEAVER, ALRD 20, (REAR DUMPER)

4. **DEFENCE VEHICLES :** AZAD SUPER, TOPCHI, STALLION (4 × 4), STALLION (6 × 6)

5. **ENGINES :** INDUSTRIAL, MARINE

The decisions concerning product mix entail decisions about the width (breadth), length, depth, and internal consistency of the product mix.

The *breadth or width* of the product mix refers to the number of types of products i.e. product lines offered by a company. The breadth of Ashok Leyland's product mix is 5 lines *viz.* buses, trucks, heavy vehicles, defence vehicles, engines.

The length of the product mix refers to the total number of items in the product mix. The length of Ashok Leyland's product mix as shown above is 29 (twenty nine). The *average length* of Ashok Leyland's product mix is 29 divided by 5 i.e. 5.8.

The *depth* of the product mix refers to the number of variants within each product type or product line. For example, Ashok Leyland offers nine kinds of buses and eleven kinds of trucks.

The *internal consistency* of a product mix refers to the extent to which various product lines and their variations are related to one another or complement one another in respect of use, production, distribution channels.

For example, Hindustan Unilever Limited produces and markets a number of product groups or lines such as soaps, detergents, personal products like tooth pastes and hair oils, cosmetics, chemicals including aroma chemicals, fertilizers, seeds/plant growth nutrient. They also export these products. It is evident that the first four groups namely soaps, detergents, personal products and cosmetics can be marketed through same distribution channels and so have much internal consistency. Again the last two groups have also substantial consistency as same group of customers would need them.

A company can expand its business by increasing the width, length, depth or consistency of the product mix, or taking recourse to expanding more than one of these dimensions.

12.5 PRODUCT LINE

As a company expands its production activities and engages itself in producing more and more goods, marketing managers need to group the products to be marketed on the basis of some common relationship among them. For example, some products may satisfy same customer need and being alternatives may perform a similar function, or there may be some products that maybe needed by same target markets or group of customers, or it may produce products that could be marketed through same distribution channels. Thus the products produced and to be marketed need to be organised on the basis of some common relationships into a number of types of products called product lines. We can define the product line as follows :

A product line is a group of products that are related in someway as serving the same customer need, being sold to same target market or group of customers, marketed through the same distribution network or outlets or falling in a common price category or given price range.

It is the management that has to decide the grouping of the products on the basis of some common relationships into a product line. Different managements do it differently depending upon their judgements.

EXAMPLES: Hindustan Unilever Limited treats soaps and detergents as a single group or product line. It includes three categories —

  (i) Personal wash category comprises Lifebuoy, Breeze, Hamam, Moti, Jai, Rexona, Lux, Liril;
 (ii) Fabric wash segment includes brands like Surf, Wheel;
(iii) Home Care Business includes Vim all these three categories are grouped into Soaps and Detergents Product Group.

Tooth paste Close-up, Fair and Lovely Fairness Cream, Pears Soap, Dove Beauty Bar, Tata Nihar Hair Oil are grouped by Hindustan Unilever in another product line termed *Personal Products.*

On the other hand, Dabur India Limited has grouped Amla Hair Oil, Vatika Hair Oil and other Oils-Shampoos, Tooth powders and Talc into a single group as *consumer care business segment.*

Length of the Product Line

The length of a product line refers to the total number of items in that product line.

For example, Dabur markets seven types of hair oils, namely Dabur Amla, Dabur Jasmine, Dabur Bhringraj, Dabur Brahmi Amla, Dabur

Anmol Coconut, Vatika & Special Hair Oil, three types of tooth pastes, three types of tooth powders, two types of shampoos, two types of luxury talc. Thus this product line comprises 17 items, i.e. it has a length of 17.

Each product or item has different depth e.g. Dabur Amla Hair oil is available in 5 sizes (50 ml, 100 ml, 200 ml, 300 ml, 500 ml). Its depth is 5.

Colgate tooth paste comes, at present, in five formulations Colgate dental cream, Gel, Total, Herbal and Calciguard, each in three sizes, the number of variants being 15. Colgate toothpaste has a depth of 15. As the companies grow over a time, they do on adding the variants and depth of their products.

12.6 PRODUCT LINE DECISIONS

In this context, the management is required to make decisions regarding the length of the product line i.e. the decisions involve determination of the length of the product line and how does it compare with competitors' product lines. Both of these types of decisions involve evaluating product performance or analysis of product-line.

Product-line Length

An important decision marketing managers have to make is to determine whether to stretch or shorten the product line. It would mean evaluating the performance and the contribution of each item in the over-all performance of the product line and determining whether such contribution justifies its continuous production and whether there is scope of increased profits through addition of more items.

A product line is said to be *long* if the dropping of an item can result in increasing the profits and a product line is said to be *short* if addition of items would lead to better performance or profits of the product line. Thus the managers have to make product-drop and product-addition decisions.

Evaluating Product Performance

Evaluation of product performance necessitates analysis of sales and market share trends, profit contribution, financial requirements to support the product, cost and price trends. Two most important criteria of measuring product performance are percentage share of the product in the total sales volume of the product line and percentage of profits contributed by the product in the product line. A product with a very low relative contribution or performance in terms of sales or profits in a product line may need to be dropped out if stakes in doing it are not high. It should also be seen that too much concentration is not laid on a very few high performance products as entry of a few new competitors in the product-market could adversely affect the performance of company's products any time. The lesson is that the performance of the product should be continually monitored and necessary measures should be taken in time to correct the situation, if need arises. The best situation will be that each product contributes consistently and equally towards results to be accomplished by the product line as a whole.

As stated above, the length of the product line can be stretched by adding new items or it can be shortened by dropping those items which are consistently lacking in performance despite boosting efforts and/or are termed as weak or deadwood. Both types of these decisions need careful study, analysis, planning and workout as considerable resources are at stake in both cases.

(a) **Lengthening a product line:** Product manager can stretch the product line by adding new items to the product line. These new items may cater for the needs of upper end customers or lower end customers. Thus the stretching or lengthening may be *upward* or *downward*. Again a company may introduce new items meant for both ways and there may be two- way stretch. With the introduction of Maruti Esteem Car in Indian market, Maruti stretched upward.

While stretching a product-line without careful study or examination, a company may face great danger. The new item may be another version of an existing product and it may lead to danger of what is called *product cannibalisation.*

Product Cannibalisation means one product in a line eating away at the sale of another product. This needs to be avoided lest it should make the company worse off. The product line manager should strive to capture new customers with the line stretch and see that existing products do not get cannibalised.

(b) **Shortening a Product Line:** Shortening the product line implies dropping of weak and low performance items from the product line. Moreover, there may be some products that might have reached the ultimate end of their life cycles. A product will not survive for ever. It has to reach its end one day. Such products need to be identified and dropped after careful study and in a way that the resources released are diverted to higher margin products or other marketing priority areas and thus to make more profitable contributions.

Product Positioning

The other type of the decisions to be taken by product line managers is that of positioning products in the line. The length of a product line is optimum when the number of items or products offered to target markets is just enough to yield optimum profits. This would need proper positioning of the product in the product line.

We have referred to product positioning as the final step in target marketing in the chapter on marketing segmentation and marketing targeting. We have discussed market segmentation and market targeting in details in chapter 8. A brief explanation of the concept of product positioning is given here.

Positioning a product in a product line means giving the customers a reason to buy the product from the particular company rather than from another and designing the company's offer so that it occupies a distinct place in the target customers' minds.

Obviously, this would require effecting changes in features, quality, price, design and other attributes of the product. This may require augmentation of benefits — one, two, three or more, product improvements, alteration in product life cycle and many other decisions in repositioning the product. This will mean heavy investments in R&D. Product repositioning can help in making the target market understand how that firm's product stands vis-à-vis competitors' offers. This calls for introducing differences referred to above in the product so that it may stand out as distinct offer with a good image. And herein lies the importance of *branding* and *packaging*. These decisions also form an integral-part of the firm's product policy. Product positioning helps the manager in designing marketing mix strategy. Again it must be ensured that the product positioning does not result in a major cannibalisation and such positioning helps in retaining existing customers and attracting new customers. Lifebuoy Gold, Alteration in shape of Lux International, Surf Excel, Nirma Super are some illustrations of products repositioning.

12.7 BRANDING

As we have seen, in modern business, branding of a product is of critical importance in positioning a product and developing effective product policy. It goes without saying that building a brand will need corresponding packaging, advertising, promotion etc. In modern world, branding decisions may affect the scope of success or failure of a product. Before studying this aspect, we should become familiar with a few terms associated with branding.

Brand: It is name, term, sign, symbol or design, or a combination of them which is intended to identify and distinguish the product of one firm from that of the other. It is the means to differentiate product of one producer from that of the competitor.

Brand Name: It is that part of the brand that can be vocalised. For example Maruti, Lux, Dalda, Colgate, Philips, Nikon, Sony.

Brand Mark: It is that part of the brand that can be identified and recognised but cannot be vocalised or uttered such as symbol, design, colouring or lettering.

Trade Mark: It is a brand or a part of brand that is given legal protection for exclusive use by the firm. A trade mark presents company's exclusive right to use the brand name and /or brand mark. New trade marks are registered with country's patent offices.

Copy right: The sole legal right to reproduce or sell a literary, musical and artistic work.

Need for Branding (Importance)

Branding involves heavy costs in the form of packaging, labelling, advertising and promotion. Obviously unbranded products will be cheaper in the market. Still the firms or sellers prefer to brand their products. There are several important reasons for branding the products as it gives the company several advantages:

1. **Market Identity:** A brand helps the seller to position its product in the market, in terms of quality, service, prestige, price. It gives the seller a market identity and a market image. For a large number of customers a brand stands out for quality, durability and reliability.

2. **Legal Protection:** A brand name provides legal protection as to the special features of the product from the competitors and as to the huge promotional expenses incurred on building it.

3. **Customer Loyalty:** A brand name helps in building customer brand loyalty and attracting loyal and profitable customers towards other products of the company as well besides repeat sale orders.

4. **Increased Profit Margins:** Brand loyalty helps in increasing the sales and profit margins of the sellers.

5. **Market Segmentation:** Branding helps the producer in segmenting markets and offering variations of the product to cater to the respective needs of those segments. For example, Hindustan Unilever is offering six or seven brands of soaps in *personal wash category*. Different brands may serve different market niches.

6. **Processing of Orders:** Branding makes it easier for the seller to process orders and solve customer's problems.

7. **Bargaining Position:** A strong brand image enhances the producer's bargaining position and strengthens his influence in channels of distribution. It is easier to get distributors for the product.

8. **Corporate Image:** Strong brands i.e. brands known for quality and performance standards build a corporate image that has long term advantages.

It is for these reasons that today companies increasingly prefer to brand their products though a lot of business in products of generic nature is carried out without branding the product e.g. grocery items and things of daily consumption. However most of even the grocery items like salt, cooking oils, rice, wheat flour etc. that were earlier sold unbranded, are being branded for marketing now-a-days.

Producer's Brand and Distributor's Brand

Branding i.e. the task of giving a brand name to a product and building brand name and image is usually done at the producer level or the distributor level. Who will sponsor the brand is a matter also requiring management decisions.

(a) A large number of big manufacturers or producers brand their products at their level and spend huge amounts in advertising and publicity to build them. These brands are also termed as national brands or global brands. Some producers give their brand names on license to third parties—manufacturers, distributors and retailers for royalties (rent) to be paid to the rightholder of the brand name. This is known as *name and character licensing*. A large number of Indian manufacturing companies have made such arrangements under *license* or *franchise*. Again most of manufacturing companies get their goods produced in accordance with their quality standards from other manufacturer-suppliers in order to meet their increasing demand under their own brand names. Such companies include Bata, Hindustan Unilever, Whirlpool, Pepsi Cola, P&G etc.

(b) Branding is also done at distributors' level. Many distributors or marketing firms (channel members) obtain their goods from manufacturer-suppliers and market them under their own brand names. A few examples are Nanz grocery products, Shakti Bhog Atta, Lal Kila Basmati Rice, Kohinoor Basmati, Aishwarya Sarees etc.

In our country, in wake of recent liberalisation and transfer of technology, a number of products are being marketed in the combined brand names of manufacturers and marketers or two manufacturers e.g., Wall's Kwality ice cream, Hero-Honda.

As brand helps a company in increasing its market share, therefore the battle of brands has become a continuing affair. No wonder, many unscrupulous persons are selling spurious goods under reputed brand names especially under those brand names which are market leaders.

Individual Brand and Family Brand

Whatever may be the level, manufacturer or distributor on which branding is done, the other decision to be made by the management is whether to apportion individual brand names for different products or an overall family brand name for each of the products.

Individual product branding means giving separate brand names for each product. This practice is followed by Hindustan Unilever, Procter & Gamble, Marico Industries, Brooke Bond Lipton etc. Some examples of individual brand are Sunlight, Lux, Camay, Ariel, Dalda, Parachute Coconut Oil, Taj Mahal Tea etc. An advantage of individual branding is that the customers are offered a variety of brands, though the company has to incur huge costs on positioning these brands. In case of a company offering many competing brands, a failure of one brand would not make much difference. But it must be ensured that too many brands do not result in cannibalisation.

Multiple Branding involves giving more than one brand name to the same product category. This is done to achieve greater market share. For example, Kelvinator, Gem & Leonard refrigerators.

Family product branding involves giving a blanket family name for all products, product groups or product lines. This may take the form of use of corporate name with individual products. This policy is followed by Godrej, Tata, Nirma, Philips, BPL, HMT, Colgate, Dabur, Amul, T-series.

Family branding has an advantage of economy as compared to heavy costs of individual branding. The reputation of family brand can easily help in launching and positioning new products in market on the basis of collective image perceived by the customers. The disadvantage is that, sometimes, a failure of even one item can adversely effect the entire image of family brand. Again, a single family brand for all products can cause confusion. Moreover, customers have different needs and may have liking for different individual brands of the needed products.

Decisions concerning individual or family branding have considerable impact on the development of marketing strategy for various products. Therefore branding must facilitate promotion of the product concept and help in attracting more and more customers for the product. In order to achieve this objective, a brand name should have the following qualities:

1. It should give a notion of the product's benefits.
2. It should reflect product quality.
3. It should be unique and distinctive.
4. It should be easily recognisable and pronounceable.

12.8 PACKAGING DECISIONS

Physical products need to be presented or sold in some container or package. Therefore packaging of products is necessary and required at least for protection of the product and for holding it conveniently. Again good packaging can attract customers towards it and has a promotional role. As stated earlier, packaging has a big role in positioning the product and building a brand image. It is an important component of the strategy of creating a product concept and is a powerful marketing tool. Packaging decisions thus form an important part of the marketing strategy.

In simple terms, packaging is the process of producing the wrapper or container/containers for the product.

Packaging is referred to as the activities of designing and producing the package i.e. wrapper or container for a product.

There are three levels of physical packaging namely:

First Level—Primary Package: It is the materials that envelope the product and hold it i.e. it is the immediate container of the product. In case of Colgate Dental Cream, the tube containing the tooth paste is its primary package or the bottle or can holding Pepsi Cola or a match box holding matches is the primary package.

Second Level—Secondary Package: It is the material that holds or protects the primary package. This is generally discarded and thrown away when the product is about to be used. The card board box containing the toothpaste or liquid tonic bottle or a paper or packet wrapping a dozen match boxes serve as secondary package.

Third Level—Tertiary Package or Shipping Package: It refers to bulk packaging i.e. material that holds the secondary package for shipment. This refers to packaging for storage, identification or transportation. The corrugated card board box containing six dozen of secondary packages of tooth paste or plastic cartons each containing 24 bottles of Pepsi Cola are examples of shipping packages.

Sometimes, the primary, secondary and shipping packages are combined. However, usually at least two levels of packaging are used depending on the nature of the product.

Besides the above levels of packaging, labelling is also a part of packaging.

Labelling

Labelling consists of printed information pertaining to the product appearing on or with the package.

Besides describing the product, it may also contain printed statements concerning safety or statutory warnings against products' misuse or hazards as a legal requirement. Moreover, companies are also required to print some specific information like date of manufacturing, date of expiry, batch number, name and address of manufacturer, maximum retail price and other information under some Food, Drug, Cosmetics Acts or Labelling laws. Again, labelling has a great promotional value. For example, Colgate Dental Cream tube contains this printed information:

FOAMING NON-FLUORIDATED TOOTHPASTE

STOPS BAD BREATH .... FIGHTS TOOTH DECAY

CLOSE-UP tube contains the information

TOOTH PASTE WITH REAL MOUTHWASH.

Why Packaging?

Packaging has become an important marketing tool. As seen above, packaging makes it convenient to handle the product. It protects and promotes it. There are a number of reasons for the growing use of packaging;

1. **Increased store efficiency:** It takes relatively less labour and lime to package items at producer's end rather than retailer's end.

2. **Self-service:** Packaging makes it easier to identify and market the product. It attracts customer's attention. As customers started buying products at super markets, malls and stores on self-service basis, the packaging became imperative.

3. **Customer's convenience and paying capability:** It was convenient and looked good to carry out the product in a package and the Customers were willing to pay a little more to buy this convenience.

4. **Brand image:** As an attractive package could attract better attention of the customers and help in building the brand image, the producers saw in packaging a great promotional opportunity. It helps both producers and consumers.

5. **Shelf-space battle:** With limited shelf space available with the retailer, how to stock and stack product posed a great problem. Packaging offered a solution to this problem.

6. **Strict statutory guidelines and safety laws:** Government regulations and laws containing guidelines and safety instructions have also been a contributory factor to the increasing use of packaging.

7. **Better opportunities:** Packaging offered greater and innovative opportunities for increasing the sales volume of the products and hence the profit of the company. There was a realisation on the part of the producers that innovative packaging could yield better results.

Thus it was the need for protection, convenience, economy, efficiency, and promotion that led to the development of concept of packaging.

Functions of Package

Besides creating benefits like protection, convenience, economy, or promotion, packaging performs many other functions.

1. An important function of packaging is that of supporting a new product strategy. In fact, packaging is considered as an integral part of the product. For example, coffee in a glass jar, paint in a plastic bucket or dry fruits in a gift pack add to the value and become a part of marketing strategy.

2. It makes distribution channels accessible. Good packaging comes to the aid of distributors as identification, stocking and stacking and pricing become easier.

3. Packaging aids in devising a sound price strategy. The quality, design and material of packages go into determination of the price of the product.

4. The most basic function of packaging is containing the product in specific quantities for their shipment. It facilitates the physical distribution of products.

5. Another essential function of packaging is protection of the product — protection against damage, spoilage, pilferage. Sealed packaging can assure quality and purity and provide protection against imitation.

6. The design, size, shape, colour, text or brand mark on a package promotes the product and helps in image-building.

7. Packaging provides important and necessary information as to the product's uses, ingredients, price, date of manufacturing, warranty, cautions and warnings.

Designing and Developing Packages/Containers (Packaging Decisions)

In order to fulfil the above mentioned functions, the packages need to be designed and developed very carefully. It must be ensured that packaging decisions support the overall product policy and marketing strategy.

One way to look at it is to make such decisions regarding packaging that the package may look attractive enough to capture customers' attention. Such decisions involve —choice of package material, package aesthetics, size, shape, design and colour etc. These decisions will necessitate the following steps.

1. **Coordination and Cooperation:** Packaging decisions involve participation of high level managers and personnel of manufacturing department and various divisions of marketing department within the organisation. Successful packaging requires coordination and cooperation of people not only within the organisation but also of the concerned people outside it, such as those from advertising agencies, middle men and packaging specialists.

2. **Market Research:** These decisions hinge on marketing research in fields of advertising, image building and packaging material. It may involve technical and engineering testing. Packaging research facilitates these decisions.

3. **Physical Package and Graphics:** The material, design, shape, colour, finish, graphics, illustrations, labelling, copy wording, printing—all contribute to making the package appealing and attractive. These must be consistent with the image desired. While making these decisions legal and environmental issues, concerns and obligations must be kept in mind. Indian laws require producers to print some particular information on the label and package for information and protection of consumers.

4. **Design Trial:** Before introducing a design in the market, it will be worth the effort to conduct a trial or test to gauge the reactions of distributors, dealers and consumers. This work is to be done very carefully, rather in a subtle and indirect manner.

Package designs also need to be periodically evaluated and reviewed in the light of reactions of consumers and distribution channels. No efforts should be spared to take advantage of innovations and inventions in this ever changing field.

EXERCISES

1. What is the concept of product in marketing management? Explain the levels through which a product passes.

2. What is meant by product mix? Enlist various decisions concerning product mix.

3. Explain the terms—the breadth, the length, the depth and internal consistency of a product mix. Give examples

4. What is meant by product line? What are usual product line decisions to be made by a marketing manager.

5. (a) How can you evaluate product performance?
   (b) What does a manager do:
      (i) to lengthen the product line?
      (ii) to shorten the product line?

6. What is meant by branding? Explain its importance.

7. What decisions need to be made while branding a product?

8. What do you mean by packaging? Explain its levels giving illustrations.

9. Why is the packaging necessary in modern marketing? What other functions does it perform?

10. Packaging has a great role in positioning the product and building a brand image. Discuss.

11. Write notes on:
    (i) Expected product and extended product
    (ii) Packaging
    (iii) Customer Product Perception
    (iv) Product Positioning
    (v) Product policy
    (vi) Labelling.

12. What is meant by customer perception hierarchy? Explain its most significant managerial implications in context of:
    (a) car cassette tape decks,
    (b) high quality vegetarian hot meals.

13. There has been a major change in packaging of hydrogenated oil from tin can to plastic containers for domestic use. Keeping in view the packaging functions in the chapter, discuss the problable advantages of this change.

13

New Product Development

In the customer-oriented marketing, product development assumes greater significance. The producer is to regularly strive for modifying and improving his products to cater to the varying needs, habits and mores of the customers. He needs to go on adding some distinct features and benefits to the existing product to meet the challenges of competitors' offerings. Development and launching of new products are also necessary to earn more profits and hence for the survival and growth of the company. A product may be relaunched with a new variant or with redesigned marketing strategy. Thus for the sake of our study, a new product could be original product, a modified or improved version or new addition to the product line or even one belonging to another product group. All these variants would require designing or redesigning of marketing strategies.

13.1 NEW PRODUCTS: DEFINITION AND CONCEPT

We can define the term "new products" as follows:

New products are those products which offer such change for the customers that may necessitate designing or redesigning of marketing strategies.

It is evident from the above definition that it is the customer's perceptions, attitudes and behaviours that decide whether a product may be regarded as new or not. A product can be called *"new only"* when there is a change in customer perception, attitude and behaviour as a result of redesigned or new marketing strategy. It suggests that a product should be new not only to the producer but it should be perceived as such by the customer as well.

Booz, Allen & Hamilton, Inc., an international management and consultancy firm, identified six categories of new products in terms of their newness both to the company and to customers in the market place. These categories are:

- **Additions to existing product lines:** Products that supplement a company's established lines
- **Improvements in/revisions to existing products:** Products that provide improved performance or greater perceived value and replace existing products.
- **New product lines:** Products that allow a company to enter an established market for the first time
- **Cost reductions:** New products that provide similar performance at lower cost.
- **New-to-the-world products:** Products that create an entirely new market.
- **Repositioning:** Existing products that are targeted to new markets or market segments.

Generally, managements tend to adopt lower-risk new product development strategies. Obviously *"new-to-the-world products"* and *"new product lines"* categories contain relatively higher risk, though they may yield higher profits.

13.2 FACTORS HINDERING SUCCESSFUL NEW-PRODUCT DEVELOPMENT (WHY NEW PRODUCTS FAIL?)

As stated above new-product development involves very high risk. In case of failure of a product, the company will have to incur huge losses and it may not be possible for it to recover its investments. High cost of capital, government regulations, shorter product life, and growing international competition enhance the risk. Studies show that failure rate of new-products is considerably high. There are many forces at work that alter consumers' needs and wants, life styles and preferences and thus affect the success of a new product. Some important factors that hinder successful new-product development are given here.

1. **Paucity of new-product ideas:** There are areas in which scope for improving some basic products is very limited and very few good ideas are forthcoming, especially in case of detergents, toothpastes, steel etc.

2. **Environmental and governmental factors:** The companies have to comply with certain governmental regulations concerning

protection of consumers' interests, maintenance of ecological balance and safety requirements. These have constraining effect on the development of new products.

3. **Competition and Fragmented Markets:** Due to competitors' offerings to the same market segments, the market share of a product can go down as the market gets fragmented. Usually, new products are also addressed to same or similar market segments. In presence of strong competition, the product's success is hampered. Again competitors' similar offerings shorten the product life's cycle.

4. **Paucity of Funds:** Development of a product and its introduction in the marketplace involves high costs and requires huge funds for R&D, manufacture, advertising etc. Shortage of funds thwarts all these efforts. Inadequate budgeting hampers the development.

5. **Time Interval:** It takes time to convert an idea into a marketable product. The time between generation of an idea and its introduction in marketplace may spread over several years. In the meantime, there may be substantial changes in socio-economic conditions, customers' needs, attitudes and behaviour and the whole effort may come to naught. Again the competitors might bring out similar products in the marketplace before the company introduces its product. Speed is the decider in such circumstances.

6. **Inappropriate Organisation Structure:** The success of a new product is also hindered if the organisation structure is not suited to the new product. Same organisation structure that may be good for existing products may not prove efficient for successful introduction of the new product. Thus managerial problems like poor timing of introduction, purchasing, inadequate controls over performance, failure to position the product also lead to product failure.

7. **Product Features:** The success of a new product largely depends on matching of its features to target markets. The success of a new product is hampered if there is no effective combination of product's features with target markets. Thus failure to establish a competitive market position also accounts for new product failure.

All these deficiencies and difficulties need to be removed for successful development of a new product.

13.3 NEW PRODUCT DEVELOPMENT PROCESS

Broadly speaking, new product planning should start with the development of overall new product strategy. New product development decisions are generally taken by the top management of the company at Corporate Headquarters level. Therefore any new product plan must be drawn in accordance with the guidelines provided and criteria set up by the corporate management or company management. In other words, it is the corporate management which is to provide guidelines for developing new products so that the new product development process becomes an integrated part of corporate new product development strategy. Formulation of corporate or business new product development strategy should, in fact, be the first step in new product development process.

Moreover, for successful new product development, an effective organisational structure needs to be built to manage new product development process and monitor consumer adoption processes.

New-product development process consists of nine steps — new product development strategy, idea generation, idea screening, concept development and testing, marketing strategy, business analysis, product development, market testing, and commercialisation. We shall explain them briefly in the following pages.

1. New-Product Strategy Development

New product strategy is corporate's basic plan of action that provides guidelines and overall direction for the new product development process. The new product development programme must fall under the ambit of the objectives of company and contribute to fulfillment of its mission. The new product strategy will link the corporate objectives to new product development programme and effort. Development of new product strategy involves identifying corporate growth role for new products, scanning external environment and identifying new product opportunities, analysing industry to determine the growth potential, assessing previous new product experience, assessing company's internal capability, appraising corporate work culture and evaluating position of existing products in the product life cycle. This will help in identifying the markets for which new products need to be developed. Thus new product strategy, in fact, prepares ground for initiating the actual new product development process.

2. Idea Generation

The next stage in the new product development process is the search for new ideas in the light of the corporate new product strategy. Some practitioners consider idea generation or discovery of new product ideas as the first step in the new product development process. The main purpose of this step is to identify new ways to satisfy customers' needs and wants and to understand evolving technologies. Idea generation means creation of or having a large number of ideas to this end. Idea generation or search for ideas will involve use of various sources for new-product ideas and creativity (or idea generation techniques). A number of new product ideas may come by accident or intuition and a few of them may prove successful also, but search for ideas is too important a matter to be left to chance.

(a) Idea Sources

Major sources of ideas are — customers, inventors and scientists, competitors, marketing personnel, distributors and middle men, consultants, investors, advertising agencies and top management. Ideas may come from any source, what is needed is to "foster and manage the flow of ideas uninterruptedly and seriously".

Customers: Study and analysis of customers' needs and wants and their reactions can lead to generation of many ideas. It can be done through customer surveys, group discussions and interviews. Marketing research has a great role in the discovery of ideas. Customer complaint letters may contain ideas for improving products.

Competitors: Another important source of ideas is the competitors' products themselves. A keen study and an intensive analysis of features of a competitor's product can yield new product ideas. Again reactions of customers toward the competitor's product and distributors' evaluation of competitor's product can be helpful in search for ideas.

These can be obtained through market intelligence. Some companies hire employees of competitors. However ethical and legal issues are involved in it.

Inventors and Scientists (R&D):   Inventors and scientists are a major source of technological improvements and new sources. Most of ideas regarding product involving use of technology and scientific formulations come from scientists and inventors. Most of the inventions and discoveries are the result of Research and Development Effort. Yet the marketing concept suggest that the ideas should come from analysis of customers' needs and R&D should determine what products are technologically possible. Products are to be developed to fulfil the needs of the customers.

Marketing Personnel or Employees: Marketing personnel or employees are the better judges of company or product as they are in direct touch with the customers and channel members. In fact, they are a privy to their reactions about the product. They can provide many tangible ideas. They need to be trained for the purpose and given due incentive for giving new ideas.

Distributors and Middle Men: Distribution channels are another source of new product ideas. On one hand, they are exposed to customers' needs and wants while on the other hand they face the competitors' challenges and are familiar with strengths and weaknesses of competitors' products.

(b) Idea Generation Techniques

Techniques which are used in marketing research can also be used in creating new product ideas. These techniques are summarised below:

Direct observation: Direct observation can help in searching an idea while analysing and solving the problems of customers. The deficiency in a product may give an idea that may lead to modified and improved product.

Listing of attributes: In this technique the features and atrtributes of an existing product are listed and modified in search of an improved product. For example, earlier Colgtate tooth brush had flat head and it was thought that it did not brush deeper. This gave an idea, that the tooth brush would brush deeper and reach all corners of the mouth if the head is diamond shaped and the Tynex, rounded bristles, are V-trimmed. This resulted in launching of Colgate Plus Zig Zag Tooth Brush.

Brain Storming: It is a technique in which small group of customers, say six to ten are asked to give as many ideas an they can. Brain storming would mean filling customer brains with all sorts of thoughts and thinking about a specific problem and encouraging the participants to toss out new product ideas. The session may last one to two hours and customers may contribute about a hundred or so new ideas. No criticism or evaluation of ideas advanced is allowed. The session is largely unstructured and freewheeling. Brain storming is characterised by total absence of any criticism or evaluation of the presented idea, freewheeling, in presentation, pooling of a large number of ideas and encouragement to participants to advance still newer ideas. Sometimes, the problem remains unspecified and a secret and a general discussion of related issues is held. The proceedings can be tape recorded. Questionnaires can also be used to elicit information.

Focus Group Discussion and Interviewing: This is another technique similar to brain storming. Here the sessions are general and free-wheeling but ideas presented and related issues are evaluated and discussed critically.

Forced Relationships: Here each object is studied in relation to other objects and ideas are invited regarding how to provide all or more benefits.

Customer Preference Testing: This is proper research technique which is used to understand customer decision processes or customer perceptions. This would involve:

(1) determination of the specified product-market.

(2) identification of attributes that customers usually compare to select out of the product alternatives.

(3) knowing customer's perception of benefits being provided by each product in terms of identified attributes.

(4) developing a model predicting customer's preference, and

(5) developing a method for searching new product ideas about an ideal combination of these attributes in accordance with customer's preference.

Environmental Forecasting: A changing environment implies changing markets and hence new product opportunities. Environmental forecasting will also provide many new product ideas.

3.   Screening New Product Ideas

The third step is to screen and evaluate the pooled ideas to select those which looked feasible and worth development. The purpose is to prune the pool of ideas obtained as a result of idea generation exercise using valid criteria and to select those ideas that could become a successful product and reject those which lacked potential. It would be desirable at this stage to drop as many ideas as possible so as to avoid wastage of resources.

Screening new product ideas is the process of determining and using criteria to evaluate the potential of new product ideas.

In this process, the management should avoid two types of errors:

(1) It should not reject an idea which could become a successful product later i.e. which is otherwise a good idea. This is *Drop-error.*

(2) It should not accept an idea that later fails i.e. which is otherwise a poor idea. This is *Go-error,* meaning accepting poor idea.

Screening Criteria

There can be two categories of criteria for screening the new product ideas, namely those determining their correspondence to the corporate goals and objectives and secondly those reflecting company's ability and resources to take advantage of those ideas.

The first category of criteria is to determine whether a particular idea is consistent with company's mission in terms of satisfaction of similar customer needs, with its image and reputation in terms of quality of the product and its superiority over competitors' offerings or new-ness, with the expected financial performance of the company in terms of market share, return on investment, pay back period and in terms of market opportunity potential. These criteria determine whether a new product idea is compatible with company objectives.

Again, the company has also to judge whether the firm has the requisite capabilities and resources to convert the idea into a successful product. Ideas need to be evaluated in terms of company's resource capabilities. This category includes investment commitment, length of existing product life cycle, technical skills required, marketing programme requirements (advertising, direct selling, distribution channels, service facilities), probable price, patent protection etc. The company must have these capabilities before accepting the idea. These criteria evaluate whether a new idea is compatible with company's resources.

Those ideas which do not meet the above criteria at any stage need to be dropped.

Evaluating New Product Ideas

In some cases, product managers are required to describe the new product idea in terms of its likely target market, size of the market, estimated costs, probable price, profitability etc., so that a sound decision could be made.

In order to arrive at the correct decisions, ideas are also evaluated using some qualitative rating and scoring criteria e.g., culling or qualitative criteria may be — Will the new product get more shelf place or get more dealers? Will the product require new technology? if the idea does not fulfil the set qualitative criteria, the idea can be dropped. Again idea can be rated against product success requirements like company goodwill, finance, location and facilities, marketing etc. There may be scoring criteria as well to evaluate the idea e.g. growth rate of sales-volume, length of product life cycle etc. etc. Scores are apportioned for fulfilling of each of the scoring criteria. Again those ideas that do not meet the set rating or scoring criteria (set score) will be rejected.

It must be kept in view that acceptance of a new product idea at screening stage will cost heavy expenditures at later stages. Therefore, this stage is crucial in decision' making and every care should be taken before selecting an idea for conversion into a new product.

4. Concept Development and Testing

A. Product Concept

A product idea symbolises a possible product that a company thinks it would offer to the market. The product idea needs to be transformed into a *product concept* so that it has a meaning for a consumer. Idea is an abstraction whereas *product concept symbolises something concrete which can be perceived by the consumer.*

For example, in case of tooth brush, the product idea is that there is need of a tooth brush which should reach every corner of the mouth. The idea can be elaborated in the form of a product concept when it is conceived that a diamond shaped head with V-trimmed bristles would make the tooth brush reach every corner of the mouth. Thus a tooth brush with diamond shaped head and V-trimmed bristles would form the product concept. This will have a meaning for the consumer and he will easily understand it. The matter has been further elaborated in the case study, given at the end of this chapter.

Product image presents a picture, a consumer's perception of a potential product. Product concept is, in fact, producer's conception of the proposed product.

Again, it is concept that can determine the degree of competition it would face in the market. But before making decision for developing the concept into real product for marketing, a product concept needs to be tested in the target market.

B. Concept Testing

Concept testing aims at seeking customer reactions to a proposed product before developing it actually. Concept testing helps in finding out whether consumers actually need the product and will be willing to buy the proposed product when offered. The concept that is presented to consumers for their reactions may be in a symbolic form consisting of some statements about the product or its picture, its features and benefits offered, or it may be presented physically. The product manager contacts a sample group of the target market and records their responses and reactions to a questionnaire prepared for the purpose. Thus in this step, potential customers are asked to evaluate a product concept rather than the actual product. Concept testing helps the management in assessing the feasibility of the proposed product, and in planning and developing marketing strategy for the proposed product, if the concept is found feasible and practical as a result of the concept testing.

It must be noted that it is not the product idea that a consumer buys. It is a product concept that a consumer buys. Again, all selling is concept selling, therefore concept development and testing are of critical importance. They work as assurance against unnecessary risks. Concept testing helps in evaluating the potential market acceptance of the proposed product.

5.   Marketing Strategy Development

After the product concept is tested and passed, the next stage is that of preparing a preliminary marketing strategy report for introduction of the new product into the market. This report should contain the following information — (1) the size, structure, preferences of target market, (2) proposed product's price, distribution strategy, advertising and promotion budget i.e. marketing mix offering for first year, (3) sales and profit objectives and required marketing mix offers over a time. The next step will be to assess whether these preliminary projections correspond to the company's objectives.

6.   Business Analysis

The next step is to assess the likely profitability of a new product and hence to find economic feasibility of new product idea. Business analysis is an indepth study of estimated financial feasibility of new product ideas and hence to determine whether or not to develop the new product. It would involve projecting sales, costs and profits. Business analysis involves use of a number of methods to estimate these figures and study economic feasibility. The accuracy of these estimates will depend on accuracy of demand forecasts and other related market factors. Such information may include estimated demand for the new product, seasonal patterns of consumption, competitor's market share of similar product, price elasticity of demand, volume-cost-profit study at sales levels etc. An important method used for business analysis is described here.

Break-Even-Analysis

Break-even point is that level of sales-volume at which total revenue is equal to total cost. It is that volume of sales that will provide revenue to cover the costs of production, distributing and selling a new product. Volume less than this level will result in loss for business and the level beyond this point will yield profit. Break-even analysis does not help in finding actual volume of sales. It determines that level of sales that is necessary to recover costs. It is the break-even-point. The management here is to assess whether or not the estimated sales would be more than the break even volume.

The break-even volume is determined by the formula

$$S_{BE} = \frac{\text{Total Fixed Costs}}{(p_u - vc_u)}$$

Where S_{BE} denotes break even sales volume, p_u the price of new-product per unit and vc_u, the variable cost per unit of the product.

Estimation of the break-even volume is not possible unless forecasts of production, distribution and selling costs are available. This would mean devising before hand the production, distribution and selling strategies.

As prices, revenues and costs are likely to change over time, the future revenues and costs will have to be discounted by opportunity cost of a company's capital, usually represented by rate of interest on debt.

These forecasts can be spread over a period of time giving due allowance for change in price level and a discounted cash flow is estimated. Business analysis will show whether or not introduction of the new product will be profitable. If the business analysis suggests favourable response of the market, then the new product is worth development. In case the analysis suggests otherwise, the idea can be dropped at this stage.

7. Product Development

After making sure that the new product concept is economically feasible, the next step is translating the concept into actual functioning product i.e. developing a prototype functioning product. Product development is a technical and engineering task usually entrusted to R&D Department. A prototype represents a physical version of the product concept. The prototype working model provides all the benefits preferred by the consumers and can be manufactured on the budgeted cost. At this stage costs of manufacturing, packaging and distribution are also estimated. This step tests functional performance i.e. product use and ensures technical and commercial feasibility.

The prototype product may require functional testing or product use testing. This may have to be done in a laboratory or in field. In some cases, products may need testing and approval by governmental agencies. *Consumer preference testing* can also be used to evaluate features and benefits. It offers scope for further addition of benefits. Some companies give samples of consumers goods to consumers for use in their homes and ask of their reactions.

After making it sure that the product performs the expected functions, the next step is to present it in marketable form for market testing.

There are some high risk goods like books, films etc. of which development of prototype is not feasible due to the high costs involved and will need to be introduced direct to the market. These would need careful scrutiny and editing before they are introduced.

8.   Market Testing

After the prototype model passes the performance test, the next step is to produce the product and embellish it with brand name and proper packaging and present the product as it is to be offered to the consumer in the market place. Before it is introduced in the market, it needs to be tested in the actual market situation. Market testing helps the management to assess the reactions of the consumers and members of distribution channels towards the new product, and to find the size of the market. Different methods are used for test-marketing depending on the types of products i.e. consumer goods and industrial goods. Market testing is more costly than the concept testing or product performance testing.

As stated earlier, some companies distribute free samples to consumers for trial and use in their homes. This can also be done to test consumers' acceptance of the product. Such a course may take time before a new product is introduced. To quote an example,

---

Hindustan Lever Limited, now Hindustan Unilever Ltd., distributed in April/May 1996, 5000 free samples of 500 gms. each and thousands of 20 gm. sachets of their detergent International Surf Excel to consumers for trial before it was introduced in the market. Now Surf Excel has become a popular brand in the market.

Some companies use it as a promotional effort to get their new product through in the market. If the product passes the marketing test, the stage is ready for its actual introduction into the marketplace with total commitment of the management.

9.   Commercialisation

Commercialisation is the last stage of the product development and involves actual introduction of the new product into the target market. This would require commitments of all resources and related decisions.

For example, for this stage, manufacturing facilities will have to be created and set into operation, raw materials and supplies will have to be arranged, distribution outlets will have to be decided and sales personnel must be hired and a marketing mix strategy will have to be devised to support the introduction of the product. All this will require huge investment. Decisions regarding timing of introduction, geographical area to be covered, segment of target market to be served and budgets to be apportioned to marketing mix components become crucial.

Timing

In marketing a new product commercially, timing is a crucial factor. The company that introduces the new product before others think to do so reaps better advantages. Innovative products are often introduced quickly lest others should initiate. Again there can be *crash introduction* or *rollout introduction*. It depends upon product market situation and likely competitive entry.

Crash Introduction: Crash introduction means introducing the new product to the entire target market quickly and intently with full support of marketing mix elements. It requires deployment of all resources in the marketplace to promote the product for adoption. Such an introduction allows little time to competitors to prepare and is advisable when there is apprehension of competitive entry. However it involves high risk.

Rollout Introduction: This is slow marketing of the new product involving division of the target markets geographically and introducing it initially in one area, region or more. In case the product achieves success in that area or region, marketing is extended to other areas to serve the entire target market. This marketing strategy is very useful where competitors are not quick enough to respond and bring out their products to compete. The resources can thus be deployed in stages. Rollout introduction is advisable in case of those companies who have not resources enough to go in for crash introduction. However slow rollout introduction can give substantial leverage to the competitors.

13.4   NEW PRODUCT DEVELOPMENT ORGANISATION

As stated earlier, development of a new product necessitates cooperation and coordination of efforts of almost all functional departments. The process can be facilitated if this function is entrusted to individuals or a group of people who may be capable enough to elicit cooperation of all and integrate various functions. This work can be entrusted to a product manager who may be responsible to look after existing product and develop new products or a manager may be appointed exclusively of developing new products (New Product Manager). Again large companies have a separate new product department with R&D facilities. In several companies, this function is performed by a group of people entrusted with this job. They work as *new product committees* consisting of top level managers from functional areas within the organisation. Another type of organisation is what is called *Venture Committees* organised as a separate organisational unit and established exclusively to develop new products. Specialists and experts can be appointed in this organisation.

13.5   CONSUMER-ADOPTION PROCESSES

The success of new product or innovation lies in its early adoption (purchase) by the consumers and diffusion throughout the markets. It poses a challenging task for the marketing department. It is, in fact, this all important task to which all marketing efforts need to be directed. Consumer-acceptance involves following considerations.

A.   Adoption Process

Studies show that a consumer goes through the following five stages of adoption process:

(i) *Awareness:*   The consumer comes to know the existence of a new product or innovation

(ii) *Interest:*   The consumer evinces interest in the product and seeks information about it.

(iii) *Evaluation:*   The consumer considers and weighs whether or not to try the new product.

(iv) *Trial:*   The consumer tries the product once or twice to find out its value.

(v) *Adoption:*   The consumer purchases and repurchases the new product and becomes a regular user.

The task of marketing manager is to take steps as may facilitate and quicken consumer movement through these steps.

B.   Consumers' Individual Differences

Again there are marked differences in individual consumers as to their responses to innovations. Some consumers adopt new product more readily than the others. Some are innovators or consumption leaders or pioneers and adopt innovation immediately. Others take their own time to do so.

Everett M. Rogers has categorised adopters on the basis of relative time of adoption of innovations (new products) into five groups.

(i) *Innovators:*   Consumers who immediately are willing to try a new product even at some risk. They form a small proportion of adopters.

(ii) *Early adopters:*   They are consumers who try the new product early but after giving due consideration. They are opinion leaders.

(iii) *Early Majority:* This is a group of consumers who try innovation on their own but are not opinion leaders. Such a section may form one third of all eventual adopters.

(iv) *Late majority:* They represent the consumers who try the new product after others have tried the new product. Again they form about one third of eventual adopters.

(v) *Laggard:* These are the consumers who take considerable time to try the product. They adopt it when the use of product has become almost a tradition. These are tradition bound and form about one sixth of the eventual adopters.

Again, it must be kept in view that diffusion of the product depends on the potential buyers perception of product's attributes and benefits and it is the responsibility of the marketing personnel to make the consumer aware of the product's strengths and remove the misconceptions, if there are any. The rate of adoption is also influenced by personal influence and opinions of other persons.

13.6 PRODUCT FAILURE-FURTHER CONSIDERED

We have already explained in an earlier section of this chapter the factors that hinder the development of new product and cause its failure. We have also referred to Go-error while discussing screening of new-ideas. It is generally the Go-error that results in product failure. The product failure is of three types:

(i) *Absolute Product Failure:* It occurs when sales do not cover even variable costs, resulting in heavy loss of money.

(ii) *Partial Product Failure:* It occurs when sales cover all variable costs and a part of the fixed costs, resulting in loss of money.

(iii) *Relative Product Failure:* It occurs when sales cover all costs (variable and fixed) and yield a profit but not as expected or targeted by the company.

A case study of product failure is appended to this chapter.

APPENDIX TO CHAPTER 13*

*CASE STUDY 13.1:
BOMBAY OIL INDUSTRIES LIMITED:
PRODUCT FAILURE

*(For MBA and advanced students)

Marico Industries Limited is a wholly owned subsidiary of Bombay Oil Industries limited. It took over the consumer products business and organisation from Bombay Oil in April, 1990. Mr. Harsh Mariwala is the managing director of Marico Industries Ltd., Prior to this, Mr. Harsh managed Bomaby Oil Industries Ltd., which produced and marketed well reputed brands like **"Parachute Coconut Oil"** and **'Saffola Refined Oil'**. In 1980s, Bombay Oil introduced a number of new products in the market which proved failures, despite good images attained by their brands 'Parachute' and 'Saffola'. The following description tells the story in brief.

Product Failures

Every thing looked to be going well for the Consumer Products Division on the surface. However, the decade of the eighties also saw some failures of products in the market place. Harsh wanted to diversify and widen his portfolio of consumer products. He did not want to be over reliant on 'Parachute' and 'Saffola' in the long run. A tooth powder by the brand name 'Whistle' was launched in select markets in 1981. The tooth powder market was dominated by Colgate with a host of small regional brands. In launching 'Whistle' it was hoped that shares could be grabbed both from Colgate and the local brands. Pricing was competitive but Bombay Oil had not quite reckoned with Colgate's intrinsic strength in the market. A distinct image was not built for 'whistle' alongside frequent changes in quality based on inadequate market feedback. The feedback that came from the field seemed to be different from market to market and changes in the product were not well researched. The volumes never really grew and it was withdrawn from the market in 1983.

A groundnut oil under the brand name of 'Parachute Filtered Groundnut Oil' was launched in select markets in 1984. Though the product created a small market for itself, it could not become a truly national brand. Significant investments in terms of advertising and promotion were not available for this brand and the margins were not substantial and the markets were very localised, this brand was also withdrawn in 1986.

Packaged pulses under the brand name of 'Parachute' were also launched in Bombay in 1984. This product did not last very long in the market as the retailers resisted the entry of this brand. To them selling loose pulses was far more profitable than selling packaged pulses.

Why did all the new products fail? Was umbrella branding the right strategy? Did all the new products receive the necessary resource back-up? While some of these questions were being echoed, Harsh summed up these failures. He said, "We can always attribute these product failures to some phenomena outside of ourselves. But I think the truth of the matter is the poor quality of thinking we managers then did. There was little attention given to product development and quality. We were not willing to commit big resources behind any of those products because, I suspect, we were not sure of our thinking. Just as in the past when we moved over from the bulk business to the consumer business, what we needed was a major transformation in the orgnaisation's thinking, to begin looking at even products outside the oil trade. We were caught napping. "These products failures each close on the heels of the other, led the Consumer Products Division into a three year hibernation at least as far as new products were concerned".

Source: The Marico Story

EXERCISES

1. What do you mean by new product? Give some broad categories of new products.
2. What are the major factors which cause failure of new products?
3. Explain the process of developing a new product
4. Describe the process of consumer-adoption of new product.
5. Write notes on:

   (a) Product failure

   (b) Break-even analysis

   (c) Concept development and testing

   (d) Idea generation techniques

   (e) Screening new product ideas

   (f) Product development.

6. A firm wants to introduce a new pack of a set of chocolates of different designs. Subject the new product proposal to your screening process and identify possible problems.

14

Stages of Product Life Cycle

As stated earlier, marketing strategy is greatly influenced by the distinct stage of the product life cycle through which a product is passing. Different stages of product life cycle pose different competitive challenges and offer different types of threats and opportunities. An understanding of the concept of the product life cycle is crucial for devising an effective marketing strategy.

14.1 PRODUCT LIFE CYCLE (PLC): CONCEPT AND STAGES

Studies of sales of products over a time show that products have definite i.e. limited lives and the sales of each product go through different stages over the time signifying a life cycle. Each product, that is introduced in the market, has a limited life span and has to go out of the market in its present form one day.

Product life cycle signifies different stages through which the sales of a product move during its life. It is fluctuations in the volume of sales of a product in time-sequence that determine the distinct stages of its life. A typical product is thought to move through four stages as measured by sales over a time, namely *introduction, growth, maturity* and *decline.* These stages representing different periods showing changes in growth of sales in the sales-history of the product form what is known as *product life cycle.* These stages of product life cycle are explained here.

Stages of Product Life Cycle

1. **Market Introduction Stage:** As the name suggests, it is the

introductory period when the market is acquainted with the new product. During introduction stage, the management incurs heavy expenses on acquainting the market with the product. *The product sales are low and grow slowly.* Profits do not exist and may even be negative. Though there is little competition from similar brands but the product has to surmount teething troubles due to its newness and because its use necessitates change in established customer behaviour. Moreover, the typical need of the customer is already being met by other product types. The customers need to be convinced that the new product being introduced is superior to other products in use.

2. Growth Stage: This is the period when more and more customers accept the product leading to increase in demand. *The product sales grow rapidly resulting in increasing profits.* Though the product type competition decreases, the similar brand competition increases due to entry of new competitors. This poses new threats and the marketing strategy needs to be adjusted accordingly. In this stage, sales and profits go on increasing.

3. Maturity Stage: This is the stage when the product has reached almost all the potential customers and achieved their acceptance. At the maturity stage, *the product sales grow slowly and the profits stabilise or are declining after reaching the peak.* During this period, the similar brand competition becomes stronger and the firm has to incur increased marketing expenses to keep the product in competition and devise a strategy aimed at assuring the buyers about the superiority of its brand over the brands of competitors. At this stage, demand tends to reach a saturation point.

4. Decline: This is the stage when the demand for the product falls rapidly as improved substitute products or new products appear on the market scene. *At this stage, the sales and profits shrink swiftly.* No product can last for ever and this is the stage when the product may have to be phased out. This stage may be due to:

(i) advance in technology leading to introduction of new products in the market and intense competition,

(ii) change in demographic environment,

(iii) changes in needs or tastes of customers,

(iv) loss of customers' purchasing capability,

(v) increasing costs of supplies leading to depressing margins and erosion of profits.

Figure 14.1 illustrates various stages of life cycle of a typical product.

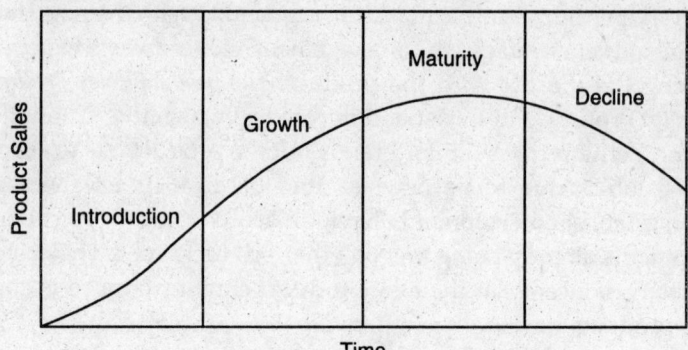

Fig. 14.1 Stages of life cycle of product.

14.2 MARKETING STRATEGIES FOR DIFFERENT STAGES OF PLC

As stated above, different stages of product life cycle require different financial, production, personnel and marketing strategies. We shall describe here briefly various marketing strategies that can be developed in different stages *viz.* introduction, growth, maturity and decline stages of PLC. What is required is to match the marketing strategy with the distinct stage of PLC.

A. Marketing Strategies in the Introduction Stage

In the introduction or pioneering stage, the main task is to acquaint the customers with the new product and its superiority over other similar products available in the market. In order to achieve this purpose, arrangements will have to be made to make the product available to the customers in the marketplace through dealers and other distribution channels. Introduction of a new product also involves special and concerted promotional efforts. The pricing of the product would also need careful consideration.

Introduction stage of PLC is marked by low sales and slow growth in sales-volume. The profits are low and may be negative. The management has to incur heavy expenses on promotion of the product. In the initial stage, competition from similar brands is missing as other firms would take time in developing a similar brand. But the new product faces competition from other product types. In this stage, the management can adopt one of the following four strategies as far as pricing and promotion of new products are concerned. The pioneering or initiating process may be slow or rapid depending upon the degree of intensity of competition and resources available with the company for promotional effort.

(i) **A rapid-skimming strategy:** This strategy involves setting a high launching price supported by high promotional effort to skim first profits from price-inelastic customers. The company incurs heavy expenses on promotion of the new product and sets a high price of the product to gain as much profit as possible before actual competition appears. This strategy helps in recovering the investment rapidly and building the brand image, though the high margins of profits would attract competition. This strategy is suitable when a large segment of the potential market is not aware of the new product.

(ii) **A slow-skimming strategy:** This strategy involves setting a high launching price supported by low-promotional efforts to skim first profits from price-inelastic customers. The company incures low expenses on promotion of the new product to reduce its marketing cost but sets a high price to gain as much profit as possible in absence of any immediate competitive threat. This strategy is suitable for a small market most of whom are aware of the new product.

(iii) **A rapid-penetration strategy:** This strategy involves setting a low initiating price supported by high promotional effort to acquire and establish a large market share of price-elastic customers, most of whom are not aware of the product. This strategy is suitable when potential competition is strong and a firm can reap the advantages of large scale production and thus can capitalise on production economies.

(iv) **A slow-penetration strategy:** This strategy involves setting a low launching price supported by low promotional effort to establish market share in a large market of price-elastic customers, most of whom are acquainted with the product. In presence of some potential competition, the company keeps the launching price of the product relatively low. At the same time, it also keeps its promotional expenses down. While on one hand, customers accept the product readily, relatively low margins discourage or preempt the entry of competitors. This strategy is appropriate if the company can effect production cost economies and retain an initial foothold.

B. Marketing Strategies in Growth Stage

In this stage, the product has achieved acceptance of a large number of customers. This stage is marked by high sales-growth and increasing profits. This is the stage when similar brand competition is increasing. The pricing aspect is important here, but there is shift in promotional effort. Here promotional effort is aimed at convincing the customers about the superiority of company's product over similar brands of competitors, promotional expenditure remaining at same level or a bit more. It strengthens its competitive position. Besides, the company's strategy will consist of developing and adding more new and improved features and searching new markets to increase sales and profits over a long period. In fact, the devised strategy should help the company in prolonging this period. Developing new product features is an essential part of the marketing strategy in this stage. The marketing strategy in growth stage generally aims at expanding market and may consist of adding new features and improving product quality, expanding distribution network, entering new markets, making additional promotional efforts and if need be, even lowering prices to reach segments of price-elastic customers.

C.   Marketing Strategies in Maturity Stage

This stage is marked by slow sales growth and declining profits. There is strong presence of competitors. The objective is to maximise profit while keeping the existing marketing share. The marketing strategy would involve improving the product, expanding the market or number of brand users or modifying the marketing mix i.e. price-distribution, advertising mechanism so as to increase sales volume. This stage is longer than the earlier stages but now the product has reached the maturity stage and the situation is ripe for planning new products that may replace the existing products when their end is near or there is time for brand switching. This is a stage when a company confronts tough competition and profits erode.

D.   Marketing Strategies in Decline Stage

This stage is marked by decreasing sales and declining profits. Sales of the product are confined to laggards. As the profits erode, the number of competitors is small but there are few to buy the product. There is need of reducing the expenditure and to encash whatever can be got out of the brand. In this stage, the strategy would involve cutting drastically the marketing mix as far as products in decline stage are concerned, and making necessary replacements by new products. In this stage, the marketing strategies may involve brand differentiation or phasing out weak products, price reduction, reducing distribution outlets and minimising promotional effort and diverting resources to replacements. It must be kept in view that post-sales service continues to be maintained for past buyers. However, every attempt should be made to build sales and market share by taking advantage of production economies, if situation permitting.

14.3 UTILITY OF PRODUCT LIFE CYCLE

It is evident from the above study that the relative importance of various elements of marketing mix would shift at different stages in the product life cycle. This makes the product life cycle a key marketing planning tool and it can be successfully used in making sound and effective decisions. PLC can help the company in many ways:

1. It helps in preplanning the new product launch as it would be possible to predict the market situation, the competitor's challenges and threats, and market opportunities likely to be available.
2. The PLC concept can help in prolonging the growth and maturity stages of the PLC through carefully designed marketing strategies.
3. It helps the management to follow a proactive approach, in the sense that it helps in anticipating the competition the product is expected to face and how to meet its challenge effectively. It is useful in analysing competitive conditions and devising measures to surmount them.
4. It helps in making decisions regarding long-term investments and deployment of resources to company's product mix.
5. It helps in timely phasing out of the products that are not wanted and thus reducing the losses to minimal level.
6. It helps in reading and assessing changes in customer behaviour and thus in adjusting the market strategy to meet various situations as they arise.

14.4 MARKET EVOLUTION

Almost analogous to the concept of product life cycle is the broader concept of market evolution in which new needs, new competitors, new attributes and new technologies play crucial parts. Markets also evolve through four stages, namely market emergence stage, market growth, market maturity and market decline.

Market Emergence Stage: New need leads to creation of a new product and hence a new market. Large pioneering companies usually create a product that may satisfy a particular need of a large number of potential buyers i.e. pioneering firms devise what is called *mass market strategy.* Pioneering firms use skimming price strategy to skim the cream and take advantage of innovations. They may use penetration price strategy to establish a larger market share in the pioneering stage until competitors enter the market with similar brands.

Market Growth Stage: As the demand of the product grows and more and more competitors enter the market with similar brands, the size of the market grows. In this stage, the pioneer strives hard to keep its market share in tact and improve it and adjust its marketing strategy accordingly, while new entrants try to achieve a foothold. On account of stiff competition and declining profits, the market growth slows down and the market reaches the stage of maturity.

Market Maturity Stage: With the entry of a large number of competitors and mass production of similar brands, each having established its own market share, the market growth slows down as the profits tend to decline. This is the maturity stage of the market evolution. In the growth stage, almost all major market segments are covered. But in maturity stage, companies try to acquire a niche of brand loyal customers and the market gets fragmented. But, in order to maintain their market share, companies have to add new attributes to the products and this results in consolidation of the market. Thus market passes from a period of fragmentation to a period of consolidation and again to fragmentation and so on till it enters the stage of decline.

Most of the large companies like Hindustan Unilever, Procter and Gamble, Colgate-Plamolive add new attributes like Fluoride, Perfumes and other ingredients and add Plus, Super, International, Excel, etc. to their brand names to make their brands more appealing and to consolidate the market.

Decline Stage: The market reaches the decline stage when the demand for the product falls substantially or ceases because it is not needed any more or a new technology replaces the old technology which becomes obsolete. New technologies, with added benefits and services, catering to new needs of the customers, will lead to creation of new markets as old markets decline. New markets will move through the same stages as referred to in the above discussion and the process goes on.

EXERCISES

1. Define product life cycle. Explain the stages of product life cycle.
2. Give some marketing strategies for different stages of PLC.
3. What is meant by market evolution? Describe its stages.
4. Write notes on the following:
   (a) Rapid skimming strategy
   (b) Rapid penetration strategy
   (c) Utility of product life cycle.

15

Managing Services

In chapter 1, we have referred to non-material goods — goods that have no shape and cannot be seen, touched, felt, tasted or transferred. These goods are intangible but satisfy needs and wants. These intangibles are offered and purchased like physical products and are called services or service products, e.g. services of a doctor i.e. medicare, education, orchestra's music, shoe repairing, stitching suits, hair cut, consultancy, business goodwill etc. Services are provided by the government sector (law and order, employment, justice), the non-business or social sector and the business sector (airlines, banks, telephone services courier services, hotels, joy rides, establishments like clubs and entertainment parks).

These services also need to be marketed like physical products and special marketing strategies need to be devised to that effect depending on the nature of these services and production-consumption blending. Again an important feature of services is that their production and consumption takes place simultaneously and involves face-to-face producer—customer (server—served) interaction.

15.1 SERVICES AND THEIR CHARACTERISTICS

As stated above, service is not a tangible product or object like a book, pen, car, aeroplane, it is an act, a deed, a performance that is offered by a producer to another person who needs it. It may or may not be concerned with the marketing and performance of a physical product. It is essentially intangible and cannot be possessed. However, it can be purchased and consumed.

A service is a deed, act, performance or effort that is offered by its producer to a consumer who needs it or target market. Unlike physical products it is intangible. It is consumed but not possessed. Its production and consumption takes place simultaneously.

Services include health care, hospitals, teaching (schools, colleges, institutes), religious ceremonies, advertising, travel, entertainment, face lift, consultancy, insurance, communication, charitable acts of social and religious organisations like Red Cross, sports bodies and organisations. Some services are also tied to a physical object such as pre-sale and post-sale services for TV, car, machines, computers and durables.

In modern times, service sector represented by services organisations has assumed predominant significance in a country's economy. It employs a substantial portion of non-agricultural workforce and makes substantial contribution to the national income.

Again, there are some characteristics that make services different and distinguishable from physical products and so different marketing strategies are required to be designed to effectively market them. The characteristics that make services or service products different form tangible physical products are intangibility, inseparbility, variability and perishability. We shall briefly explain them and their marketing implications here.

1.   Intangibility

The most important characteristic of services is that they are essentially intangible and comprise performances, deeds, efforts which fall under the realm of abstractions. When a buyer or consumer purchases a service, there is nothing tangible or concrete to show for it.

Marketing Implications and Strategies

Intangibility or abstractness of services has important marketing implications for marketing managers. It poses mainly two challenges, namely (i) identifying and defining attributes of the service that may be of benefit to the customer and fit in his perception and (ii) turning abstract attributes into something concrete of importance to customers i.e. bringing life to identified service attributes in such a way that their importance may be clearly perceived by potential customers. This challenging task is what Theodore Levitt called *"to tangibilise the intangible"* i.e. showing tangible evidence of importance of attributes of the services offered to the customers.

"Tangibilising the intangible" is a challenging task and the marketing manager will have to design special marketing strategies and employ separate marketing mix to achieve the goal. Here the process is the reverse of what is employed in case of a tangible physical product. In case of a physical product, images for customer perception are drawn from the very tangible product, i.e. efforts are made to show the attributes of the product in the form of abstract images or intangibles whereas in case of marketing services, the task is to turn abstract attributes into concrete form and show concrete benefits of the marketed services.

At the end of this chapter, we are giving a few examples and cases how service organisations tangibilise the intangible and market their services.

2. Inseparability

Production and consumption of services take place simultaneously. The provider and receiver of a service are both present at the time service is rendered. It cannot be stocked as inventory. The producer and production of service is present at the time of its receipt by the customer. Production of service cannot be separated from the producer. The quality of service will depend on the ability and capability of the provider. The marketing strategy will involve *producer-customer interaction or production-consumption blending*.

Marketing Implications and Strategies

The inseparability of production of a service from producer and its simultaneous consumption puts dual responsibility on the producer. He is both a producer and a marketer in this customer- oriented market. The producer may be satisfying the specific need of a customer and thus rendering a customised service. Both the production and marketing functions need to be taken care of.

Another implication of this characteristic is that standardisation of quality of service poses a great challenge. In production-consumption interaction, customer's preference for a provider of service has an important place as quality of service cannot be separated from one who renders it.

Another challenge in the production-consumption blending is to match the supply of the service with the demand for it. For example, in a seminar, more people may come than the number for which arrangements were made. If the demand exceeds supply, then producers may have to increase their productivity. Thus marketing strategy will be designed accordingly. To illustrate, a teaching tuition may involve teacher-single student interaction or a teacher may teach a group of two or more or a larger group of students at the same time. The same strategy can be employed in conducting employees training and rendering other services to adjust supply to the demand.

3. Variability

The quality of service provided by different persons may differ from person to person. Again the quality of service provided by the same producer is bound to vary if rendered at different places or times.

Marketing Implications and Strategy

Maintenance of standardised service quality is a great challenge for the marketing manager. The quality can be ensured by employing competent persons adequately trained in the job. Educational and professional institutes, hospitals, hotels, and airlines select competent and experienced people and spend huge amount of money on their training and develop them for providing quality service. They take timely measures to detect deficiencies and make timely corrections. It would involve quality control.

4. Perishability

Services cannot be stocked or stored in an inventory and transferred therefrom to the customer at the time of need. The production and consumption of service take place simultaneously. Its perishability poses important marketing implications and poses problems of production-consumption blending.

Marketing Implication and Strategies

Fluctuations in demand for services are not easy to tackle with. A service may be in much demand at a particular time whereas at another point of time it might not be wanted. Service producers use different marketing strategies at the periods of peak demand and at off-peak hour demand.

Airlines, railways, cinema houses, hotels usually have a set reservation system to blend production-consumption. Railways increase their train services in peak seasons to meet peak-period rush whereas they curtail them during off-season periods. Hotels at hill stations charge less than normal rates during off season period.

15.2   SERVICE INDUSTRIES AND MARKETING CONCEPTS

Gone are the days when public spirited individuals and philantheropists, religious, denominational and charitable organisations ran schools, professional and technical institutes and hospitals to provide free education and medicare respectively to the people at large. Thanks to the winds of change in the wake of market economy, increasing materialism and liberalisation, they have now turned them into lucrative businesses rather than service institutes. May it be D.A.V. Public Schools, Delhi Public Schools, Christian schools, Sanatan Dharm or Agrasen schools and institutes or in case of hospitals Sir Ganga Ram Hostpital, Apollo, Escorts, Batra or others—all have become business ventures and it has become extremely difficult in their case to make a distinction between a business organisation and non-business or non-profit organisation i.e. between business sector or social sector. Anyhow, whatever be the motive, profit or social, the basic marketing concepts and techniques need to be applied to both if they are to be run efficiently and effectively. Both are producing and providing services to the people. Both have to face marketing problems and need marketing strategies and skills to tackle them. First, we shall discuss service industries in business sector.

Service Industries in Business Sector

Some important service industries in business sector are — hotels, airlines, railways, roadways, water transport, communication (telephones, other postal services, couriers, overseas communication), banks, non-banking financial institutions, credit institutions, insurance (life and non-life), real estate, dwellings, professional teaching and training institutes, clinics and nursing homes, recreation and entertainment, consultancy firms, repair and servicing, legal firms, laundries, beauty parlours, saloons etc. etc. They exist both in public (government) as well as in private sector.

This list is expanding day by day and new types of service industries are emerging on the scene, thanks to the policy of liberalisation of economy and increasing cultural influences from the west. Some recent additions are— resort country clubs, health and nature care centres, beach resorts and hill resorts, tourism and real estate corporations offering services like time sharing and holiday plans in these resorts and hotels, cabarets, discotheques, night clubs, astrological consultation centres etc. One of the most recent entries is that of placement firms and agencies or Personnel Search Service (PSS) organisations who search personnel for middle, senior and top level management positions for placement in Indian and foreign companies. Some foreign firms have also entered in this field. Some of them charge as much as $25000 (Rs. 12 lakhs or more) from their clients for a single placement at senior level in MNC in India or overseas.

Of late, tertiary or service sector has assumed greater prominence in the economy and consumers' lives. In India, this sector contributes about 54 percent of the total net domestic product. Besides, there is a large number of workers in primary (agricultural) and secondary (manufacturing) sectors who are rendering intangible services to these sectors.

What is important in this context is that *like physical products, services also need to be managed, marketed and delivered efficiently and effectively, if the service industries are to survive and prosper.* Non-application of marketing concepts and marketing principles in delivery of services adversely affects the functioning of the service industries. It is due to this very reason that a number of service organisations especially in public sector are in a mess.

Marketing Concept in Service Industries

In chapter 2, while discussing marketing philosophies, we have explained the four tenets of the marketing concept, namely a market focus, customer-orientation, integrated and coordinated marketing, and profitability or productivity. We have also referred to the need for internal marketing i.e. need for proper hiring, training, developing and motivating marketing personnel of the organisation so that they can contribute their best towards meeting and satisfying the needs of customers, and accomplishing the organisational objectives and its integration with external marketing. The same principles hold good for service industries as well.

An essential element of the marketing concept as applicable in case of service industries is delivering *quality service*. A service industry can compromise with the quality of service it delivers, only at its peril. Therefore maintenance and enhancement of service quality should form the essential ingredient of its marketing strategy. It should go on adding further benefits and satisfactions or what is called value to its services, if it is to survive and grow in the face of increasing competition in the field. *It would involve continuing addition of secondary service features to primary service package to distinguish its quality of service from that of the competitors and reap early advantages. Consumer satisfaction through service excellence should be its chief motto.* Again service quality as perceived by the consumer also depends on consumer-seller interaction or on the service skills and competence acquired by the deliverers of service.

15.3 MARKETING STRATEGIES FOR SERVICE INDUSTRIES

As stated above, service industries also need to devise their marketing strategies and offer specific marketing mixes to serve their customers through quality service. These strategies are to be devised after analysing the marketing environment and situation existing at a particular time. This suggests that alongwith other elements of marketing strategy, a service firm must convince and assure the customer of the quality of service to be provided by it and it should match the customer's perception of it.

Christian Gronoos presented a service quality model in which he has suggested two other elements namely *internal marketing* and *interactive marketing* in addition to 4Ps (Product, Price, Place and Promotion) of the marketing mix offer.

Internal marketing means first hiring competent persons with positive attitude and then training, developing and motivating them and all those service personnel who deal with customers. This is very necessary to provide high quality of service.

Interactive marketing aims at creating service skills in those who provide services. It suggests that quality of service is highly influenced by the quality customer-provider interaction. *Interactive marketing* puts emphasis both on technical quality as well as functional quality. The service provider must not only be technically competent, he must also show concern for the customer and inspire confidence in him.

The service marketers must focus on consumer's perception of the quality of service expected, on consumer perspective and on efficient communication to all concerned.

To sum up, an effective marketing strategy for a service industry should aim at *consumer satisfaction through high service quality*. In order to achieve this objective, the management should go on:

 (i) distinguishing its message and service from that of the competitors
 (ii) inducting high service quality i.e. service excellence into the organisation
(iii) finding ways and means to increase productivity or profitability, without jeopardising the quality of service so as to accomplish the objectives for which the organisation has been set up.

And the marketing strategy should reflect all these aspects.

(i) Meeting Competitors' Challenges

It must not be forgotten by the service marketers that the market place is usually rampant with fierce competition and the competitors would most probably be offering the same, if not higher level of service. So the firm should go what is called beyond "the point of encounter", distinguishing its message and service from that of the competitors. As stated earlier, this can be done by continually adding value i.e. by adding

Services

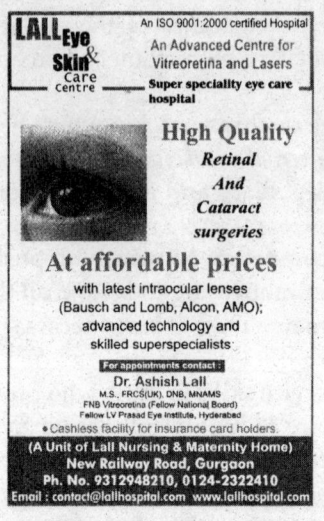

new features, benefits and satisfactions to the service and thus creating competitive differential, and building credibility and image. In this context, communicating all improvements and augmentations through advertising, and sales promotion is of critical importance. The focus here must be on consumer perspective and on efficient communication to all audiences, internal and external. There is need to convert the service story into a "buy message".

(ii) Enhancing Service Quality

Another important task of service marketers is to strive for providing higher level of service to the customers. What is required here is the acquisition of service skills and hence improvement of not only technical quality but of functional quality of the service. As already stated, internal marketing and interactive marketing can contribute a lot towards this end.

Consumer satisfaction through service quality is the result of expectations and experience. Besides, researchers have developed a number of lists of factors that determine service quality. One such list has been developed by A Parasuraman, Vilarie A. Zeithamal and Leonard L. Berry. According to them, the ten major determinants of service quality are:

(1) easy *access* to service (in terms of location and time)

(2) accurate and effective *communication*

(3) *competence* of the employees

(4) *courtesy* shown by the employees

(5) *reliability* i.e. consistency and accuracy in performance of the service

(6) quick and creative *responsiveness* to customers' requests and problems,

(7) *security* i.e. safe and risk free service,

(8) proper *understanding* and knowledge of customers' needs and

(9) service *tangibles* correctly reflecting the service quality.

(10) customer's satisfaction with the service rendered.

It is clear from above discussion that in order to ensure excellent service quality, the marketing strategy should involve

1. finding out the needs, interests and problems of the customers/consumers.

2. searching alternatives available to meet the needs and problems of the customers. For it, the service marketer heeds to put himself in consumer's shoes.

3. finding out marketing opportunities available to tackle those problems to improve services and determining *"thrust areas"*.

4. devising plan to take advantages of the opportunities identified.

5. allocating funds for thrust areas.

6. hiring competent and skilled employees with a positive mind, who may empathise with clients and may have customer's best interests at their heart and genuine belief in-quality performance.

7. ensuring continuous training of personnel of all levels of management, particularly in areas of thrust.

8. communicating effectively and correctly, especially in consumer's language, all improvements, offered to customers. This information needs to be supplied to all employees working in the organisation.

9. monitoring the delivery of services and correction mechanism.

10. follow up to ensure that service is delivered to the customers according to latter's expectation and perception.

Above all, what is required is the total commitment and dedication of the top management and employees to the concept of quality. Alongwith consumer satisfaction, employees satisfaction is of paramount importance. It goes without saying that delivery of good quality of service necessitates maintenance of healthy management-employees relations. In this context, employees satisfaction is as important as consumer satisfaction.

(iii) Profitability or Productivity

Service industries also need funds to provide services. Often, huge investments are made in them. Being labour-intensive, costs of services go on rising. Again being business ventures, they must get a reasonable return on investment in the form of profit. Therefore, service firms must be always on look out for ways and means to increase service productivity without jeopardising its quality. They should seek profit-making opportunities and strive to find profitable ways to satisfy customer needs through service quality. In this context, employees need to work with determination and dedication and undergo continuous training to increase the productivity. It is not without reason that the yearly bonus of many employees even in service industries is linked with productivity.

15.4 MANAGING SALE-LINKED SERVICES (MANUFACTURING INDUSTRIES)

As stated above, service quality is of vital importance for the survival and growth of service industries. The same holds good for services delivered to the customers by the manufacturing industries selling physical products. These services are usually linked with the sale of their products and have a great bearing on the performance of the firms and their products. These services are known as *"Product Support Services"*.

In modern business, companies manufacturing airbuses, machines, equipment, computers, photostat machines, electronic gadgets, refrigerators, televisions, music systems, furniture, watches, washing machines, cars, appliances and products like those, are offering a number of pre-sale and post-sale services to their customers to convince them of the prompt service and quality performance of their products, to gain competitive advantage and hence to build a credible positioning image for themselves and win the loyalty of the customers.

Pre-sale services may include prompt quotations, easy contact, reliable timely delivery, product warranties, replacement guarantee, technical advice, credit, test facilities etc. In case of highly priced machines and equipment, companies offer architectural services, installation services and training services for the staff and operators for efficient operation of the machines or equipment.

After-sales services include equipment maintenance and repair services, replacement guarantee, training services etc.

For example, Panasonic Television **Certificate of Warranty** contains the following condition.

"Within one year of the date of purchase, any replacements for component(s) found to be defective due to faulty workmanship or defective material will be made, and repairs executed free of charge by National Panasonic India Pvt. Ltd. Parts repaired or replaced are under warranty throughout the remainder of the original warranty period only.

Repairs under warranty may also be carried out by the Authorised Service Centres of National Panasonic India Pvt. Ltd. or the local National Panasonic dealers."

Some multinational companies like Minolta and Canon (Cameras), Rolex and Sieko (watches), National Panasonic (TVs and VCRs), Pioneer (Music Systems) etc., provide *International Warranty Certificates* and they have appointed authorised service centres in various countries to carry out replacements and repairs. Companies are now offering more and more value-added services to attract customers. Again what is important in this context is not the quantum of promises made by the manufacturers but the quality of service and actual performance of the equipment or product to the best satisfaction of the customer.

(a) Pre-sale Service Strategy

In order to arrive at proper product service package, the product marketers need to find out additional services expected and valued by the consumers of the product. This survey would help the company to augment its product service offer and acquire competitive advantage. Moreover the product needs to be designed in such an innovative way that the service cost is minimum. The best strategy would be to design the products in such a way that they do not give any trouble and if there is any trouble, the customer may himself rectify it at little expense.

To cite an example, Pioneer Electronic Corporation, Japan has designed its Stereo Multi-play CD Cassette Deck Receiver (Music System) in such a way that trouble shooting is often possible at home without a visit to centre, if its instructions are followed

(b) After-sales Service Strategy

As seen in the example given above in connection with' National Panasonic TV, the manufacturers can serve the customers in the following ways.

1. They themselves may provide these services (replacement, repairs and training).

2. They may appoint or franchise authorised service centres to provide these services. (Like Maruti Cars Service Centres).

3. They may carry out these tasks through distributors or local dealers.

4. Third parties or independent units may provide these services. Such an arrangement offers relatively cheaper and more prompt service, especially after the warranty period.

Whatever may be source of various product support services, a customer expects a prompt and satisfactory action on his problem and complaint and the best strategy is to provide fast quality service to him. This can only ensure his satisfaction and loyalty.

15.5   NON-PROFIT OR NON-BUSINESS MARKETING (SOCIAL MARKETING)

Thus far we have focused our attention on marketing of services provided by service industries and product-based industries in business sector. The one main objective they pursue is profits. There are a number of non-profit or non-business organisations and institutions, both public and private, that are providing services to the general public, their members or to some carefully identified target groups or are contributing to social causes through their services. Most of these non-business organisations are run by religious, charitable, public spirited and governmental bodies. They include:

1. Educational and research institutes (universities, colleges, schools etc.)
2. Health care organisations (hospitals, dispensaries, nature- cure centres, health care organisations etc.)
3. Cultural organisations (museums, art associations, arts groups, organisations like Sahitya Akademi, Sangeet and Natak Akademi, Kala Kendras, All India Council of Cultural Relations etc.).
4. Professional and trade associations (trade unions, trade associations, industry groups, FICCI etc.)
5. Religious organisations (religious foundations, denominational bodies, temples, churches, mosques, gurudwaras etc.)
6. Public Interest groups and social organisations (environmental groups, antinuclear groups, animal protection groups, anti pollution campaigners, groups campaigning for sanitation and beautification etc.).
7. Charitable and philanthropic institutions (social welfare organisations, orphanage, old age homes etc.)
8. Government organisations (governmental agencies responsible for public administration and community services etc.)
9. Political organisations (Political parties and their front organisations etc.)

The list can be expanded

The non-profit organisations by and large provide their services free, at nominal prices or at highly subsidised rates. To run these services, they need money and have to raise the necessary funds through donations from public (donors) and government grants from public exchequer. Some organisations organise charity balls, film stars nights, cricket matches and music concerts etc. to raise funds. Such shows for charitable and public causes are generally exempted from tax-payments.

Some Characteristics of Non-business Sector

On the basis of above study we can enlist some unique characteristics of non-business marketing. These characteristics are —

1. **Dependence on a number of groups or publics (Multiple Publics):** Business organisations are mainly responsible to three publics, namely shareholders, employees and customers, but non-profit or non-business sector has to respond to many groups or publics depending upon the nature of organisation. For example, charitable organisations run exclusively on donations have only two main publics to respond to—donors and beneficiaries or clients.

Whereas a university or college may be responsible to many groups or publics—students, faculty, parents, alumni, donors or contributors, legislators and many other groups. Thus non- business organisations need to address and direct their marketing programmes to several publics having their own needs and aspirations. Thus in case of non-profit organisations, environmental analysis (analysis of needs of various publics) becomes more important.

2. **Variety of Objectives (Multiple Objectives):** Business organisations are set up to pursue rather a single objective namely profits, but non-business organisations are set up to fulfil a variety of objectives from providing education to prohibition, healthcare to elimination of drug abuse, environmental protection to soul upliftment, rehabilitation of the helpless and the destitutes and so on. The pursuance of a variety of objectives would require specific marketing strategies for attaining specific objectives.

3. **Social Sensitivity and Controversiality:** Most of the services provided by non-business organisations concern social causes and practices and aim at changing social behaviour. Often they are controversial in nature and are opposed e.g. campaigns for anti-alcoholism and vegetarianism, abortions, protection of environment (Chipko Andolan, Campaign against Sardar Sarovar Valley Dam Project etc.). Generally they operate in matters of public interest. Most organisations pursue strictly social objectives. They need what is called *"social marketing"*.

4. **Intangibility:** Most of the non-business organisations provide services comprising ideas, causes and practices which are more often intangible. Their marketing thus needs application of principles of services marketing.

5. **Subjectivity to Greater Public Scrutiny:** As stated earlier, most of non-profit organisations depend for their existence on donations from public grants, from public exchequer, contributions from members and volunteers, funds raised through funds raising campaigns and enjoy tax exemptions. They are often more subject and liable to public scrutiny and environmental pressures.

6. **Dual Management:** Most of non-business organisations have dual managements in the form of *generalists* (administrators or professional managers) and *specialists* (who are qualified, experienced and skilled and actually provide the specific services). This situation tends to breed conflict. The success of such organisations depends on avoidance of the conflict.

Though non-business marketing cannot be said to be different from business marketing and the basic marketing concepts, principles, tools and techniques are applicable to the non-business sectors as they are applicable in case of business organisations, yet the above characteristics of non-business (non-profit) organisations, namely responsibility to multiple publics, multiple social objectives, subjectivity to greater public scrutiny etc., suggest that non-profit marketing has several unique aspects. For example non-profit or non-business marketing would involve relatively greater environmental analysis, greater application of principles of services-marketing etc. Again, many non-profit organisations are set up exclusively for fulfilling certain social objectives as is evident from the above list of non-profit organisations. Their efforts may be addressed and directed to general public, their members or well defined target group. The solution lies in what is called *"Social marketing"*.

"Social marketing is the design, implementation and control of programmes seeking to increase the acceptability of a social idea, change or practice in a target group."

The marketing process for services is basically same in non-profit marketing/social marketing as in profit marketing and marketing strategies of non-profit organisations should also include all four Ps, internal marketing and interactive marketing in order to provide high quality of service to their clients and beneficiaries. However keeping in view the unique aspects of non-profit organisations, the social marketing may have to place relatively more emphasis on advertising and promotion than on distribution channels. It has been observed that the channel of distribution components of non-profit organisations often overlap with promotional and communication media elements.

CASE 15.1: AUGMENTATION OF SERVICES (HARYANA ROADWAYS)

Haryana Roadways, a state owned passenger road transport organisation, has a fleet of more than 5000 buses. It is providing travelling facilities, services to about 20 lakhs (2 millions) passengers daily on the average on about 1700 state and inter- state routes. Besides, it is providing buses to elite or public schools for carrying students to and fro at highly concessional rates. It also provides buses to marriage parties during the season, subject to their availability.

Augmenting Benefits

Haryana Roadways bus services are well appreciated for their punctuality, safety and comfort. It has been consistently adding additional benefits and features to its services to cater to the needs of different groups of its consumers. For example, it introduced *Express Service* to ensure speedier journey charging 25 per cent more than the normal ordinary fare and *Deluxe Bus Service* to provide additional comfort through luxury seats and more space. In order to cater to the varying requirements of different categories of passengers on different routes, it is running different types of buses e.g. Ordinary, Express, Semi-deluxe, Deluxe, ACC buses, Video-coaches. In video-coaches, passengers are entertained with movies during the journey. In summer season, Haryana Roadways serves cold drinks like Frooti, Pepsi Cola, Campa Cola etc., on payment to its passengers on some buses. Passengers in some Deluxe buses enjoy food and other facilities at Haryana Tourist Department's resorts and restaurants during the journey. Recently it has introduced low-floor and semi-low floor delux buses and AC buses on some routes for which passengers have to pay higher fares.

Scope for improvement

Yet, there is a vast scope for improvement in the services being rendered by the organisation. On some routes, there is excessive and dangerous overcrowding and frequency of bus service is dismally inadequate. Again, maintenance, cleanliness, drainage and sanitation at most of the bus stations are in a mess and need substantial improvements. In rainy season, passengers have no dry ground to wait and board the bus at most of the stations. Even bus stands at important district headquarters turn into ponds of water in rainy season. At bus stations, the food and snacks provided by private contractors or shopkeepers are of very low quality while charges are exorbitant. The consumer passengers are totally at their mercy and whims. Here, Haryana Roadways has not kept passengers' (consumers') interests in view at all and have rather sacrificed their interests. This is totally against the marketing concept. The present system needs to be overhauled. Passengers at large must get food and snacks of good quality and at fair market prices, at these bus stations. Haryana Roadways Authorities must realise that conditions like this are adversely affecting its image. Haryana Roads can add value to its services by taking timely measures in this direction and providing additional benefits and facilities to its passengers.

EXERCISES

1. What is meant by services? Explain their chief characteristics.
2. Can marketing concept be applied in service industries? Discuss.

3. Explain some marketing strategies for service industries.

4. What is meant by social marketing? Describe its characteristics.

5. Write notes on the following:

    (a) Augmentation of service

    (b) After-sales service

    (c) Product support service

    (d) Non-business marketing.

6. Using a specific non-profit organisation with which you are familiar, draw a marketing plan to promote the services it offers.

16

Pricing Decisions

16.1 PRICES: MEANING AND ROLE

In simple words, price means the value which a seller sets on his goods and services. In this sense, money paid in exchange for various commodities, rewards paid for services of various factors of production such as land's rent, labour's wages, capital's interest, entrepreneur's profit, travelling fair, commissions and fees charged by people for services rendered - all are prices.

Literally, pricing means setting prices on products and services. In marketing context, managers are required to make two types of pricing decisions:

(i) Setting a price on a new product on its introduction or on an existing product on its introduction *in a new marketplace* as a part of company's market development strategy.

(ii) Resetting or adjusting the prices of current products in the face of varying costs, competitive and marketing conditions.

As discussed in chapter 6, price is one of the essential elements of marketing mix and marketing strategy. In fact, it is the only element that generates revenue, the other three elements, namely product, place (distribution) and promotion incur costs. Obviously, the price of a product has great bearing on its demand, sales-volume, revenue, profits and company's market share. Again, pricing decisions have an important role in meeting the competition.

Moreover, prices play an important role relative to other marketing mix variables. Pricing decisions may affect the quality of product and resource allocation to other elements of marketing mix especially promotional and distributional activities. Pricing decisions are intertwined with promotional decisions. A change in the former would lead to change in the latter. For example, a change in price may require changes in labelling, advertising message, packaging etc. Pricing decisions, in fact, should be consistent with other marketing mix variables, it is evident from the above study that the price has an important role in building and enhancing the image of a product.

16.2 PRICING — THE PROCESS

In context of our present study, we can define pricing in the following way—

"Pricing is the process of selecting the pricing objectives, determining the possible range of prices (or pricing flexibility), developing price strategies, setting the final price, and implementing and controlling pricing decisions."

The steps given in the above process are explained here.

Step 1 :   Selecting the Pricing Objectives

The first step in the Pricing Process is that of setting pricing objectives. *Pricing objectives are the specific quantitative and qualitative goals or targets that the management seeks to attain or accomplish through its pricing decisions or pricing strategy.* These goals reflect the role of pricing in the marketing plan.

The companies may have short-term objectives and long-term objectives. Some companies select a set of pricing objectives alongwith primary pricing objectives like profitability or growth. They may set some secondary or collateral pricing objectives like promotion of new products, building image of the firm. Whatever be the objectives, they should be expressed in clear, specific and concise terms so that they are understood by all those who are involved in making and implementing pricing decisions. Moreover, these objectives must be consistent with the mission of the organisation and should, in fact, emanate from the corporate mission and marketing objectives. Some important pricing objectives are given below:

(i) **Growth Objectives - Sales Growth and Market Share Growth:**   Some companies would like to increase the sales-volumes of their products through their pricing decisions to earn long-term profitability, while others may set such price as may contribute to increase their market share.

(ii) **Profit Based Objectives:**   Some companies set the price that may maximise return on investment. Some companies have maximisation of profits, both short run and long run, as their objective and they would set the prices towards this end. For example, see the following statement:

> We are interested not just in short-term profits but a balance of short-term profits and long-term profitability. We aim at making optimum profits rather than maximisation of profits. We aim for profit based on all-round efficiencies, increased volumes and deserved price premiums.
>
> **Marico Industries**

(iii) **Competitive Objectives:**   These objectives include maintaining price leadership, discouraging new entry, establishing market position, discouraging price cutting and maintaining price stability etc. Some companies use prices to achieve these objectives so as to meet competitive challenges.

(iv) **Maximisation of Revenue and Cash Flow:**   Some companies would like to maximise the sales revenue and cash flow so as to rapidly recover the costs incurred on development of new products.

(v) **Maintenance of product-Quality leadership:**   Some companies want to maintain product-quality leadership in the market. Their pricing decisions contribute to accomplishment of this goal. For example, Godrej refrigerators and furniture (especially cupboards and cabinets) are priced much more than competitors' products i.e. refrigerators and furniture respectively.

(vi) **Collateral Objectives (Enhancement of other strategy elements:**   These objectives include promotion of new products, building image of the product and firm, maintaining full product line.

Step 2:   Determining the Possible Range of Prices or Pricing Flexibility

After selecting the pricing objectives, the next step is to determine the possible range of the prices i.e. the lower and upper limits of the price or what is called the extent of *the pricing flexibility.*

In order to earn profits, price must exceed per unit costs. Even for bare survival of the firm, revenue recovered must at least cover the cost of production and other costs. In other words, the price must at least cover per unit costs. Thus per unit costs determine the lower limit of the price (floor price). The upper limit of the price (ceiling price) is determined by demand for the product and competition. Again, prices are subject to legal constraints. This suggests that the lower and upper limits for prices are determined by costs, demand competition and legal and ethical considerations. They are briefly discussed here.

1. Determining Costs

Production requires inputs of various kinds e.g. raw materials, labour, management etc.. The money required for purchasing the inputs, together with certain other items is called cost of production. It includes wages and salaries of workers, expenditure on buildings, machines, raw materials, rent of land and building, interest on investment and borrowed capital, depreciation charges on fixed capital, expenses on light, fuel, electricity, advertising, freight and insurance premium and taxes paid to the government and local authorities.

A. Costs are categorised as follows:

(i) **Explicit Cost:** These are costs which a producer pays directly to factors of production as a reward of their services. They include rent of land and buildings, wages and salaries, depreciation charges and replacement and repair of machines and equipment and marketing costs like expenses on advertising, promotion and distribution of the product in order to increase sales.

(ii) **Implicit Cost:** These are the costs that a producer pays to himself as rewards for factors of production owned by him e.g., wages for the work performed by the owner of the firm, interest on the capital supplied by him, rent of land and buildings belonging to him and used in production. The estimated value of these services i.e. rewards are also to be accounted for in the determination of the costs.

(iii) **Normal Profit:** Total costs of production also includes a fair return for firm's effort and risk i.e. normal profit. Normal profit is the reward that prompts an entrepreneur to engage himself in business and bear risk. In economics, normal profit is considered as a part of production costs and is included in their determination.

B. Costs of production are also divided into two parts *viz.* fixed costs and variable costs.

(i) **Fixed Costs:** These are the costs that a firm incurs on use of fixed factors or inputs e.g., rent of land, buildings, interest on capital, bonds, expenditure on machines, depreciation, salaries of permanent staff. These costs are fixed and do not change with increase or decrease in quantity of production. These are the costs which a firm will have to bear even if the production is stopped temporarily. These are also called *supplementary or overhead costs*.

(ii) **Variable Costs:**   These are those expenses of production that vary with change in the output. Variable costs are expenses that are incurred on purchase of variable inputs. Increased amount of output (production) would require increased amount of labour, raw materials, energy, fuel and a firm will have to spend more. When the output is reduced the variable costs are also reduced. With the stoppage of production, variable costs cease and do not exist. Variable costs are also called *prime costs*.

Total costs are the sum of the total fixed costs and total variable costs at a given level of production and in the short run, the price set by the company must yield revenues that may at least cover the total production costs. Only then the company can survive. Per unit cost, thus, determines the floor price of the product.

Understanding Cost Behaviour

(a) Understanding cost behaviour vis-à-vis level or scale of production is essential for estimating actual production costs. It has been seen that average cost or cost per unit varies with the scale of production. In the short run, the average cost curve is U-shaped, it slopes downwards initially and then rises upwards. It suggests that the average cost is high in the initial stages of production i.e. when a few units are produced. As the scale of production is increased and more and more units are produced, average cost falls. It goes on falling till the level of production that can be maximally produced per period according to the capacity of the equipment or plant

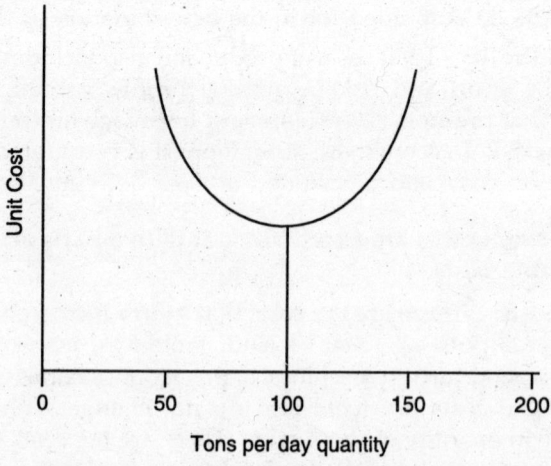

Fig. 16.1   Cost Behaviour (Fixed Size Plant).

installed. The reason for this fall is that now fixed costs are spread over more units. Beyond this limit, the average cost will begin to rise. If the scale of production is expanded beyond the maximum capacity of the equipment and plant, the breakdown of machines, the increased difficulties of coordination and management and diseconomies push the average cost upward.

Thus in the short run, as output is increased, the average cost falls up to the point of maximum capacity of the equipment and scale used in the firm and rises thereafter.

Average cost of production is lowest when the firm produces the maximum amount it can with its existing equipment and scale at a particular time. For example, a dairy plant with a capacity of 100 ton per day will have the least cost when it just produces 100 TPD. In case, it believes it could produce 200 TPD, it could do so but at higher average cost. It would be better he installs another plant.

Thus when a producer finds that expansion of scale will increase average costs, it will be advisable that he builds another plant instead of expanding the old plant. In case the demand for the product is much more, he can build a new plant with higher capacity. However average costs increase when a firm installs a plant with a capacity more than what is known as "*optimal size*", due to diseconomies of the scale and resultant inefficiency.

(b) Again, it has been seen that as a firm produces more and more units over a time, it gains what is called *cumulative production experience* and as a result average production costs decline. In other words, with the increase in accumulated production (i.e. in the

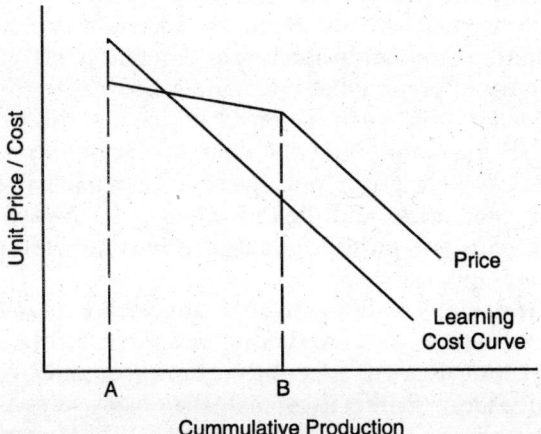

Fig. 16.2 Learning/Experience Cost Curve.

cumulative number of units produced), costs per unit of production decline. This is shown by *experience cost curve or learning cost curve.* The decline in production costs is due to benefits of accumulated production experience, production economies, increased efficiency of workers and other improvements.

For example, in case of 100 ton per day dairy plant, the average cost when it produces 30000 tonnes cumulatively will be less than when its accumulated production was 20000 tonnes. This is the reason that some companies set their prices sometimes at lower level when a new product is produced as they feel that they will make up the difference as the volume of production and hence sale increases.

2. Evaluating Demand

Whereas costs determine floor price or the lower limits of the price, the upper limits of the price are determined by demand and competition.

Demand for a product is the quantity of product that a consumer is ready to purchase at a given price at a particular point of time. There are a number of factors that determine and influence the demand for a product. These factors are — the price of the product itself, the prices of other related products including substitutes (competitors' prices), level of income of the consumers, consumers' tastes and preferences, seasonal changes, size, composition and level of population, distribution of income etc.

The law of demand states that other things being equal, when the price of a good is raised, less of it is demanded and at a lower price consumers will purchase more of it. In other words, if other factors are held constant, people will buy more at lower prices and less at higher prices. Thus there is an inverse relationship between price and demand, provided that there is no change in other factors enumerated above.

Estimating market demand for a product is a difficult task as it involves estimating customer responses to different prices of the product. Companies do try to prepare their demand schedules for their products.

A demand schedule is statement or table showing different quantities or amount of a firm's commodity that would be demanded by the market at different prices in a given time period. Generally, it shows lower quantities at higher prices and the curve has a downward slope but in case of luxury items and prestige goods, this may not hold good and the curve may have upward slope.

Forecasting demand at different (alternative) prices involves gathering marketing information and marketing research. Firms employ time analyses of historical prices and sale-volumes or directly contact the consumers in the target market to estimate their likely response at different price levels to prepare their demand schedule. The demand schedules may look like.

Table 16.1 Market Demand Schedule.

Price (Rs.)	Quantity Demanded (Units)
25	44000
30	40000
35	35000
40	32000
45	24000

Marketers are not only concerned with the absolute levels of demand at different prices (demand schedules) but are also concerned with how the market demand responds to a change in the price of the product or what is called in economics the *price elasticity of demand.*

Price Elasticity of Demand

Price elasticity of demand is a measure of the relative change in quantity demanded in response to a relative change in price. *It is defined as the ratio of the percentage change in demand to the percentage change in price.*

$$E_p = \frac{\text{percentage change in quantity demanded}}{\text{percentage change in price}}$$

For example, if a 40% increase in price reduces the demand by 25% then the elasticity of demand

$$E_p = 25\%/40\% = 0.625 \tag{<1}$$

On the other hand if there is 40% increase in quantity demanded as a result of 20% fall in price, then

$$E_p = 40\%/20\% = 2 \tag{>1}$$

According to Mrs. Robinson, "the elasticity of demand at any price is the proportional change in quantity purchased in response to a small change in price divided by the proportional change in price, then coefficient elasticity,

$$E_p = \frac{\text{proportional change in quantity demanded}}{\text{proportional change in price}}$$

If p_o, q_o are the original price or initial quantity demanded respectively, Δp, Δq as absolute changes in price and demand respectively, then according to the above formula

$$E_p = \frac{\Delta q}{q_o} \div \frac{\Delta p}{p_o} = \frac{p_o}{q_o} \times \frac{\Delta q}{\Delta p}$$

For example, at the price of Rs. 8, the quantity demanded is 400 units, but when the price is reduced to Rs. 6, the quantity demanded is 800 units.

p_o = Rs. 8, q_o = 400 units, Δp = Rs. 8 – Rs. 6 = Rs. 2

Δq = 800 units – 400 units = 400

$$E_p = \frac{p_o}{q_o} \times \frac{\Delta q}{\Delta p} = \frac{8}{400} \times \frac{400}{2} = 4$$

In the above examples, we have seen that price elasticity of demand can be less than 1 (<1), greater than 1 (>1) and equal to 1 (=1)

(1) When the coefficient of elasticity is greater then 1, it means that the proportional change in demand is much greater than that in price or in other words demand is very responsive to price changes i.e. the demand is elastic. In such cases, a small increase in price will result in great decrease in demand and a small cut in price will yield much higher demand and revenues.

(2) When the coefficient of elasticity is less then 1, the proportional change in demand is much less than that in price or in other words, demand is not very responsive to price change i.e. demand is inelastic. In such cases, even a large change in price may not make a discernible change in demand. In such cases, an increase in price will result in increase in total revenues if $E = 0$, then demand is perfectly inelastic and product demand is not at all responsive to the price changes.

(3) When coefficient of elasticity is equal to 1, the proportional change in quantity demanded will be same as in the price. This is called "*unit elasticity of demand*".

The price elasticity of demand is generally different at different price levels or at price points in a demand schedule.

There are many factors that govern elasticity of demand such as nature of products e.g. necessaries, comforts, luxuries, (the demand is inelastic for necessaries), availability of substitutes, fashions, habits, level of increase, standard of living of consumers etc.

The elasticity of demand is very helpful in determining the price of products. The firm can increase the price of those products which have inelastic demand. In case of price sensitive or price responsive consumers, a small decrease in price will result in great demand and yield higher total revenues. The price elasticity of demand also helps in setting discriminatory prices for different segments of the market. The firm can charge higher prices from customers with inelastic demand whereas it will charge lower price from consumers with elastic demand.

One important aspect of constant and heavy advertising is that it effects a change in fashions, attitude and behaviour of the consumers and make the demand of advertised product inelastic so that the firm can charge higher prices. A very high price could reduce the demand substantially. Thus while unit costs determine lower limits (or floor prices) of the product, elasticity of demand helps in determining the upper limits (or the ceiling prices of the product). The price to be set must be somewhere in between these limits, depending upon existing conditions.

3. Evaluating Competition

In order to know the extent of pricing flexibility and determine price, it is very essential to study and analyse the competitive structure of the industry and competitors' offers and prices. Competition directly affects the price elasticity. The price elasticity will be higher if substitutes are made available to the customers by the competitors. Again, before setting a price, management must also gauge the possible price reactions of the competitors. Competition thus also helps in determining the extent of pricing flexibility.

Information about competitors' prices and offers can be procured from information services supplying price lists and such information, through routine shopping, or by purchasing actual products manufactured by competitors. This will also help the firm in positioning its product in the market.

Price competition is often likely to receive strong reaction from the competitors and may lead to price war and may sometimes backfire, especially when a consumer is accustomed to use a particular brand of a product. It would be much better for the firm to meet the competition by improving its offer in terms of quality, improved benefits and services i.e. through non-price competition and value addition.

4. Legal and Ethical Considerations

Pricing decisions are also subject to legal and moral considerations. We have already referred to, in the appendix to chapter 3, a number of legislative enactments in this country to safeguard the interests of the consumers against black marketing and use of unfair trade practices. The limits of the price are thus also influenced by legal and ethical constraints. While setting the prices, legal obligations are also to be adhered to.

Step 3. Developing Price Strategies (Policies)

After selecting pricing objectives and determining the range of the prices, the next step is to develop and formulate price strategies to be pursued by the firm. *Price strategies are the guidelines and policies to develop specific plans for setting the price of a product or line of products.* Price strategies are the guidelines that help matching pricing decisions with the market conditions. These strategies include price discrimination, price lining, maintaining price levels and price stability, life cycle pricing, psychological pricing etc.

A. Price Discrimination: Price discrimination is a price policy or strategy that determines price variability.

Price discrimination means charging different prices from different buyers of the same product or service.

Discriminatory pricing can take the following forms:

  (i) **Personal discrimination:**   Here different prices are charged for the same product or service from different segments of consumers. For example, Indian Railways charges 30 per cent less fare from senior citizens than the normal fare for adults. Again marketers may charge higher prices from buyers whose intensity of desire is large or whose ability to pay is greater. Usually it is done in a disguised way e.g. railways charge from upper class passengers much more than the cost of providing the additional facilities.

 (ii) **Local or Geographical Discrimination:**   Different prices may be charged from different localities or locations. Prices in fashionable shops or in shops located in fashionable areas are often higher than those in ordinary shops. Again movement of products to some distant locations may involve higher costs due to transportation or taxes imposed by the local authorities. For example, Food Corporation of India sells wheat and rice relatively at higher prices in Tamilnadu and J&K state than those in Haryana or Punjab.

(iii) **Trade or Use Discrimination:**   This occurs when different prices are charged for the same product or service for its different uses. For example, electricity boards charge different rates for electricity for factory consumption, domestic consumption and agricultural consumption or for electricity used for commercial purposes. Again the electricity boards charge at higher rates from those domestic consumers who consume units beyond a fixed consumption.

(iv) **Discrimination on basis of Timing or period (Time Pricing):**   Different prices may be charged at different times or period (season) for same service. Hotels located in Hill stations charge relatively low accommodation tariffs in off-season period and charge higher tariffs in summer season.

 (v) **Product-form Pricing:**   This discrimination means charging significantly different prices for only marginally different products. For example, the price of library edition of the books are relatively much higher than paperback edition. The prices are set at significantly higher level than additional costs of production.

Discounts

One important advantage of price discrimination or variability strategy is that it allows the company to grant quantity, cash, trade (functional), geographic, seasonal discounts and allowances to buyers and make price adjustments. These discounts are briefly explained here.

(i) *Quantity or Volume Purchase Discount:* As the name suggests, quantity discount is reduction in price for bulk purchase by the buyer at the time during a given period of time. This discount is a form of incentive to the buyer for purchasing large volumes.

(ii) *Cash Discount:* It is allowing reduction in price, say 2 per cent in the bill amount, if payment is made in advance or in cash at the time of purchase or within a specified period after the purchase, say within 10, 20 or 30 days. A large number of companies follow this practice. In Haryana, municipal committees allow 10 percent discount in the amount of House Tax Bill, if it is paid within 10 days of the receipt of the bill.

(iii) *Trade Discount:* This discount is also called functional discount as it is allowed by the producers to the members of distribution channels for performing certain functions concerned with sale and marketing e.g. to C & F agents.

(iv) *Seasonal Discount:* As the name suggests, it is the reduction in price of products or services during off-season period. Hotels, manufacturers of refrigerators and woolen cloths offer this discount during off-season periods.

(v) *Geographic Discount:* It is the price reduction allowed to buyers in certain zones or areas. Generally, this practice is followed to retain the existing market share or as a market penetration strategy in that zone or area.

(vi) *Trade-in and other allowances:* These allowances are allowed to promote and increase trade. Companies manufacturing durables like refrigerators, scooters, TVs adjust the price or resale price of old items against the purchase of brand new item. Thus resale price of old item is reduced from the list price. Kinetic Honda, Godrej, Whirlpool, and many other companies run such schemes in India. Some companies make payments in the form of price reductions towards promotional efforts made by the channel member and toward their participation in advertising or sales campaigns. Such reductions are called *promotional allowances*. (See page 246)

B. Price Lining: Price lining is a strategy of setting a number of price levels within a product line to cater to the preferences of different customers. A large number of companies produce many products in a

product line rather than simple products. For example, Maruti Udyog Limited manufactures several types of passenger cars such as Maruti 800, Maruti 1000, Omni (ST), Omni (HT), Gypsy, Maruti Esteem (in 3 varieties) with different features and sets price levels to match with the needs and preferences of different customers. Price lining helps to serve various market segments without competing with other products in its product line.

Prices of a product at each level of marketing channel are also set by what is called inverted pricing. This helps in determining the maximum manufacturing cost of a product at which the manufacturer has to produce the product. We shall explain the method in the next section.

C. Price Level and Price Stability Strategy: Before setting the final prices, the firm has to decide about the general price level for its products. To make decision, the management has to take into consideration factors such as the existing and desired image of the firm, product line objectives, strengths and weaknesses of the competitors and economic environment. It has to decide whether it has to build a low-price image or high price-quality image like Godrej refrigerators, Ray Ban glasses, Rolex watches etc. Depending on this decision or determination of general price level for the company's products, the firm will devise a general strategy of low prices or higher prices.

Again, the firm has also to devise such strategy that may ensure resale price stability and discourage price competition among sellers. This can be done by effecting resale through selected distributors and dealers and taking measures to ensure that they sell the products at suggested prices.

D. Life Cycle Pricing: This strategy aims at effecting changes in the prices of the products as they move through different stages of product life cycle so as to reap best advantage in face of increasing competition. We have already referred to decline in per cent costs of production in case of accumulated production or learning experience. The prices can be adjusted as the product passes through different-stages of PLC and the company can enjoy competitive advantage during each stage of the product life cycle.

E. Psychological Pricing: Psychological pricing is a strategy which uses price as an instrument to appeal to consumers' minds and influence consumers' perceptions of the product quality. The management can take advantage of psychological factors in setting prices for those who are influenced more by price perceptions rather than law of demand. Cravens, Hills and Woodruff define psychological pricing *"as a strategy based on customer price perceptions so as to have special appeal in certain target markets."*

Psychological pricing strategies include:

(1) *Odd-even pricing:* e.g., Bata Shoes price their shoes Rs. 599.95, Rs. 799.95, Rs. 1099.95 as they may look in the Rs. 500, Rs. 700, Rs. 1000 ranges instead of Rs. 600, Rs. 800 and Rs. 1100 ranges respectively.

(2) *Prestige Pricing:* It is a pricing strategy to derive benefit from what is called snob appeal. This is true of high-hat-shopping. A customer is willing to pay the high price, not that the quality of the product is superior but because he has bought it from New Delhi Connaught Place Shopping Centre or from Oberoi Hotel Arcade which serve affluent shoppers.

(3) *Customer Price Perceptions of Product Quality:* Many customers perceive higher price of a product as connoting higher quality. In their case, price acts as indicator of product quality. This is true in case of expensive cars, perfumes, women's apparels in boutiques in posh areas.

Step 4.   Setting the Prices

After having selected the pricing objectives, determining the possible range of prices and selecting price strategies, the stage is ready for setting and establishing prices. In setting the price, cost, demand and competition factors need to be taken into consideration. Usually, the following methods or approaches are employed in establishing prices of the product.

1.   Cost-oriented Methods

We know that costs determine the lower limit of the price (or floor price) of the product. Generally, the established price should be more than the floor price so as to earn some profit. This necessitates adding something to per unit cost. There are three approaches in this context—cost-plus pricing, markup pricing and break-even analysis.

(a) Cost-plus Pricing: Cost-plus pricing involves adding a certain percentage of the cost to the per unit total cost to set the price. For example, *in case of a pen,*

EXAMPLE 16.1

Say,	unit variable cost	= Rs. 11.00
	unit fixed cost	= Rs. 02.00
	unit total cost	= Rs. 13.00
	Plus 35% of the cost	= Rs. 04.55
	(covers expenses and profit)	
	Set price or selling price	= Rs. 17.55

(b) Markup Pricing: Markup pricing involves adding a certain percentage of the selling price (or return on sales) rather than percentage of the cost as in cost-plus pricing explained above. Markup price is determined by using the following formula.

EXAMPLE 16.2

for example, say

unit variable cost	= Rs. 11.00
unit fixed cost	= Rs. 02.00
unit total cost	= Rs. 13.00
desired markup percentage	= 35% (0.35)

to cover secondary expenses like insurance, handling costs, handling cost, pilferage plus profit.

$$\text{Markup Price} = \frac{\text{unit costs}}{(1 - \text{desired markup percentage})}$$

$$\text{Markup price or selling price} = \frac{13}{\left(1 - \dfrac{35}{100}\right)} = \frac{13 \times 100}{65} = \text{Rs. 20}$$

In practice, there are different markups for different products or categories of products. They are most often determined by industry tradition, firm's strategy, operating expenses, expected sales and other factors.

EXAMPLE 16.3

As stated above manufacturers add percentage markup to their costs to set their prices for different products. Distributors and retailers add their fixed markup percentages to their costs or purchasing prices. For example in above illustration distributor's markup is 20 per cent and retailer's mark up is 33-1/3% then,

Distributors per unit cost = Rs. 20.00

$$\text{markup} = 20\% \quad (0.2)$$

Distributor's selling price or Retailer's purchase price

$$= \frac{20}{(1 - 0.2)} = \frac{20 \times 100}{80} = \text{Rs. 25}$$

Retailer's per unit cost = Rs. 25.00

Markup percentage = 331/3% = 100/300

Retailer's Markup price or selling price

$$\frac{25}{\left(1 - \dfrac{100}{300}\right)} = \frac{25 \times 300}{200} = \text{Rs. 37.50}$$

A Note on Inverted Pricing

In the previous step, while discussing price lining, we have referred to inverted pricing. In this strategy, we set a target market selling price or retailers' selling price for the target market and with the help of given markups, we determine the costs of retailer, distributor and the manufacturer. In order to get unit costs we use the following formula

Unit cost = markup price (1 – markup percentage)

EXAMPLE 16.4

In the above exercise, the required markups are 33 1/3 % per cent for retailer, 20% for distributor and 35% for the manufacturer.

Retailer's Markup Price = Rs. 37.50
Retailer's Cost or Distributor's markup price

$$= 37.5(1 - 100/300) = Rs. 25$$

Distributor's Cost or Manufacturer's markup price

$$= 25(1 - 20/100) = Rs. 20$$

Manufacturer's Cost = 20(1 – 35/100) = Rs. 13

It suggests that the manufacturer has to produce at a cost of Rs. 13 or less, so that the retailer may sell it at Rs. 37.50 in the target market.

(c) Break-even Pricing: Another cost-oriented approach to set the price is to find out the break-even point or level. Break-even point is that level of output or sales at which the money cost of production per unit (variable cost + fixed cost) is just equal to the market price. In other words, break even point is the level of sales that covers all relevant variable and fixed costs.

EXAMPLE 16.5

If fixed costs are Rs. 200000, unit variable cost is Rs. 1600 and the price per unit is Rs. 2000, then the volume of sales at break-even point must be Rs. 200000 ÷ (Rs. 2000 – Rs. 1600) or Rs. 200000 ÷ Rs. 400 = 500

$$\text{Break evern output/sale} = \frac{\text{Total fixed costs}}{\text{Price} - \text{Variable Cost}}$$

because at this point,

Total revenues = Rs. 2000 × 500 = Rs. 1000000

Total costs = Rs. 200000 + 1600 × 500 = Rs. 1000000

This is volume of sale/output at break even point

(d) Target Return Pricing: This is another cost-oriented method based on break even analysis. It is setting prices at a target percentage return over and above the break even point. In this approach, a target percentage return on total costs of a break even volume, or a standard level of output or sales or on total investment is added to the total costs and price is arrived at by dividing the total revenues by the standard level of sales.

EXAMPLE 16.6

For example, in the previous example, the desired percentage return on total costs is 20%,

then Total Costs	= Rs. 1000000
Return percentage @ 20%	= Rs. 200000
Total revenues	= Rs. 1200000
Volume of output or sales	= 500
Set Price = 1200000 ÷ 500	= Rs. 2400

EXAMPLE 16.6

Suppose the total costs of standard output level of 800 units is Rs. 15,00,000 and 25 per cent rate of return on costs is desired, then

total revenue management needs to receive

= Rs. 1500000 + 25% of Rs. 1500000

= Rs. 1500000 + 375000 = Rs. 1875000

Standard output / sale level = 800 units

Set Price = Rs. 1875000 ÷ 800 = Rs. 2343.75 per unit

Prices of supplies are likely to fluctuate. The costs will rise if there is inflation. While setting the price, the firm should build an adjustment mechanism to meet inflationary challenges. Inflation is one of the critical factors and it needs to be incorporated in price setting methods.

2. Competition-oriented Methods

A large number of firms set their prices in relation to the prices of their competitors. The service providers like Insurance, transport, railways, health care organisations also follow this strategy. Firms also bid for contracts in competitive bidding. We come across two approaches, namely going-rate pricing and sealed-bid pricing.

(i) Going-rate Pricing: Going-rate pricing is a method of establishing price in relation to the prices of the competitors. In this approach, firms set the same prices as their competitors, or charge more or less prices than the prices of the competitors. Firms charge a small percentage above or below competitor's prices. In this approach little attention is paid to the cost or demand factors. Most often, small firms follow the large competitors, the price leaders, as they are in a better position to set more suitable prices. They would change their prices when leaders do the same. They often rely on what Philip Kotler calls *"the collective wisdom"* of the industry which they feel would yield them a fair return on their

Trade-in Allowances

investment without subjecting them to the risk of a price war. Most of the retailers follow this approach as it is easy for the customers to make comparisons.

(ii) Sealed-bid Pricing: A large number of firms and government agencies invite sealed bids and tenders from competitive suppliers (vendors) for supply of a large number of goods and services. Many contracting firms have to submit bids to get contract work. In it, they need to follow a competitive bidding approach. The lower the bid price submitted, the better is the chance or probability of its being accepted and higher the bid price, the lower the chance of its being selected. In case of low bid pricing, the profit margin will also be low.

No firm would bid a price lower than the cost as it would entail loss. However, it would try to price its bid in such a way that may offer superior chance of being accepted than competitors' offers or in other words it would price its bid lower than its expectations of the competitors' price. The firm would like to have a reasonable profit and in competitive bidding it cannot set its price very higher than its cost as it would mean lower chance of its acceptance.

Firms that make many bids can take the risk of playing odds and price the bids using expected profits criterion but the firms that want that their bids are accepted have to price them at lower percentage above the costs without considering the expected profits they are likely to get. Anyhow we can estimate the expected profits by using the following formula.

Expected profit of a bid $(A) = p(A) \times z(A)$ where A is the actual bid, $p(A)$ the probability of the bid A, being accepted; $z(A)$, the profit if bid A is accepted.

For example, Let the cost of the product = Rs. 1000

Table 16.2 Expected Profit on Different Bids.

Firm's Bid		Firm's profit (Price — Cost)	Probability of a acceptance of the bid (assumed)	Expected Profit
(i)	Percentage of Cost	(ii)	(iii)	(iv)= (ii) × (iii)
Rs. 1100	110	Rs. 100	0.9	Rs. 90
1300	130	300	0.7	210
1400	140	400	0.2	80
1500	150	500	0.1	50

The above table shows that a bid of 130 per cent of the costs (i.e. 30 per cent above the cost) are expected to yield the highest value to the bidder but it has less chance to be selected than the first bid price of 110 per cent of costs.

3.   Demand-oriented Methods

(i)   Perceived-value Pricing

It is an approach of setting the price on the basis of the customers' perceptions of the product's value. As the name suggests, the price of the product is determined by what a customer perceives as the value of the product to him in terms of its potential benefits and uses. In this method, the price of product is set in a way as to match the perceived value of the product built in customer's mind. A company can set the prices of its products at higher level if customers perceive that the products offered by the company possess relatively more potential benefits, namely superior quality, more durability, greater reliability, service quality etc. They will be willing to pay more for better perceived value or offer.

Sometimes perceptions may be far from reality and the company may overestimate or underestimate its offer's perceived value. In this context, importance of market research cannot be overemphasised. Market research can help in determining the real perceived value of the product. The perceived value can be determined by computing a weighted index of attributes of each product offer as rated by the buyers according to their perceptions. Perceptions have great impact on price elasticity. The product will not sell if the set price is higher than its perceived value. The price must be set either equal to or below the product's perceived value.

(ii) Demand Modified Break-even Pricing

It is an approach to set prices that may yield maximum profit over the break-even point, taking into consideration the estimates of market demand at each feasible price. In other words, it is a method to set price at that point of demand schedule over break-even point that may yield the highest profit. The main problem in this method is to get accurate and trustworthy demand schedules or estimates of price-demand relationships.

EXAMPLE 16.8

The fixed costs of a company as Rs. 500000, the unit variable costs are Rs. 15, the estimates of quantities demanded at prices of Rs. 25, 30, 35, 40, 45 are given in table 16.2. We have to determine the price that will give maximum yield.

We have already given the formula to determine break-even point in example 16.5

Break even sale = Total fixed costs/(P – V)

First, we have to find break-even points at prices Rs. 25, 30, 35, 40, 45. They are shown in table 16.3

Table 16.3 Demand Schedule and Break-even Analysis.

Price	Demand	Break-even Points	Total Revenue	Total Costs	Expected Profits
(Rs.)	(units)	(units)	(Rs.)	(Rs.)	(Rs.)
25	44000	50000	1100000	1160000	– (60000)
30	40000	33333	1200000	1100000	100000
35	35000	25000	1225000	1025000	200000
40	32000	20000	1280000	980000	300000
45	24000	16667	1080000	860000	220000

In the above table, five prices viz. Rs. 25, 30, 35, 40, 45 have been considered. The Rs. 40 price yields the highest profit of Rs. 300000, the firm can set the price of the product at Rs. 40/- to get highest profit.

Step 5: Implementing and Controlling Pricing Decisions

The primary responsibility of implementing price decisions lies on the marketing department, its salesforce and members of distribution channels and there should be perfect coordination and cooperation among them to this end. But an effective and successful implementation of pricing decisions also requires cooperation of other departments in the organisation such as manufacturing, materials, financial departments etc. In fact, their consultation is necessary while making pricing decisions as its success is dependent on their contribution as well.

Again, pricing objectives are determined in consonance with the company's mission, objectives and philosophy, therefore it is very necessary that the salespeople and resellers (members of channel) are fully acquainted with these matters so that they could understand the logic and rationale behind the price decisions and the price changes. It is, again, the sales people and distributors who have to explain and justify the pricing decisions to the buyers and convince them of the corresponding product value. Messages need to be communicated to salespeople and distributors promptly and effectively.

As stated earlier, pricing decisions affect company's and distributors' performance, competitors' business and customers' response. All these need to be monitored to determine the effectiveness of pricing decision. Controlling pricing decisions would involve assessing the performance of the company and channel members and the reactions of the competitors and the customers to the price decisions and to see that the pricing objectives are achieved as planned. In case of low performance and adverse reaction of the consumers, immediate remedial measures will have to be taken to correct the situation. This would need collection of necessary accurate information and data on these aspects.

16.3   SOME OTHER PRICING POLICIES

In chapter 14, we have discussed a few marketing strategies used in the introduction stage of the product life cycle. The same strategies can be devised in case of new products as well. A brief description of a few additional pricing strategies is given here.

   1. Skimming Pricing Strategy: This strategy advocates setting a high price of the product initially so as to reap profits from price inelastic customers and skim the cream off the market and rapidly recover the new product development costs in absence of competition. As the competitors enter the market with their products, the prices are lowered for price sensitive customers.

   2. Penetration Pricing Strategy: This strategy involves setting a low price on the product so as to penetrate into the market and rapidly acquire a large market share. This pricing policy discourages competitors' entry as profits are low. The firm concentrates on deriving the benefits of production economies to keep the costs lower.

   3. Loss-Leader Pricing: This is the strategy, generally employed by middlemen and retailers, of deliberately selling a product or brand below cost or giving it free in order to attract customers for other goods. Such practices arise whenever the competition is keen. The product which is sold at a loss or given free is called *"Loss-Leader"*. Manufacturer: do not approve of this practice as it may adversely affect the image of their brand. It is a sort of promotional pricing.

   4. Some other Promotional Policies: May take form of special event pricing, cash rebates, low-interest credit or financing, free warranties and free service facilities etc.

16.4   REACTIONS TO PRICE CHANGES

As stated earlier, price changes not only affect the company's and distributor's performance, they also affect customers and competitors. Their reactions to these prices also need to be monitored and considered if pricing objectives are to be achieved.

   (a) Customers' Reactions to Price Changes: The reaction of the customers to a price change can be advantageous or disadvantageous depending on their price elasticity and their perceptions. As discussed in previous section, the customers' responses to the change in price (hike or fall in price) will depend on their price elasticity. For example, a reduction in price will result in increase in the demand in case of price elastic or price sensitive customers. Again, the response of customers to the price changes will depend how they perceive the pricing decisions of the firm.

Their perception of firm's product value, product quality, brand image, financial position of the firm etc. will also dictate their response to the price change.

(b) Competitors' Reactions to Price Changes: One response of competitors' can be what is called *"following the leader"* and other responses will depend on what the competitors think is in their self interest. There may be no response at all or there may be a limited response.

(c) Price Competition and Non-price Competition: Generally companies compete against one another by reducing prices. But under monopolistic competition, a price reduction may not have much effect. Again, manufacturers do not like price war or price cutting as it may lead to a general fall of prices of similar goods thus *"spoiling the market"* for all producers.

Now-a-days, companies resort to non-price competition. Competition takes place rather through—

  (i) improvement of the quality of the product,

 (ii) provision of additional benefits and services and

(iii) gifts or

(iv) advertisements.

What is important is to know why price changes are being made and what are the pricing moves of the competitors. The importance of marketing intelligence cannot be overlooked. The marketers should have some contingency plan to meet the challenge. A proactive approach in this context will be very helpful.

EXERCISES

1. Define Pricing. Describe various steps involved in the pricing process.
2. What do you mean by price strategies? Explain some important price strategies.
3. Explain major cost-oriented methods of setting the prices.
4. How do the customers and competitors react to price changes?
5. Write notes on
   (a) Pricing objectives
   (b) Experience cost curve
   (c) Price elasticity of demand
   (d) Price lining
   (e) Going rate pricing

(f) Penetration pricing strategy

(g) Discounts

(h) Pricing flexibility

6. What are the uses of pricing decisions a marketing manager is required to make?

7. Identify and discuss major factors to consider in pricing a new soap to be distributed through a conventional channel of distribution.

8. What role does price occupy in the marketing mix (programme) for a new cola soft drink company?

9. Under what circumstances would penetration pricing for a new product be most appropriate?

CHAPTER

17

Managing Promotion Programme (Marketing Communication)

17.1 PROMOTION: DEFINITION

After a product is developed and priced, the next important task for the company is to inform and make its prospective customers aware of the product, its features and benefits, convince them of the product value, quality and superior performance over competitors' products through repeated contacts and communication, and persuade them to purchase the product. This important task of influencing the buyer behaviour and making him purchase the product is called *promotion*. Not only existing and potential customers, the company has also to inform about and acquaint other audiences like middlemen (distributors), investors, suppliers, employees and general public with the product/products to build the brand image and corporate image.

As referred to in chapter 6, promotion is an essential component of marketing mix and strategy, the other ingredients of marketing mix being product, price and place (distribution). All the above activities concerning promotion i.e. informing, acquainting the customer with the product, reminding and convincing him into buying it neccessitate communication. In other words, communication is the life blood of promotion. Therefore, promotion is also called *marketing communication*. We can define the promotion as follows:

Promotion refers to all those activities that are designed by the producer to influence buyer behaviour through communication.

Communication takes place through advertisements, brochures, direct mail, personal and on-line contacts etc. Before discussing the tools of promotion, we shall briefly explain the general meaning, nature and process of communication.

17.2 COMMUNICATION: MEANING, NATURE AND PROCESS

(a) Meaning and Nature

In simple, words, communication means the process through which two or more persons exchange ideas and understanding among themselves. According to Haimann, "communication means the process of passing information and understanding from one person to another." It involves sharing with and transmitting to others ideas, facts, thoughts and values.

Allen defines communication as "the sum of all the things one person does when he wants to create understanding in the mind of another. Communication is a bridge of meaning. It involves a systematic and continuing process of telling, listening and understanding." Thus communication is the transfer of meaning from one person to another and is reflected in attention, understanding, acceptance and results or response. It is evident from the above definitions of communication that the process of communication is complete only when the message transmitted by sender (firm) is correctly understood by the receiver i.e. the customer after its receipt and the sender receives necessary feedback or the receiver's response to that effect. Only then the use of message can be effective and meaningful. Transmission of information or meaning necessitates use of several means or devices such as oral or written words, actions, facial expressions, audio-visual devices etc.

(b) The Communication Process

The communication process is the method by which the sender transfers information and understanding to the receiver. It consists of the following six steps:

Step 1: Developing an Idea or Thought: The first step is to develop the idea, thought or message to be communicated. It is the subject matter of communication and may be in the form of opinion, feelings, views, suggestions, orders. This is a key step in the sense that it determines the worthwhileness of the ideas or message to be communicated.

Step 2: Encoding: This step consists of converting the idea or message into form of words, symbols, pictures, actions for transmission.

Step 3: Transmission through Channels: Encoded ideas or directives are to be transmitted through certain channels or media known as communication channels like radio, telephone, wireless, mobile, fax,

TV or any print medium etc. This step allows the message to reach the receiver.

Step 4: Receipt by the Receiver (Customer): At this point, the message is received by the receiver (customer) i.e. the person for whom the message is meant.

Step 5: Decoding: In this step, the message received is decoded to understand it. The message received is analysed and interpreted so that it is correctly understood. The communication is not effective or complete unless the receiver (customer) understands the message. This step gives meaning to the message.

Step 6: Use of Response: After the message is understood, the receiver is to make use of the same., This step involves response or reaction on the part of the receiver (customer) regarding his understanding of the message. The communication is complete with this step. But it remains to be seen whether the receiver has understood the message in the same sense as the sender wants. This can be ensured only when the receiver sends feedback to the sender. It is very essential for the communication to be effective. This would involve two-way communication.

Feedback: Feedback ensures exchange of information and understanding between the receiver and the sender. It ensures that the message has been correctly understood. This gives rise to two-way communication which results in a communication circuit and takes the form of a circular flow of information as shown in the Figure 17.1.

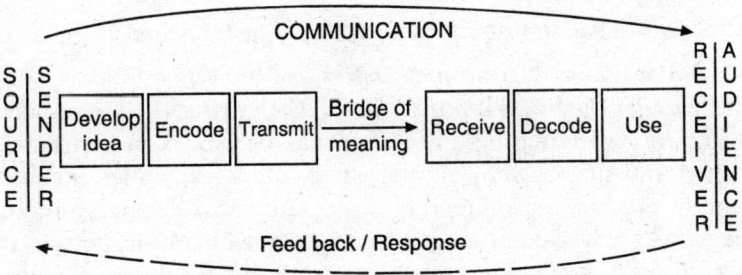

Fig. 17.1 Communication Process.

(c) The Objectives of Communication

From the above study we can sum up the objectives of sound and effective communication. These objectives are:

   (i) to transmit the right information at the right time to the person who needs it.

(ii) to get the message accepted, understood and acted upon.

We shall elaborate it further later in the chapter.

17.3   PROMOTION MIX (MARKETING COMMUNICATION MIX)

A company has to communicate not only with its customers but also with its other *audiences* — distributors, suppliers, investors and other publics as they affect the performance of the company. Moreover, companies spend huge amounts on publicity and public relations campaigns to establish a corporate image. As stated above, a company uses advertisements, brochures, letters, personal contacts and other tools etc., to communicate with its audiences. All these tools contribute to influencing the behaviour of the customer and are called *promotion tools*.

Major tools of communication are—advertising, sales-promotion, publicity, personal selling and on-line selling and form promotion mix or marketing communication mix. We define promotion mix as follows:

Promotion mix is a particular combination of promotion tools used by a company to communicate with its audiences.

Promotion mix consists of:

(1) Advertising

(2) Sales Promotion

(3) Publicity/Public Relations

(4) Personal Selling

We are briefly describing these four major tools here.

(1) Advertising: *It is a presentation and promotion of ideas, goods or services paid for by an identified or identifiable sponsor.* It is non-personal mass communication programme and has become a potent means of education and mass selling. It consists of all the activities involved in presenting product information to target audiences through advertising media such as newspapers, magazines, catalogues, booklets, posters, radio, television, cinema, novelties, calendars, cards, boards, logos, sky writing, posters, transport advertising.

(2) Sales Promotion: It consists of a number of incentives, mostly short-term, designed to stimulate quicker and/or greater purchase of a particular product by consumer or trade. *It is referred to as a number of non-personal non-media incentives to increase and boost the sales of goods and services.* The incentives may include discounts, gifts, gift coupons, distribution of free trial samples, exhibitions, displays and demonstrations, contests, games, prize lotteries, trade-in allowances etc. The audience

clearly perceives the organisation which is the source of these incentives/ messages.

(3) Publicity/Public Relations (PR): According to UK Institute of Public Relations, PR is "the deliberate planned and sustained effort to establish and maintain mutual understanding between an organisation and its public". It is referred to everything that is conducted to improve mutual understanding between an organisation and target audiences with the aim of building goodwill and good image, *It is referred to as non-personal stimulation of demand for a product or service by planting commercially significant news about it in a printed medium or obtaining favourable presentation of it upon radio, TV, or stage, that is not paid by the sponsor.* The publicity activities include print media news stories, speeches, seminars, annual reports, donations, publications, lobbying, sponsorship etc. Media are the source of messages.

(4) Personal Selling: *This is referred io as oral presentation in a direct conversation with one or more prospective purchaser or customer for the purpose of making sales.* It is face to face selling. It is done by salesforce i.e. paid employees of the company, the organisation being the source of messages. The activities include in-person sales presentations, telemarketing, on-line marketing, trade fairs.

Managing promotion mix means planning, organising, directing and controlling promotion tools, namely advertising, sales promotion, publicity and personal selling. We shall discuss management of each specific tool in details in the next three chapters. In this chapter we shall give a general explanation of how to manage communication flow and the promotion mix as a whole.

17.4 EFFECTIVENESS OF COMMUNICATION (COMMUNICATION FLOW)

Communication is regarded as the basic skill of management as it is an essential part of everything the manager does. Communication always involves two parties, one who transmits or sends the message i.e. the sender or the source and the other who receives the message i.e. the receiver or in our present context, the customer or the target audience. Only sending a message is not communicating. It is not one-way street. It is a two way affair. For communication to take place, the message must be received by one for whom it is meant. Again communication is transfer of information, understanding and meaning from one person to another. A significant feature of effective communication is that the message or information sent is properly understood by the target audience and evinces a proper response.

(1) The beginning of all communication is an idea or problem and it

must first be clearly formulated in the mind of the sender. As John Dewey has stated "a problem well stated is half solved". The clear conception of message in the mind of the communicator is essential for effective communication. This would help in designing a clear and effective message that may capture the attention of the target audience. We shall discuss this aspect of communication in the next section.

(2) A climate of trust and confidence is essential for effective communication. Keeping faith with the target audience, transmitting an information and facts honestly and positively, responding to their views and grievances by the management are necessary conditions to win the confidence and trust of the target audience. The management must see that there is no variance between what it says and what it does. Effectiveness of message thus depends on the credibility of the sender. The credibility of the sender or source will depend on how the customer or receiver (the target audience) perceives it. In other words, the effectiveness of message or communication depends on customer's or audience perception of the source. We shall study this aspect further in the next section.

(3) Even when a sender makes good efforts to develop a message and transmits it correctly and the customer on his part receives it intently, there are many factors that may hamper or block the flow of communication resulting in misunderstanding, distortions, or break downs in the communications or what is called the *"Noise"*. *Noise* in environment implies misunder-standing or distortion of message in the communications system. In modern enterprises, the senders of the message and the audiences have separate words of their own with their own feelings, emotions, views, abilities, prides and prejudices and interests. *The factors, conditions or interferences which limit the audience's understanding and block the communication flow or, in other words, create noise are known as barriers to communication.* They include personal, physical, semantic (language) and other barriers which hamper flow of communication.

17.5 DEVELOPING A PROMOTION (MARKETING COMMUNICATION) PROGRAMME

Development of an effective promotion programme involves the following steps:

(1) Identifying the Target Audience
(2) Determining Promotion/Marketing Communication Objectives.
(3) designing the Message — Content and Format
(4) Selecting Message Channels/Promotion Media
(5) Allocating Promotion Budget
(6) Determining the Promotion Mix

(7) Evaluating the Performance

We will discuss these steps in the following pages.

Step 1:   Identifying the Target Audience

The nature and type of promotion programme will depend on the nature and type of the audience for which the programme is to be designed. Therefore, the first task is to identify the target audience and its characteristics.

As stated above, marketing communication is not restricted to the company's customers. It has to communicate with many other audiences such as distributors, shareholders and investors, government officials, company employees and general people at large to build its image and reputation and promote its products. A promotion programme directed to general public or middlemen i.e. influencers will be different from that which is directed to customers i.e. actual purchasers or deciders. Often, a company needs to communicate with a mix of these audiences.

Again, the promotion programme needs to be tailored to the target audience's perceptions of the company image and its products. The members of the target audience may have a good or bad perception of the company's image and its products. Their response to company's products may be favourable, unfavourable, indifferent. Thus buyer's behaviour toward the company and its products greatly influences the devising of the promotional programme and its objectives.

Identification of the target audience involves *seeking and analysing responses* from its members to know their perceptions of the company and its products. The responses can be gathered and analysed through market research.

Moreover, among the target audience, there may be persons who may not be knowing the company or the product-brand. Again, some may have tried the brand while others might have not done so. Otto Ottesen, a Scandinavian researcher has shown that within a target group there are knowers and nonknowers. Among knowers, there are persons who might have tried the product and those who have not tried the product. *Even among knowers-triers, they have different attitudes. Some are preferers, some are indifferents and some rejectors.* This suggests that it is the audience's attitudes and perception of the product and the company that would determine the objectives of the promotion programme. Therefore a marketing manager must first identify the target audience and study and analyse its characteristics and perceptions. This would help in determining the communication objectives. Identifying the target audience is the central promotion decision. Other promotion decisions are based on this.

Step 2: Determining Promotion Objectives

After identifying the target audience, the next step in planning a communication programme is to determine the objectives of the programme i.e. pinpointing the objects and goals that the communication should aim at achieving. This would involve studying the members of the target audience and their behaviour as buyers. Moreover, a promotion programme necessitates effecting change in attitude and buying behaviour of people in the target audiences. It is the study and analysis of the buyer behaviour of the target audience that would give us the objectives of the communication.

As stated earlier the main objectives of an effective communication are:

(i) to transmit the right information at the right time to the person identified.

(ii) to get the messages understood, accepted and acted upon.

Thus communication objectives have three main components:

(1) Cognitive i.e. knowledge component (learning),
(2) Affective component which pertains to feelings and emotions (feeling), and
(3) Conative or behavioural component i.e. getting the consumer to act (doing).

In our present context, the ultimate objective of communication is to influence buyer purchase decisions and help the people in the target audience in making their purchase choices. What is important in the determination of communication objectives is that they should match the purchasing behaviour of the people in the target audience. Only then the communication can be effective and meaningful.

A buyer passes through a number of stages before he makes decision to purchase a product. We have already referred to them as cognitive, affective and conative or behavioural stages. Each stage influences and has an effect on the buyer which leads to another stage and effect, thus forming a *"hierarchy of effects"*.

We are giving here two models of "hierarchy of effects" showing various stages through which the buyer passes. These give us the objectives of communication or promotion programme as a whole.

A. The AIDA Model

The letters of the term "AIDA" denote attention, interest, desire, action. This model of marketing suggests that the buyers pass through stages of attention, interest, desire and action before making actual purchase. This model advocates that the communication be so designed that the message should capture attention of the consumer, hold his interest, stir up desire and elicit action on his part to purchase the product. According to this model, the objectives of communication are:

(i) to get the attention of the customer
(ii) to hold his interest in company's product
(iii) to stir up desire in the customer for the product
(iv) to make him purchase the company's product

Attention belongs to conative field, interest and desire belong to affective field and action is the conative or behavioural component. Attention leads to interest, interest leads to desire and desire results in action.

B. "Hierarchy of Effects" Model

A. Elrick and Robert J. Lavidge, American researchers have developed a communications model for advertising known as *"Hierarchy of Effects"* model. According to this model a buyer progresses through following stages

(i) *Unawareness:*   When the buyer does not know the existence of the product.
(ii) *Awareness:*   When the buyer knows about the availability of the product.
(iii) *Knowledge:*   When the customer knows the product and its attributes and understands its use.
(iv) *Liking:*   This is the stage which shows how the buyer feels about the product i.e. whether he likes it or dislikes it.
(v) *Preference:*   It is the stage when a buyer has to decide his preferences out of several alternative choices i.e. it is the stage of developing preference for a product to others.
(vi) *Conviction:*   When the customer is convinced that the product is a good buy.
(vii) *Purchase:*   When the customer takes the final step and purchases the product creating sale for the company.

In this model, cognitive stage comprises awareness and knowledge, affective stage consists of liking, preference and conviction and purchase represents the behavioural stage. Awareness leads to knowledge, knowledge to liking, liking to preference and then to conviction and purchase in that order, making the model an hierarchy of effects.

According to this model, the objectives of communication are:

(i) to build awareness among the target audience about the company's product

 (ii) to provide information about the product, its attributes, and benefits and services it would deliver

(iii) to make the customer to have a liking for the product or fostering favourable feelings

 (iv) to build customer preference

  (v) to convince the customer to buy the product

 (vi) to make him purchase the product, through additional information, if need be

It is evident from the above study that communication objectives heavily depend on the identified target audience and their response i.e. on the stage of the market. We have already referred to Otto Otteson's study on responses of the target audience. Within the target audience, there are knowers and nonknowers, triers or non-triers. The communication objectives for knowers will be different from those for the non-knowers as there would be no need of rebuilding awareness among the former. Again, if the number of nonknowers among the target audience is very small, then why to waste money and time on building awareness among them. Again, the objectives for knowers-triers will be different from those for knowers-nontriers. In case of a new product, when the number of knowers and triers is small, the objective of communication will be to increase their number. In case of triers-indifferents, the objective may be to turn them into preferers. In short, the communication objectives will be determined on the responses of the target audience or in other words, the stage of the market or the stage through which a particular segment in the target audience is passing through. We shall discuss the objectives further in the next chapter on advertising.

Step 3: Designing the Message

After identifying the target audience and determining the communication objectives in consonance with their responses, the next step is to design a message that may appeal the most to the identified audience. We have already referred to in the previous step the AIDA model which suggests the desirable qualities of an ideal and effective communication message i.e. the message should be such as may capture the attention of the audience, arouse their interest, stir up their desire and persuade them to purchase the product.

Designing a message is one of *the most creative part of promotion* and it involves determining what to say to the target audience (i.e. message contents), how to say it effectively (i.e. message format or lay out) and who is to say it (i.e. source of the message).

(a) Message Content: *It is the information comprising facts, opinions, ideas, themes, persuasive arguments, unique selling points (USPs) provided to the target audience to achieve promotional objectives.* USPs are strong claims and advantageous position of the proposition. It is a sort of appeal addressed to the target audience with a view to inform them, capture their attention, arouse their interest and persuade them. The message content or appeal may contain information about physical features of the product or functions and use of the product. It may present comparison of the company's brand with other brands (comparative advertising). It may convey an emotional appeal evoking emotions and feelings of fear, guilt, humour etc. Often most of the message content is product-oriented and consumer-oriented. The content of the message should hit the right audiences.

A few examples of *message content* i.e. the information provided are given here.

Colgate Dental Cream— Stops Bad Breath. All Around Protection. Fights Tooth Decay. Colgate Smile.

Close-up Toothpaste— Shiny white teeth. Fresh breath. Tooth paste with Real Mouthwash.

Pond's Cleansing Bar— Enriched with Pond's Cream for Clean and Soft Skin.

Lux— The Beauty Soap of Movie Stars, Peach & Cream for Velvety and Soft Skin.

Lifebuoy— Where There's Lifebuoy, there's Health.

Dove— The Cream beauty bathing bar.

Surf— Extra Strong, Whiter than Ever Before.

SPG Agro Bond: When You Are Among Sky High, Nature is the Best Buy.

Campa: The Great Indian Taste.

(b) Message Format or Layout: It is the arrangement in which the text and visual displays pertaining to the message are presented to capture audience's attention and communicate the message effectively. A message format or layout is the presentation of various elements of message like headline, copy, diagrams and illustrations, symbols, coloured matter in a way that may enhance the message to be conveyed. A well-designed format and content together make the message interesting, attractive and eye catching. A message presented with a poor layout will not get attention to be effective. The message format should be such that the message may be easily understood and comprehended.

(c) Message Source: The effectiveness of communication also depends on the source from which a message originates. How the audience perceives the source of the message is a critical factor. The message will be more effective if the customer perceives the source as honest, trustworthy and attractive, may it be an organisation or person. An organisation or a person with a good image perception will foster positive feelings of the audience towards the message. Source credibility and source attractiveness or likeability are two important factors.

Source credibility implies the *source's trustworthiness and expertise with respect to the content of the message as perceived by the audience.* The perceived similarity of the source enhances its likeability and attractiveness. Company's reputation and image play a great part in fostering credibility. However, the real image builder is the company's product and its performance or quality of service delivered. A company can foster credibility if it can deploy trained and expert salesforce to sell its products, persons who could spell out the benefits of the product/products effectively. It is not without reason that companies and advertisers use celebrities such as film stars, sportsmen, top fashion models to build brand image of their products.

Step 4: Selecting Message Channels (Promotion Media)

However the message content and layout be well-designed and attractive, it will be of no use unless there is an efficient delivery system to carry the message to the target audience. Herein lies the importance of promotion media. *Promotion media refer to various vehicles or channels used for carrying the promotion messages to the target audiences.*

There are two types of communication channels— personal and non-personal. Personal communication involves personal contact or interaction. Non-personal communication is carried on without personal contact or interaction. The promotion media include newspapers, magazines, catalogues, booklets, yellow pages, radio, television, cable TV, car cards, novelties, calendars, boards, neon-sign boards, sky writings, displays, direct mail, door-to-door demonstration. Then there are sales-persons, opinion leaders, associates, friends to carry the message and influence the target audience. The message also reaches the entire target audience through the *"word of mouth"*. Anyhow, no single medium can reach the entire target audience. The companies use what can be called "media mix", a combination of channels to reach the target audience, depending upon the perceptions of its members. We shall study more about media in the next two chapters.

Step 5: Allocating Promotion Budgets

Promotion programme cannot be successful unless adequate budget is available for it. On the amount of money available for promotion will depend the selection of communication media and determination of promotion mix. In fact, the decisions concerning various stages, namely promotion objectives, message content and format, media budget and mix are interrelated. A marketing communication programme is a part of the overall marketing plan. The budget provision for promotion programme is generally a component of overall marketing budget determined to achieve marketing objectives. There can be a number of methods to arrive at the amount of promotion budget. These are briefly explained here

(1) Objective-and-task Method: This method involves estimation of the budget needed for achieving the identified promotion objectives. First, a company determines the tasks needed to be performed to fulfil those objectives and then estimates the total cost for performing them. That would give the estimate of the promotion budget. But the main problem here would be whether the company would be in a position to afford amount budgeted as a result of this exercise.

(2) Affordable Method: As the name suggests, according to this method, a company decides about the amount it can afford for the promotion programme. The promotion budget varies every year and depends on the financial position of the company.

(3) Percentage of Sales-volume/Turnover Method: Most of large companies set a specific percentage of their current or anticipated sales-volume or of the sales value for their promotion programmes. It is done on the basis of the past experience. However the percentage is modified per requirements of promotion goals/marketing goals.

Different companies in the same industry spend varying percentages of their turnover on promotion. Again, a company may set different percentages for promoting different products. Table 17.1 shows the expenditure being incurred by some companies on promotion (advertising, publicity, sales promotion) of their products.

If a firm wants to increase sales or market share, it may have to spend more on promotion.

Step 6:   Determining the Promotion Mix

In modern business a company cannot depend exclusively on only one of the four tools of promotion—advertising, sales promotion, public relations and personal selling. A brief description of these tools has been given in section 17.3 of this chapter. Each tool has its own unique characteristics and qualities, strengths and weaknesses. A company has to secure such combination or mix of these tools that may give the best results and achieve the identified objectives with the given total budget. Therefore, the next important task is to distribute the total budget over different tools of promotion, namely advertising, sales promotion,

Table 17.1: Expenditure on Promotion Programme.

Name of the Company	Product/ Product Group	Element	Expenditure Amount (Rs. Crores)	% of Sales	Year
Hindustan Unilever	Soaps and detergents Personal Products Beverages Icecreams Processed Foods	Advertising and Promotion	1422	10.37%	2007
Marico Limited	Refind Oils Coconut Saffola Hair Grooming Products	Advertising and Promotion	245.5	12.9%	2007-08
Dabur India Ltd.	Consumer Care Consumer Health Foods	Advertising and Publicity	248.1	11.75%	2007-08

publicity and personal selling—how much to be allocated to each tool.

As stated above, every tool has its own qualities, strengths and weaknesses and has its own role to play. We shall discuss the qualities and roles of each of these tools in details in the next three chapters. Here we shall study the criteria which help us in making decision about the structure of the promotion mix.

Generally, five criteria on the basis of selling tasks are used to determine the role of each promotion tool. They are—cost of reaching an audience member, ability to reach target audiences without leakage, ability to deliver message effectively, ability to interact with audiences, and credibility. We shall compare the roles of these tools on the basis of the above criteria.

Advertising: Advertising has a number of vehicles at its disposal such as newspapers, TV, bill boards, direct mail etc., each with its own unique role in promotion. Measured in terms of above criteria, the cost of advertising per audience member is low. Its reach to the target audience and delivery of message vary from poor to good. The interaction with audiences is totally absent. It carries only a monologue. The credibility is low.

Sales Promotion: The cost per audience member is low and ability to reach the target audience is good. It cannot effectively deliver a complicated message. The interaction with the audience is absent. Credibility of sales promotion is low.

Publicity: The cost of publicity per audience member is extremely low. Its reach is moderate, delivery poor to good, interaction low to moderate but credibility is high.

Personal Selling: In case of personal selling, the cost per audience member is very high but its ability to reach the target audience, ability to deliver even the complicated message and interaction with the audience (customers) are extremely good but credibility of this tool varies from moderate to high.

A manager has to set a promotion mix and allocate funds to each of the tools, keeping in view their likely use and potential in order to attain promotion objectives. Besides these considerations, promotional mix is also influenced by the following factors:

(1) Type of Product Market: In a number of studies, it has been shown that the relative importance of different tools of promotion in case of consumer markets and industrial markets is not same. It varies. For example, the relative importance of advertising is the highest in case of consumer goods, that of sales promotion, personal selling, publicity following in that order whereas industrial markets rate relative importance of personal selling high above sales promotion, advertising and publicity.

Studies have also shown that a mix of these tools can yield better results and increased sales over their exclusive use as a tool, whatever be the rating of their relative importance. Each tool plays a significant role and the task of the manager is to arrive at the best combination.

(2) Push and Pull Strategy: The promotion mix also depends on whether the company has devised the push strategy or pull strategy in order to achieve the promotion objectives.

(a) *In a push strategy*, the producers address their promotion activities to channel members or middlemen and induce them to place larger orders and promote the product(s) to the end users or customers.

(b) *In a pull strategy*, manufacturers address their promotion activities to the end users or customers who ask for the product from the distributors and middlemen and induce them to place larger orders with the manufacturers. Different companies follow different strategies—some depend on push strategy and some follow a pull strategy.

(3) Characteristics of Target Audience or Stage of buyer responsiveness: The promotion mix also depends on the characteristics or perceptions of the target audience i.e. stage of buyer responsiveness or what is called *"buyer-readiness stage"*. In other words the marketing mix will rely on whether the buyer is a nonknower or knower or is in the stage of comprehension, conviction, purchase or repurchase. Whereas advertising and publicity may be the most cost effective when a buyer is in non-awareness or awareness stage, sales promotion will be most effective in case of repurchasers i.e. at later stages.

(4) Product Life Cycle Stage: Promotion mix is also influenced by the stage of product life cycle. In introduction and maturity stage, all tools viz. advertising, sales promotion, personal selling, publicity would be needed in suitable measure and prove cost effective whereas in the stage of decline sales promotion is the most effective and personal selling will not serve the ends.

Now the promotion programme is ready for implementation and the decisions made need to be executed using the set promotion mix. Again, it is the responsibility of the marketing department to implement, monitor and control the promotion programme. Successful implementation necessitates cooperation of all including channel members and middlemen and coordination of various promotional activities. In some big organisations, the promotion work is handled by Communication Department or Vice-President (Marketing).

Step 7: Evaluating the Performance

After the promotion programmes and decisions are implemented, the next step is to evaluate its performance which involves measuring the results in terms of its impact on the target audience.

In evaluating the performance of the promotion programme, what is needed is to find out how far and to what extent the objectives set have been accomplished. If the objective is to increase the sales or market share then its effect on sales or market share can be recorded and measured. However most of the objectives set are in terms of changing buyer's behaviour, attitude or response and elicit certain actions among members of the target audience. Evaluation would involve measuring the frequency of the occurrence of these actions. This involves collecting information from audience members about the times the message was conveyed to them, its content and their response to the message or such information, how many of them knew the product, how many came to know of it as a result of the promotion programme, how many tried the product and how many of them were satisfied with the product. This feedback will tell the marketer how far promotion programme has been successful and in case of inadequacy, what changes are needed.

Management of specific tools of the promotion mix will form our subject of study in chapters 18, 19 and 20.

EXERCISES

1. What is meant by promotion mix? Explain its ingredients.

2. Describe the steps involved in developing a promotion programme.

3. Explain the AIDA model and Heirarchy of Effects model of marketing.

4. Write notes on
   (a) Designing a message
   (b) Push and Pull Strategy
   (c) Communication process
   (d) Message content

5. What criteria are used to determine the role of promotion mix tools? Explain.

6. Evaluate the role of advertising and publicity as promotion tools using the criteria as given in the answer to question number 5.

7. Enlist the advantages and limitations of the following promotion tools—sales promotion, personal selling, publicity.

8. For many years, Maruti India used no advertising in its promotion mix. Only recently has it started advertising its cars in the domestic market. Discuss possible reasons for the change in strategy.

18

Designing and Managing Advertising Programmes

In Chapter 17, we have briefly described various tools of promotion mix, namely advertising, sales promotion, publicity and personal selling, used by companies to communicate with their audiences. As stated earlier, advertising, one of the four promotion tools, is a potential means of communication, education and mass selling. In this chapter, we further explain the concept and role of advertising as a promotion tool and in marketing. We also discuss here how an advertising programme that may hit the audience and elicit desired response can be effectively designed and managed.

18.1 ADVERTISING: CONCEPT AND IMPORTANCE

(a) Concept

As stated earlier, "advertising is a nonpersonal presentation and promotion of ideas, products or services paid for by an identified or identifiable sponsor". It consists of all activities involved in presenting product information to the customers through advertising media. It is thus a potent tool of mass communication. We can define advertising in the following way

Advertising is the communication of a message about an idea, a product or a service to a specified audience with the objective of eliciting a desired response.

In this sense, a puppet play that communicates a message on health and sanitation to village folks or the speech of a great orator at a public meeting is as much advertising as a TV spot, a hoarding or a press ad.

(b) Importance of Advertising in Modern Business

In modern business, large companies have to produce goods on a large scale to stay in business. Mass production demands mass consumption which necessitates use of mass media i.e. advertising to provide information about the products to the consumers, i.e. right upto the very homes of the target group so as to elicit their desired response. Advertising is a potential marketing tool and it has lately revolutionised the life and times of consumers in their selection of goods and services in the market place. A market economy means advertising inevitably. Its importance cannot be overemphasised. We can enlist its benefits:

1. Advertising creates mass markets and helps the manufacturers to derive production economies, reduce costs of production and provide products to the customers at cheaper prices.

2. It creates an awareness among customers of products, especially new products and innovations resulting in better choices and improved standards of living and leading to attitude change.

3. Advertising makes the market more competitive. Competition leads to enhancement of quality and product value.

4. It helps to increase sales, market share and profits.

5. Advertising generates income and employment in advertising and media industries.

6. It helps in building the image of the product (*product advertising*), brand (*brand advertising*) and the company (*institutional advertising*).

7. It establishes direct contact between the producer and the consumer.

8. It makes the work of salesforce easy and helps them to establish more stable relationships with customers.

9. It helps furthering social and environmental causes (*advocacy advertising*).

18.2   DEVELOPING AN ADVERTISING PROGRAMME

Advertising is a potent promotional tool and as discussed in chapter 17, the first task of the manager is to identify the target audience and its characteristics (buyer's responsiveness stage or buyer motives). However, before designing an advertising programme, the marketing manager must know about what and how much role advertising is expected to play in the overall promotional strategy. In other words, he must know what job advertising is to perform as a part of the marketing strategy.

Before spelling out the steps involved in developing an effective advertising programme, it would be worthwhile to refer here to the DAGMAR MODEL.

DAGMAR MODEL

DAGMAR is a method developed by Russel H. Colley for turning advertising objectives into specific measurable goals and thus measuring advertising effectiveness. Colley named this method after his book "Defining Advertising Goals for Measuring Advertising Results" (DAGMAR). According to this model, advertising has to perform a specific communication task among a specific audience within a specified period. According to this model, advertising is not a sales task but a communication task.

According to the model, marketing managers need to make five major decisions in developing an advertising programme known as the 5Ms, *viz.* the *mission* or the advertising objectives, *money* (to be spent), *message* (to be transmitted), *media* (to be used) and *measurement* i.e. evaluation of advertising results.

Development of an effective advertising programme thus involves the following steps:

(1) Determining the mission of advertising
(2) Determining the advertising budget
(3) Designing an advertising message
(4) Deciding on the advertising media
(5) Evaluating advertising effectiveness

Step 1. Determining the Mission (Objectives) of Advertising

Determination of the mission of advertising in itself involves establishing the role of advertising in the overall promotional strategy, identifying the target audience, and setting advertising objectives.

(a) Establishing Role of Advertising

The role to be performed by advertising in a marketing programme depends on three factors—

(i) On the weightage or importance given to advertising in the overall marketing programme vis-à-vis other components of mix. This further depends on the type of the audience the advertisement is required to cater to i.e. whether it is consumer market or industrial market. Advertising is assigned a large role in a consumer market as compared to its role in industrial market.

(ii) On its relative strengths vis-à-vis capabilities of other tools of the promotional mix or in other words, the role of advertising will be determined after weighing and comparing its merits with the strengths and weaknesses of other tools, namely sales promotion, personal selling.

(iii) On coordination of roles of all the promotional tools: This work needs to be performed by marketing manager or a committee formed for the purpose.

Assigning a proper role to advertising forms one part of the advertising mission.

(b) Identifying the Target Audience

In chapter 17, we have referred to various types of audiences to which promotion efforts are often directed. Advertising in generally directed to consumers i.e. end users and the trade i.e. members of distribution channels. Identifying the specific segment of consumers or trade audience to communicate with is also an essential part of advertising mission. The main aim of advertising is to inform, remind and persuade the target audience to purchase the company's product. Thus *trade advertising* means communicating with the trade i.e.. resellers (wholesalers and retailers). We have already referred to the *"push and pull strategies"* adopted by companies to promote their products, in chapter 17. Identifying target audience also requires allocation of separate budgets for trade advertising and consumer advertising.

(c) Setting Advertising Objectives

Though the ultimate aim of advertising is to influence buyer purchase decisions and lead them to actual purchase resulting in increased sales, yet marketing managers have different opinions about the nature of advertising objectives. Some advocate setting of sales objectives for advertising while others opt for communication objectives.

Sales Objectives. These objectives can be

(i) to increase demand and turnover,

(ii) to increase market share, and

(iii) To increase profits.

Those who are in favour of setting communication objectives argue that advertising is not the only factor that influences buyers' purchase decisions and contributes to increased sales, there are a large number of other factors that help in this direction. Increase in sales or profits may not be the effect of advertising.

Communication Objectives

A. As stated above, DAGMAR model suggests that advertising is essentially a communication task that can be assessed and evaluated in terms of its accomplishment and performance. The communication objectives are—

(i) to inform the target audience about the company's products (both existing or new) or services,

(ii) to persuade the target audience,

(iii) to remind the target audience to keep them thinking about the product.

Most often, companies use *comparison advertising*, showing the superiority of their product or brand over the competitors' products or similar products called "ordinary" to persuade the audience.

B. According to AIDA model as referred to in chapter 17, the objectives of advertising are

(i) to get the attention of the customer,

(ii) to hold audience's interest in company's product,

(iii) to arouse desire in the customer for the company's product, and

(iv) To elicit action on the part of the target audience.

C. According to *Hierarchy of Effects Model* (A communications model for advertising) as explained in chapter 17, the objectives of advertising are—

(i) to increase awareness,

(ii) to impart knowledge about the product,

(iii) to instill in customers a liking for the product, and

(iv) to build customer preference for the company's product,

(v) to convince the customer for making purchase decision.

This model helps in measuring the effectiveness of advertising

Step 2. Determining the Advertising Budget

After determining the advertising objectives, the next step is to determine the advertisement budget for each product. Advertisement budget is a financial statement showing the amount of money/resources to be allocated to advertising and also how it is to be spent itemwise to achieve advertising objectives.

We have already described three methods of determining budgets, namely *affordable method, percentage of sales method* and *objective-task method* in chapter 17. These methods are also used to establish advertising budget of a product or a brand. In Table 17.1, chapter 17, we have also shown the expenditure incurred by some Indian companies on advertising and other tools in terms of percentage of sales.

Competitive Parity Method.   There is another method for estimating the cost of advertising. Some companies keep parity with competitors and spend the amount that may correspond to what their competitors are spending on advertising. It is done in the following ways:

(1) A company may set its advertising budget in relation to competitors' advertising budgets in the ratio of their market shares for the product. This is called *share-of-voice parity.*

(2) A company may spend the same percentage of its sales on advertising as competitors might be doing.

(3) A company may spend the same amount on advertising as the competitors are spending or equal to the average expenditure on advertising for the industry.

(4) A company may spend on advertising a certain percentage of market leader's advertising expenditure.

Designing and Managing Advertising Programmes

Besides, there are many other factors that need to be considered while determining the advertising budget. These factors include market-share of the company and customer loyalty, number of competitors, number of substitutes in the product-market, frequency of advertisement needed and stage of product life cycle. Marketing managers have to consider these factors also when they set an advertising budget.

Step 3.   Designing an Advertising Message

It is the message that connects the product (and the company) with the customers or target audiences. Message is, in fact, the soul of an advertising programme and needs to be designed very carefully and imaginatively. We have explained in details the considerations that should go into designing and developing an effective promotional (advertising) message in chapter 17. We shall further elaborate the theme in this chapter in context of devising an effective advertising campaign.

An advertising campaign is referred to as a set of messages communicated to the target audience in such a way as may achieve the desired advertising objectives.

Factors to be Considered in Designing Advertising Campaign (Messages)

An advertising campaign involves deciding on what to say, how to say it and when to say it effectively. While designing and developing an effective advertising campaign (i.e. messages), a marketing manager must consider the following aspects:

(i) **The Target Audience,** its characteristics and the stage of its responsiveness i.e. the customers' requirements.

(ii) **Marketing Positioning of the Product:** *Positioning is the act of designing the company's image and value offer (the product) so that target audience understands and appreciates how the company's product stands in relation to competitors' products i.e. how the company product differs from those of competitors* e.g., low price position, high quality position. It would involve assessing company's own strengths and weaknesses as well as competitors' strengths and weaknesses. Company's marketing position statement provides meaningful and truthful ideas for designing messages.

(iii) **Advertising's Unique Selling Points (Proposals):** As referred to in chapter 17, the text of the message should contain not only information, facts, ideas, themes about the product but also persuasive arguments, unique selling points i.e. strong claims and advantageous position in terms of added value, benefits and services and appeals. The message must provide information about the uses of the product and the likely resulting experiences. Most companies indulge in comparison advertising to show the superiority of their product over competitors' products. The content of the message is termed as *advertising copy.*

(iv) **Advertising Presentation (Message Format or Layout):** The advertising has also to look to how the themes or specific selling points are being presented to the audience. There is need to present various ideas and selling points in such a way as may capture audience's attention. It involves presentation of message content in an attractive format. This is the work of creative specialists.

Message format or layout is presentation of various elements of message *viz.* headline, symbols, colour in a way that may enhance the message to be conveyed. A well-designed, coordinated and integrated format and content together make the Ad copy interesting, attractive and effective.

In the present day jargon, *"advertising copy"* does not mean only written content or script in the advertisement, it covers all elements and features of an advertisement, namely text, pictures, symbols, labels, logos and drawings. Now the task of preparing and developing advertising copies is entrusted to creative specialists or advertising agencies.

Evaluating Advertising Messages

(a) **Pretesting of Alternative Ads:** Generally, companies or advertisers design a number of alternative advertising copies which are pretested to determine which copy out of the alternatives is the best to communicate selling points to the target audience and thus has the strongest appeal to the audience. The alternative that has the greatest impact is chosen as the advertising copy for communicating the message to the audience. The pretesting is conducted by market researchers or advertising agencies themselves who have the required expertise to do so.

(b) **Evaluating Criteria or Essentials of an Effective Message:** D.W. Twedt has suggested three criteria or essential features to rate an advertising message — desirability, exclusiveness and believability.

- (i) *Desirability* meaning that the message should contain some desirable or interesting information about the product.
- (ii) *Exclusiveness* meaning that the message should contain some exclusive information that may differentiate the company's product from those of the competitors and may show its superiority.
- (iii) *Believability* meaning that the message and claims made therein should be believable and stand the test of truth. *Truth is essential in advertising.*
- (iv) *Relevance* meaning that the advertisement should be relevant to the target audience.

All the alternatives are evaluated using the above criteria and the alternative with highest rating is selected for communicating with the target audience.

(c) **Need for Creative Specialists:** As stated above, designing an effective message is the most creative part of advertising. It demands originality and imagination on the part of the message designer to express and communicate the message in a fresh and interesting way. Now-a-days companies entrust the work to those who have specialised in the field. Most of the reputed advertising agencies employ creative specialists who develop and produce ads. It is a part of the services rendered by them to their clients. We shall discuss the role of advertising agencies later in the chapter.

What is important in this context is that the company must acquaint the outside creative specialists with the complete profile of the target audience i.e. audience's characteristics, values, life styles and requirements, advertising objectives of the firm, marketing positioning strategy and product's unique selling points (USPs) etc. Again, creative work needs no unnecessary fetters and specialists should be given freedom to show their creative talents provided the message does not adversely affect public interest, order or ethical standards. This is to be ensured that the advertising does not use sex gratuitously, exploits minority or oppressed groups or stimulates violence or antisocial behaviour.

In this context, Colgate-Palmolive Advertising Placement Policy given in the Appendix to this chapter is worth studying.

(d) Celebrity Advertising: Advertisers have been using celebrities like film stars, reputed models, sportsmen, athletes and famous people in other fields to promote and endorse products, services and causes. Basically, this genre came into India in the 1930s with the classic "Beauty Soap of the Film Stars" theme (Lux), when foreign film stars like Ginger Rogers or Loretta Young endorsed it. Leela Chitnis was the firs Indian actress to endorse it in 1941. Hindustan Unilever even today continues that theme with great success. Other advertisers are also using silver screen's heriones and beauties to promote their products. Advertisers have used other celebrities like Nawab Pataudi, Sunil Gavasker, Kapil Dev, Azharuddin, Sachin Tendulkar M.S. Dhoni, Gaurav Bindra and other sportsmen to promote products from Pepsi Cola and Coca Cola to Action Shoes. Most recently, the services of Menaka Gandhi and Seshan have been secured to further a social cause of vegetarianism as they endorse Saffal Vegetables. Such an advertising does elicit *"noticeability"* which is its prime objective.

It should be kept in view that, by and large, celebrities are paid for endorsing the product and the advertiser or marketer has to see whether they lead to so much of additional sales that may exceed the costs involved.

The real question for a marketer is to assess their value i.e. how far is the celebrity advertising effective? David Ogilvy, an authority on advertising while referring to "Testimonials by Celebrities" writes : "These are below average, in their ability to change brand preference. Viewers guess that celebrity has been bought and they are right. Viewers have a way to remember the celebrity and forget the product."

Celebrity Advertising

■*Creating markets, building brands*

Comparison Advertising

Advertisers have been indulging into comparison/contrast advertising as a part of persuasive superioty of their brand(s) over competitors' brands/products or other ordinary brands of same product available in the market. Sometimes, it leads to **"battle of brands"**. Comparison advertising is generally visible in case of products like soaps, detergents, toothpastes, shampoos, hair oils, automobiles etc. Captain Cook Iodised Salt, Nirma Detergent, ads are the examples.

We are giving here an advertisement published in The Statesman on June 16, 1896, more than 113 years ago to illustrate the point, besides a few other advertisements published in late 19th century. (See page 282)

Step 4.   Deciding on the Advertising Media

As referred to in chapter 17, advertising media are the various vehicles that the companies use to carry their advertising messages to the target audiences. They include television, radio, magazines and journals, newspapers, direct mail, outdoor boards etc. Each of these vehicles has its advantages and limitations. Besides, no single vehicle can reach the entire audience. There is need of combination of vehicles to reach the target audiences. Moreover, target audiences/ members need advertising exposures a number of times lest they should forget the message. Reminding is necessary. Again, it is also important to determine the timings when messages are to be advertised.

The above study suggests that an advertiser has to make three types of decisions. They comprise

   (i) selecting most cost effective media mix,

  (ii) deciding frequency of advertising insertion or exposures, and

 (iii) determining media timing and schedule.

(i) Selecting Most Cost Effective Media Mix (Specific Media Vehicles)

As there are a number of vehicles of communication available with the advertisers and no single vehicle can reach the entire target audience, the advertiser has to decide on which combination 19th Century Advertisementsof these vehicles or marketing media would most effectively achieve the advertising objectives.

Media mix is referred to as a combination of vehicles (e.g., newspaper, radio, TV, magazine, direct mail, outside boards) that can be used by the advertiser to communicate the advertising messages to the target audience.

Selecting and constructing a media mix that will achieve objectives necessitates a careful study and analysis of the strengths and weaknesses i.e. advantages and limitations of each vehicle or medium, in terms of its reach, location, impact and cost. In the following pages, we are describing important advantages and limitations of various types of media.

Some Major Advertising Media

A. Newspapers

Advantages

1. It reaches large audience.
2. It gives good local market coverage.
3. It conveys broad information on products and companies and has broad acceptance.
4. It has high believability which can benefit the advertiser
5. It takes short time to place the advertisement.

Osler's Kerosene Lamps
Substantial, Brilliant, Durable.
Safe, Windproof Cut Crystal
Glass, best quality, with Duplex
windproof Burner — Rs. 32.

19th Century Advertisements

Limitations (Disadvantages)

1. Its cost is relatively high.
2. It has a short life.
3. It has poor reproduction quality.
4. It has a nonselective audience and the advertisement may not be relevant to a large number of readers.
5. It has a limited appeal to young.
6. It does not last long and has a small pass-along audience.

B. Television

Advantages

1. It reaches mass audience.
2. It is highly visible mass medium.
3. It appeals to the senses as it has visual and audio capabilities. It uses sight, sound and motion to transmit a message and captures high attention. It provides instant exposure of pictures and ideas.
4. It provides great flexibility for use of creativity.
5. It has a short lead time to place the advertisement.

Limitations (Disadvantages)

1. It has high cost which makes it sometimes prohibitive.
2. It has little audience selectivity and cannot reach some specific target audiences.
3. It has short fleeting exposure.
4. It requires the services of promotion specialists.

C. Radio (including FM Channels)

Advantages

1. It reaches a large audience. It is a mass medium
2. It has a relative low cost
3. Its reach is high for geographic and demographic segments.
4. It has audio capability and in it use can be made of sound and music.
5. It takes short time to place the ad.

Disadvantages

1. It has short exposure time and a limited information delivery capacity
2. It gains lower attention than TV.

3. It serves fragmented audience.

4. It has no visual capability.

D. Magazines and Journals

Advantages

1. Its reach is high for demographic and geographic segments. It offers concentrated coverage.

2. The quality of production is high. Some magazines add a touch of class.

3. They have a high credibility and prestige which can benefit the ad.

4. The ad lasts longer and has a good pass-along audience or readership

5. The cost vary from magazine to magazine, from journal to journal.

Limitations

1. It has a long lead time to place and purchase the ad. The ads need to be booked in advance

2. It provides no guarantee over location of ad in an issue.

3. There is a circulation wastage and this medium has limited flexibility in capturing attention.

E. Outdoor Advertising

Advantages

1. It is highly visible mass medium

2. It has a relative low cost.

3. It evinces little competition

4. The ad gets repeated exposures.

Limitations (Disadvantages)

1. It has no audience selectivity. It cannot reach well-defined audiences.

2. The ability to convey product information is limited.

3. It is subject to regulations of local public authorities.

4. It has a very short exposure time.

F. Direct Mail

Advantages

1. It is highly selective and personal and provides flexibility in reaching the target audiences.
2. It is completely controlled medium.
3. It is easy to personalise the advertising message copy and lay out.
4. There is no competition within the same medium.
5. There is immediate response.

Limitations (Disadvantages)

1. It has a low response rate.
2. It is relatively expensive as obtaining appropriate mailing lists is a costly task.
3. It has junk mail image and is treated as such.

Besides these, there are other media like telephone, cable TV, yellow pages but they have not yet become dynamic selling media in India.

It may be interesting to no that even Newspapers, Magazines and TV Channel etc. have to advertise in other media to sell their space, time and programmes and pay for them.

Media Vehicle Costs

In order to decide on the combination of various vehicles through which message is communicated, it is essential to find out first the amount needed to buy media space or time. The cost of medium is generally calculated in terms of amount required to reach per thousand persons at one time. A popular measure of cost is calculated as cost per thousand. *Cost per thousand means* the cost of reaching per thousand persons in a vehicle audience with a single advertising insertion. For example, in case of a newspaper ad,

$$\text{cost per thousand} = \left(\frac{\text{Amount charged by newspaper for space}}{\text{Estimated readership}} \right) \times 1000$$

The cost of ad per 1000 persons can be compared to that of other vehicles with similar audiences to find out which vehicle is the most cost effective.

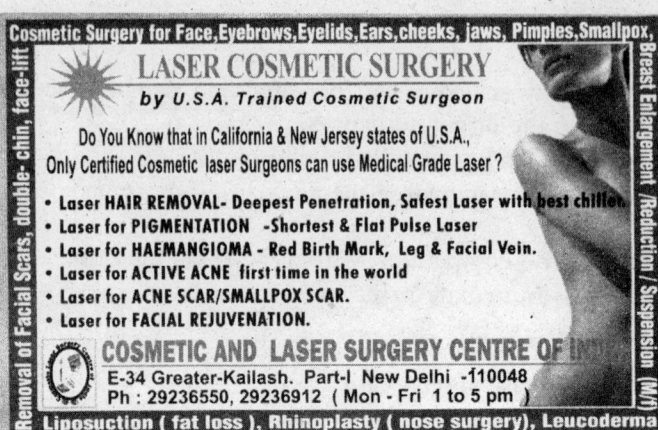

Table 18.1 Costs Involved in Various Ad Media.

Medium	Tentative Cost (Black & White)
News Papers	
English	Rs. 3,00,000 to Rs. 6,00,000 Full page per insertion
Indian Languages	Rs. 1,00,000 to Rs. 1,60,000 Full page per insertion
Television	Rs. 60,000 to Rs. 1,00,000 per 10 seconds
Magazines	
English	Rs. 50,000 to Rs. 1,00,000 Full page per insertion
Indian Languages	Rs. 2,000 to 15,000 Half page

* Rates for quarter page ad and half page ad can be calculated accordingly.
** Rates vary within the above range depending on particular region or coverage or commercial purpose.
*** For colour ads - the rates are higher.

Vehicle Audience Matching

Again, in order to select ad vehicles it should also be seen how the audiences served by a vehicle match with the target audience. In other words, the advertisers should study the characteristics of the vehicle's audience i.e. information on the type of people reading the newspaper or viewing TV/TV channel, and select the vehicle having audiences matching or overlapping the company's target audience. Now-a-days, news-papers, magazines, TV and other media supply information about their audiences, their characteristics, age-group, life styles, circulation and such information as of interest to their clients i.e. advertisers.

(ii) Deciding Frequency of Advertising Insertion or Exposures

Other important decision on advertising that an advertiser is required to make concerns the extent to which target audience is to be exposed to the advertisement during a specific period. It must be noted that the entire target audience is not exposed to the advertisement. Only a percentage of it is exposed. An advertiser has to decide the following:

(a) Reach of the advertisements i.e. the percentage of the target audience or the number of audience members that are exposed to the ad at least once during a specified period, in other words, the number of persons a particular ad reaches.

(b) Average frequency of exposure i.e. the number of times a target audience member is exposed to the advertisement during a specified period.

(c) Total number of exposures among the audience: An advertising exposure means the times a person actually sees the advertisement e.g., five advertising exposures implies that an audience member has actually seen the advertisement five times in a specified period. The repetition of advertising has a reminding function as audience members are likely to forget the message if they are exposed to the ad only once or twice.

Herbert E. Krugman advocates *three exposures* to make advertising effective.

To arrive at appropriate decisions, an advertiser should be well aware with

   (i) the characteristics of target audience,

  (ii) the characteristics of vehicle audiences,

 (iii) characteristics of different media and

 (iv) the resources available with the company for the purpose.

(iii) Determining media timing and schedule

After having decided the frequency i.e. the number of insertions to be placed in a vehicle during a specific time period say a budget year, another important decision is related to the determination of timings when the advertisements are to be inserted into various media for delivery to target audiences. The schedule will comprise specific times of ad insertions e.g. daily, weekly, monthly, off-season, at prime time or at non-prime time. These decisions are important for sequencing the insertions and to make advertising effective. Appropriate media scheduling increases the impact of the advertisements. Scheduling ensures *continuity* which necessitates spacing of insertions throughout a specific time period.

It is the selection of media mix to reach target audience together with schedule of ad insertions in these media during a specific time period and allied decisions which from what is called a *media plan.* And the stage is set for implementing and executing the advertising plan so as to achieve the advertising objectives.

Now a large number of companies take service of advertising *agencies to* design an advertising message, select the media and implement and run the advertising campaign. We shall describe the role of advertising agencies in the next section.

Step 5. Evaluating Advertising Effectiveness

Finally, the advertising managers must find out how far the advertising campaign has been effective and has achieved the advertising objectives. Evaluation of advertising effectiveness is, in fact, a part of controlling function. It is important to learn how far the deployment of resources in advertising have been meaningful and effective to promote the product and company. But it is extremely difficult to measure effectiveness of advertising and it requires serious market research. Not much headway has been made in this direction. Companies and advertising agencies spend very little on marketing research on studies relating to evaluation of advertising effectiveness or performance. However, it goes without saying that effective control demands evaluation of advertising performance vis-à-vis advertising objectives. Evaluation of advertising performance or effectiveness involves evaluation of advertising campaign *before, during* and *after* the advertising i.e. marketing research in this context involves pretesting and post-testing of the advertising activity.

Pretesting is a marketing research study used to evaluate audience reaction to an advertising activity before a decision has been made to implement the activity in full. *It is done also to evaluate alternative Ads.*

Post-testing is a marketing research study used to evaluate audience response to an advertising activity during its implementation and after it has been implemented in totality.

We have already described the communication objectives and sales-related objectives that are set by the company for the advertising campaign to fulfil. Researchers have developed alternative tests, both pre-test and post-test to assess advertising effectiveness in meeting these objectives. Thus there are tests to evaluate communication effects and tests to evaluate sales effects of advertising.

A. Tests to Evaluate Communication Effects (Communication Tests)

In this chapter, we have already referred to DAGMAR method to measure advertising effectiveness. Major communication objectives are - creating awareness of the product and company, effecting attitudinal changes (persuading), and building product preferences. A number of tests have been developed to measure awareness of the product and its unique selling points. These tests include recognition tests, recall tests and physiological tests. These tests are conducted on a sample.

(i) Recognition Tests: These tests find the number of persons or the percentage of sample who remember of having seen the advertisement in a particular vehicle (newspaper, magazine or on TV). This assesses the attention the message has received. For example, if 300 persons among a sample of 400 say that they have seen the advertisement, the recognition index is $300/400 \times 100 = 75$ (75%)

Again, percentage of those who saw the advertisement and read some particular parts of it including the name of the product and company can be known. This is called *associated score*. Further, we can know the percentage of those who said that they read most of the advertisement. This is called *most read score*. All these scores are measures of the attention paid by the audience to the Ad.

(ii) Recall Tests: As the name suggests, these tests ask the sample members to prove that they saw and read that particular Ad. They are asked to recall the text of the message and its format. Recall can be unaided (without any help given by the interviewer) or aided (when an interviewer gives a hint or clue). Here a list of products is presented to the respondents and they are asked to pick up the product they saw in the advertisement. Other questions concerning the ad and their attitudes towards the product are also asked. Comparison with advertisements of other companies is also made. These are a sort *oi Portfolio Tests*.

(iii) Physiological Tests: These tests measure physiological or bodily reactions, such as heart beat, blood pressure, eye movements, pupil dilation, brain waves etc. The degree of attention paid by a respondent to an ad results in fluctuations in bodily reactions and movements. Special equipment and devices are used to measure these changes in reaction and record the physiological data which are regarded more reliable indicator of impact of an ad. All these tests are also used for pretesting and post testing i.e. for copy-testing and effect-testing.

Attitude change and change in proposition (change in brand preference) can be assessed after respondents' multiple exposure to the ad at regular intervals.

The research techniques used for pretesting (or copy testing) include in-home tests (with the help of a projector), trailer test (in a shopping centre), theatre tests.

B. Evaluating Sales Effects (Sales Effectiveness)

Sales-related objectives include growth in sales, market-share and profits. As stated earlier, advertising is not the only factor that contributes to increase in sales or profits. There are many other factors that influence sales volume. Therefore it is extremely difficult to measure sales effectiveness of advertising. However researchers have devised a few approaches to study sales-advertising relationships and measure sale effectiveness of advertising.

(i) Analysis of Time-Series (Historical approach to sales-advertising relationships):   Companies use data on advertising expenditure and sales volume over a period of time comprising many years to find out the relationship between the advertising expenditure and sales. Statistical tools like correlation measures and regression analysis are used to find out anticipated sales as result of a given advertising expenditure. But the method is complicated as sales are also affected by carryover effect of the past advertisements. It is difficult to separate the two effects.

(ii) Experimental Approach: This approach is based on what is called *experimental design* to measure the sales effect of advertising. In this approach audiences in some areas (experimental groups) are exposed to advertising at different levels whereas in some other areas audiences (controlled group) are not exposed to advertising i.e. no advertising. The purchases made by audiences in different areas are measured and compared with those in controlled areas. The researches conducted in this context have shown that a higher advertising expenditure leads to increase in sales but at diminishing rate.

A large number of companies would be interested in evaluating the performance of advertising campaign in terms of sales effects rather than communication effects. In case a company finds during an advertising campaign that advertising's performance is not in consonance with the set advertising objectives, there is immediate need to take corrective measures. It may involve modifying advertising campaign including review of advertising objectives.

A number of companies have their own communication or advertising department to run their campaigns. However a large number of companies hire the services of outside agencies to undertake this creative work.

18.3   ADVERTISING AGENCIES, THEIR WORKING AND ROLE

(a) Advertising Agencies: Types and their Working

As stated earlier, a large number of advertising agencies have now-a-days emerged in the market to undertake specialised advertising work or run the entire advertising campaign for the companies. They employ specialists for creative work, design message, buy appropriate media, plan it, schedule it and run advertising campaign on behalf of their clients. They also undertake research studies. A brief description of various types of advertising agencies are given here.

(i) Full Service Advertising Agency: A full service advertising agency runs campaign from start to finish and offers complete range of services to the client. Before planning and implementing the advertising campaign, the agency needs to be intensively briefed by the client on matters like the target audience, the advertising objectives set to be accomplished, the estimated budget etc. The services offered by full service advertising agency comprise doing the creative work, determining the media mix, scheduling, buying necessary media space and time and overlooking the effective implementation of the campaign. Generally, they have four departments *creative department* which designs, develops and produces Ads; *media department* which selects media and buys space; *research department* which studies audience requirements, characteristics, stages of responsiveness and other research work on effectiveness of advertising, and *business department* which undertakes agency's *business activities*. For each client, the agency opens an *account*.

For their work, advertising agencies are usually paid commission by media and a fee by the client for the work done or a mix of both.

(ii) Creative Agency: This type of agency designs and develops messages and produces creative work. They produce advertisements for print media and TV commercials for their clients. They employ creative specialists on permanent or adhoc basis. Formerly they were known as 'hot shops'. They are paid a fee by the client on the basis of "work done".

(iii) Media Independents: These agencies specialise in selecting media and buy space in it. They do not undertake any creative work.

Then there are direct response agencies, sales promotion agencies, public relations (PR) agencies etc.

(b) Selecting an Advertising Agency

A large number of companies now-a-days use outside advertising agencies to run their advertising campaign. Choosing the right advertising agency goes a long way in the successful implementation of an advertising campaign and is vital for achieving marketing objectives as an advertising campaign costs vast amounts of money and resources. Companies usually use the following criteria for selecting an advertising agency.

(1) Their general background—business and financial position, turnover, organisational structure, reputation and image

(2) Their Clients — names and addresses of their businesses, number of years with agencies, accounts acquired, accounts lost.

(3) Expertise: Experience in relevant markets, media, products.

(4) Expertise and abilities in creative services and in-house services.

(5) Expected quality of work in relation to the established advertising objectives.

(6) Fees, allocating overheads and charges.

(7) Research experience and available services.

(8) Professional capability of managerial personnel and their personal compatibility.

Companies, generally, invite information and data on above aspects from different advertising agencies, compare them using the above criteria, shortlist them and select the one they feel is most suitable for running the advertising campaign effectively.

A few leading advertising agencies in India are—Hindustan Thompson Associates (HTA), Ammirati Puris Lintas, Mudra, Ulka, Nexus, Madison Diffusion-Dyer, Magic, Everest, O & M etc.

18.4 DIRECT MARKETING

A number of companies, now-a-days, are using advertising media (newspaper, radio, TV), mail and other vehicles to reach the target audience, seek audience's direct response to their messages and to sell their products directly to the consumers without the aid of any middleman or intermediate channel of distribution. This is called *direct marketing* and a fair share of advertising rupees is being spent on it. According to the American Direct Marketing Association (DMA).

Direct marketing is an interactive system of marketing which uses one or more advertising media to effect a measurable response and/or transaction at any location.

As the response is to be measured in terms of direct orders or enquiries from the customers, the system is also called *direct-order-marketing*.

Direct marketing aims at reaching the target audience, encouraging it to buy the product through a response device such as a coupon, telephone or fax. The main objective is to sell, but the context here is immediate response to the message. Direct marketing methods include use of coupon ads in the press, telephone, mail, catalogues etc.

From the advertising point of view, direct marketing can be divided into two parts—

(a) **Direct Response Advertising:** Off the page in the press or in response to an announcement on TV, and

(b) **Direct Mail** i.e. direct delivery through mail, courier, or the door drop.

A few major tools of *direct marketing* are given here,

(i) **Tele Marketing:** This is a system in which the marketer makes use of telephone to sell their products directly to the costumers who on their part make calls to place orders. With the advance in technology, automatic dialling and recorded message players play a vital role. This method is not yet popular in this country due to heavy expenses involved in making calls.

(ii) Television Direct Response Marketing: (TV Home Shopping). In this set of direct marketing, television is used for the purposes of direct marketing. It may take the form of an advertisement on TV lasting 10 seconds to sixty seconds or more i.e. TV slot, seeking orders giving a phone number or fax number to place an order. It is a form of teleshopping.

Some companies devise entire Home Shopping TV programmes or devote an entire TV channel to sell goods and services. In these programmes, they not only provide information about their products but demonstrate their performance and uses as well to persuade the customers to place orders through phone or fax or mail. Products are supplied through door drop or pay on delivery basis in big cities or by VPP mail, charging extra postal charges. In our country services of a number of popular TV artists are being used to demonstrate the utility of these products. UTS (United Teleshopping), Mumbai is a leading company organising TV shows almost daily and offering teleshopping to about 5000 customers.

(iii) Catalogue Marketing: As the name suggests, in this system, marketers mail catalogues to customers. These catalogues contain detailed information about a number of products, illustrated in multicolours, and providing order forms. In India, this system is not in vogue yet. However, in U.S.A., this is a popular system, each house-hold receiving a number of catalogues every week or month. The success of catalogue marketing depends on "the company's ability to manage customer lists, control its inventory carefully, offer quality products and project a distinctive image"

(iv) Newspaper Direct Response Marketing: Marketers use newspapers to provide information about the products, their use and price and make direct response offers to the customers providing order form or coupon in the advertisement. Otto Burlingtons Mail Order in collaboration with Times of India is one organisation engaged in newspaper direct response marketing.

(b) Direct Mail (DM):   We have already referred to "Direct Mail" as an important advertising medium and listed its advantages and limitations in an earlier section. Direct mail is the larger part of the direct marketing business. We can define it as follows:

Direct Mail (DM) is a piece of information designed for a specific product to be sent to a specific customer.

The information about the product is communicated directly to the customer through a single mail-piece letter, circular, flyer, folder. The entire concept of direct mail (DM) depends on obtaining and managing a mailing list. Now, a number of agencies compile and provide lists of specific group of customers which are available on price. Direct mail marketing involves one-to-one relationship with the consumer. DM marketers are offering a wide range of products now-a-days. The mail response orders contain a number of value adding offers and additional benefits like coupons, gifts, concessions, incentives, free services etc. to attract customers.

As stated earlier, DM has a high target market selectivity and is highly personalised. It is the only form of media advertising where response is completely measurable. Response can be analysed in terms of number of sales/enquiries per advertisement. The cost can be measured per sale/enquiry.

In USA, direct mail marketing has become very popular. Some direct mail marketers are mailing audio-cassettes, videotapes and computer diskettes to their customers. In USA 33% of all national expenditure is spent on direct mail.

As referred to in above, the success depends upon managing a mailing list and a data base containing detailed data about the target audience and its characteristics. Therefore, direct marketers need to have complete personal data of all members of the target audience. *A marketing data base is an organised collection of data about individual customers, prospects or suspects that is accessible and actionable for such marketing purposes as sale of product or service or maintenance of customer relationships.* The availability of data base is necessary for its implementation.

Now, a number of organisations provide marketing data base to their clients. For example, Datamatics Direct, Mumbai is one of India's largest "mail database and direct mail advertising provider, its clients being Hindustan Unilever, Citibank. Crompton Greaves, Godrej Group, Sterling Resorts etc.

18.5 ON-LINE MARKETING

Thanks to the rapid advancements in the fields of knowledge and information technology in recent years, the electronic gadgets like computers and electronic printers have become potent instruments of instant communication and are providing services which link the sellers and the consumers or customers and create exchanges facilitating marketing of number of goods and services. The other inventions and innovative practices and services in the fields of information, communication and banking, in the wake of liberalisation and globalisation of economy, (like global web of computer net work, availability of e-mail, fax and internet facilities, courier services, credit card facilities), have accelerated the pace of use and growth of electronic gadgets in the field of marketing as well.

Thus Online marketing has become now-a-days an important medium of creating exchanges. We can difine it in the following way:

> *Online marketing is a process that creates exchanges by linking the consumers with sellers or producers electronically through an interactive computer system.*

There are two types of Online marketing channels: commercial online services and the internet.

(a) Commercial Onlin Services: These are services which offer online information and marketing services to those clients and customers who pay a monthly or annual subscription fee for the same. These services provide information about news, views, education, travel, sports, entertainment, happenings and events, shopping services besides e-mail services. The main providers of these services in USA are America Online, Mirco-soft Network (MSN), Prodigy to name a few.

(b) The Internet: The Internet is another online marketing channel which comprises a vast and growing Web of computer networks which is not managed or owned by any central authority. There are a number of online service firms, organisations and companies which offer internet access as primary service. The use of internet grew rapidly with the development of what is known as *World Wide Web* (www) and standard browser software such as Microsoft Internet Explorer, New scape Navigator. Now, the user can obtain any information about any thing such as food, hotels, air tickets and railway tickets; send e-mails, exchange views, shop products and get services by having an access to the internet for which the seeker of information or user of internet has to pay a monthly fee or subscription to internet provider of internet connection, otherwise, internet itself is free. Only the person had to buy a PC, a modem and right software. The usual fee for getting the facility of internet service has to be paid to commercial access providers. A few internet service providers in the country are BSNL, MTNL, Tata Indicom, Airtel, Reliance Telecom who are offering internet connections under different payment schemes to the users or on fixed monthly rent. In recent years, there has been a rapid growth of Online marketing.

Electronic Commerce (e-commerce)

The growth of internet usage has ushered into a new field of commerce known as electronic commerce. *Electronic Commerce or e-commerce is a vehicle of creating exchanges i.e. buying and selling of goods and services supported by electronic means.* Like physical market places, electronic market are market spaces in which sellers or producers offer their products and goods electronically and the purchasers or customers seek information, identify the product or service they want and need and place orders using a credit card or any other mode of electronic payment. For the purpose of searching for information, the customer has to log onto the Web Site relating to product or service or web site of the seller or corporate website and search the information.

Most recently in India several Online sellers and internet portals are offering air tickets, mobile phones, MP 3 players, Cameras or airconditioners, books to online consumers at competitive prices. To quote a few examples,

(i) **e Bay India** is promising low price offers to online shoppers of popular gadgets like mobile phones, digital music players and cameras.

(ii) Yatra. com, an internet portal sells flight tickets and holiday packages which are available from its website.

(iii) Another leading online shopping portal India Times. com ran an e-mail campaign titled Branded AC's, Lowest Price Guaranteed.

(iv) Some time back, another portal Indiaplaza had run a price challenge campaign for books and mobile phones.

It is interesting to note that often, these campaigners are supported by various online as well as offline marketing initiatives and involve outdoor ads, radio spots and internet marketing tools.

18.5.1 Managing Online Marketing

These are four ways to carry on online marketing. We are describing them in brief.

(i) Getting an Electronic Online Presence: A company can create an electronic online presence by

(a) buying space on a commercial online service or

(b) by opening its own website.

The former involves either renting storage space on the Online service's Computer or establishing a link from company's own computer to the commercial Online service's shopping office. The company has to pay an annual fee to the online service for the purpose.

The latter involves opening or creating one's own website. Now-a-days, most of Corporate Houses have their own Web sites. Usually, their aim is to acquaint the customer with the mission, philosophy, values, products, product lines and their marketing details of the company to earn the customer's goodwill and supplement the efforts of other sales channels, rather than sell its own products. In the process the company also gets the necessary feedback from the customer or buyer. There is a large scope of interaction with the consumers in this process which could induce the buyers to make a favourable decision to purchase its products.

(ii) Placing Online Advertisements: This way involves placing online advertisements in the form of banners moving across the screen and full-screen ads on Web sites of other companies to attract visitors of those web sites to its products and its own web site.

(iii) Another way to establish an electronic online presence is to participate in internet forums, news groups and bulletin boards that may organise activities, both commercial and non-commercial, which might attract interested publics or groups. They may discuss some specific topics or participate in real time message exchanges in 'Chat rooms' like internet buddies.

(iv) A company may register its presence by encouraging its buyers and prospective customers to ask questions, make suggestions and lodge complaints, if any, by using e-mail to which company representatives can quickly respond besides introducing new offers and schemes. Webcasting is another service which involves automatic downloading of customised information about the company to the interested visitors who are ready to pay for the same.

18.5.2 Limitations of Online Marketing

Online marketing has tremendous uses and offers numerous promises and its expansion may revolutionese the marketing of the products and services in future but it has following limitations and challenges:

(i) It has a limited consumer exposure and reach and only a few might actually buy the product or service.

(ii) It might appeal to the elite and upper strata of society but it does not appeal to public at large, especially those consumers with less income. Only technically savvy persons are attracted to it.

(iii) The surfing, navigating and exploring web sites for information is a boring, tiring and time consuming task. It is difficult to find right information quickly.

(iv) There is every possibility of an online customer being subjected to fraud by some unscupulous visitors as it is not security-proof.

(v) It has the potential of abusing the customer's privacy and visitors may indulge in unethical practices and exploit the actual consumers.

Anyhow, it has unlimited uses and has the potential of providing innumerable services. Advance resevation of railway journey tickets is a typical example of online marketing.

Example: Advance Reservation Through Internet

INDIAN RAILWAY CATERING & TOURISM CORPORATION LTD. *(IRCTC LTD) a PSU of Ministry of Railways have developed a system for advance booking of rail tickets through internet.* **The web site for online booking is www. irctc.co.in** *The site is Veri Sign secured. Booking procedures are simple and user friendly.*

Booking of Internet Tickets (i-tickets)

- Customers should register in the above site to book tickets and for all reservations/timetable related enquiries. Registration is **free**.

- Full fare tickets, Child tickets and tickets for senior citizens at concessional rates can alone be booked through the web site. Internet tickets can be booked for journey between any two stations on the route of the train including originating station and destination.

- **Service charge of Rs. 40 per ticket in lower class and Rs. 60 per ticket in upper class (maximum 6 passengers per ticket) is leviable in addition to the train fare.**

- Tickets will be delivered at the customers address through courier at NO EXTRA COST other than mentioned above.
- A maximum of **10 tickets** can be booked in a **month** by an individual.

Payment for Internet Tickets

- Payment can be made by using all **Master/Visa/Diners club credit cards**. A service charge of **1.8%** is charged by the Banks.
- Account holders of 361 banks, including **ICICI, HDFC, State Bank of India** can use their net banking facility for making payments for tickets booked through the Internet
- Ticket on Internet can be booked from **05.00-23.30 hours on all days.**

(Excerpts from Railway Time Table
"Trains at a Glance" July 2008- June 2009*)

APPENDIX TO CHAPTER 18

CASE 18.1:
COLGATE-PALMOLIVE ADVERTISING
PLACEMENT POLICY STATEMENT

Colgate-Palmolive provides products that offer consumers real and unique values. The role of our advertising is to communicate those unique values in a manner consistent with the highest ethical standards, because our advertising creates more than a product image. It mirrors Colgate's reputation for reliability, dependability and trust worthiness.

Our advertising in its placement as well as its content and format, is sensitive to the public and its concerns, interests and sensibilities.

Specifically, Colgate-Palmolive docs not advertise in a media context:

- That includes gratuitous or excessive violence; violence which is not necessary to the development of a program's character or story line or an article's development. It also eliminates those programs where some violence is an integral part of the story line, but which feature unnecessary violent details, brutality, or suffering.
- Which it considers anti-social or in bad taste, or which could stimulate anti-social behavior through viewer imitation.
- In which schedule behaviour is grossly aberrant and/or offensive.
- That lends actual or implied support for those activities that may abuse the physical or mental health of an individual.
- Which insults, ridicules, or denigrates people because of their age, gender, sexual orientation, race, religion, ethnic origins or engages in other inappropriate stereotyping.

Colgate-Palmolive charges its advertising agencies and their media buying services with the responsibility of pre-screening any questionable media content or context. If there is any doubt about media suitability for Colgate-Palmolive advertising, it is referred to Colgate-Palmolive media management for review and discussion.

As a corporation, Colgate-Palmolive is sensitive and responsive to consumers' changing needs and values. Our advertising policy guidelines are periodically reviewed and revised to ensure their appropriateness in fulfilling both the needs of the company and those of our consumers.

EXERCISES

1. What is meant by advertising? Discuss its importance in modern business.

2. Describe the process of developing an advertising programme.

3. Explain some major advertising media and their respective advantages.

4. Write notes on
   (a) DAGMAR model
   (b) Marketing positioning of the product
   (c) Celebrity advertising
   (d) Direct Marketing
   (e) Direct Mail
   (f) Advertising ethics
   (g) Media planning

5. What is an advertising agency? Describe the types of various advertising agencies?

6. What points should be kept in mind while selecting an advertising agency.

7. "A creative advertising copy using gimmics, imagery and boasting usually does not present an unbiased description of the product". Should the company use such techniques? Discuss.

8. What are essentials of an effective ad-campaign?

9. What points are kept in mind while determining advertising budget?

10. What is product positioning and how is it useful in advertising?

19

Managing Sales Promotion and Public Relations Programmes

19.1 SALES PROMOTION: DEFINITION AND CONCEPT

In chapter 17, we have referred to "sales promotion" as one of the four tools of marketing communication (or promotion) mix, the other three being advertising, public relations or publicity and personal selling. In chapter 18, we discussed in details the role of advertising and how an effective advertising campaign can be developed. The discussion was mainly concerned with the use of various media in advertising or media advertising. The marketing communication or promotion other than media advertising is called *non-media communication* or *"below-the-line"* advertising. Thus we can define "sales promotion in the following way"

Sales promotion is referred to as a tool of communicating with an audience through a variety of nonpersonal nonmedia vehicles or short-term incentives to encourage purchase or sale of a product or service.

The incentives or nonmedia vehicle may include bonus offers, free samples, gifts coupons, discount, displays and demonstrations, patronage awards, contests, prizes etc. The Institute of Sales Promotion (UK) has given the follow definition:

Sales promotion is the function of marketing which seeks to achieve given objectives by the adding of extrinsic, tangible value to a product or service.

The methods used to add tangible value to a product or service comprise a number of incentives, mostly short-term, as listed above.

Aims (Purpose) of Sales Promotion: The aims of sales promotion can be described as follows:

(1) to boost sales in the short and long-term;

(2) to increase profits

(3) to launch a new product, if any

(4) to reinforce the brand image with the customers

(5) to encourage brand loyalty

(6) to unload accumulated inventory and to increase buffer stock held at retail level — (Buffer stock is the stock held at which level the retailer re-orders)

19.2 DEVELOPING A SALES PROMOTION PROGRAMME (SALES PROMOTION DECISIONS)

Designing sales promotion programme involves the following steps:

(i) Identifying the target audiences

(ii) Determining sales promotion objectives

(iii) Deciding on sales promotion vehicles (Incentives)

(iv) Setting sales promotion budget

(v) Evaluating sales promotion effectiveness.

Step 1. Identifying the Target Audiences

In an effort to boost purchase of a product or service, a marketing manager requires the cooperation and attention of all the three parties *viz.* consumers or the end users, the distribution channel members (i.e. the trade), and the salesforce. Thus sales promotion programme aims at three target audiences.

(i) Consumers or the end users

(ii) Trade i.e. distributors, wholesalers, retailers (channel members)

(iii) Salesforce

The company will need to study and analyse the characteristics (nature and structure) of all these three types of audiences to arrive at the marketing opportunities available in market place. An in-depth study and analysis of the nature and make up of the selected target audience will provide guidelines for determining sales promotion objectives and deciding on the incentives and inducements to be provided to the target audiences to achieve sales promotion objectives.

Step 2. Determining Sales Promotion Objectives

As stated in chapter 17, the sales promotion objectives are derived from the promotion objectives, which are derived from overall marketing objectives. Again sales promotion objectives for different target audiences will differ. For example,

(a) Sales Promotion Objectives for Consumers i.e. the end users may be

(i)   to encourage the non-user to visit the store or showroom and make an enquiry.

(ii)   to make the customer try the new product.

(iii)   to encourage the consumer to purchase more units and stock its product.

(iv)   to attract the customer using competitors' brands to its product or making consumers switch brands.

(b) Sales Promotion Objectives for the Trade may be

(i)   to elicit cooperation of the trade or retailers in the company's total marketing effort.

(ii)   inducing them to maintain larger inventory.

(iii)   encouraging off-season purchase.

(iv)   to motivate distributors to provide more shelf or floor space to company's products and push them.

(v)   to encourage them to sell the product at lower price and

(vi)   to build brand loyalty of retailers/stockists.

(c) Sales Promotion Objectives for Sales Force may be

(i)   to encourage sales force to make more calls on consumers and distributors.

(ii)   to make them put in more efforts and emphasis on certain brands or new products or during particular periods such as off-season.

It must be kept in mind that sales promotion vehicles are short-term incentives and these objectives need to achieved in a short-time, say a few weeks or a few months.

Step 3. Deciding on Sales Promotion Vehicles (Incentives)

After having established the sales promotion objectives, the next task is to select promotion tools that may achieve the set objectives. While selecting the sales promotion tools and incentives, marketers have to keep in view the characteristics of the target audience, competitors' offers and cost effectiveness of each vehicle.

A. Consumer Promotion Vehicles/Incentives

Some major sales promotion incentives to induce the consumers (end users) are given here:

(1) **Coupons or Vouchers:** Manufacturers offer many types of coupons or vouchers as tools of sales promotion. A coupon or a voucher is a certificate that entitles a customer to exchange it for some specific good or service, cash gift or a stated saving (discount) on the purchase of a product per its conditions. These coupons can be inserted in a pack of a product or in a newspaper or magazine ad or mailed directly to a customer. In most cases the redemption of coupons necessitates active cooperation of the dealers/distributors and retailers. These coupons induce the consumers and channel members to stock up on the product. They have special importance when a new product is launched or for strengthening the image of an existing product. *Gift coupons or tokens* are coupons or tokens enclosed with a product in a pack and a number of them can be exchanged for a gift as mentioned in the coupons.

(2) **Price Packs:** These are offers of "Save off' the normal price of a product (e.g., Rs. 2 Off) printed on the package itself. It is called reduced price pack. The package may contain 2 units instead of a single packet and offered at the price of one (two for the price of one) or it may be a banded pack containing an additional unit of related product at the same price e.g. a 7'o clock blade with a safety razor or a toothbrush with a tooth paste or an extra quantity of the product in the pack (20% extra).

Then another version is BUY 1, GET 1 FREE ! The offer is printed on pack meant for free distribution.

(3) **Free Trials (Samples):** Customers are given or sent free samples to try for themselves in the hope that they will buy the product. This is a very effective device in launching a new product. Manufacturers of cars, motorcycles and scooters often offer free test drives to evince interest in their products.

(4) **Cash Refund Offers:** In some cases, cash refund is made after the product is purchased. The consumer is asked to send the bill of purchase to the manufacturer or distributor and get a given amount of rebate refunded.

(5) **Point of Purchase (POP) Displays and Demonstrations (In-store Promotions):** Some marketers think that the audiences give their best attention when they are shopping in a store or a retailing shop. These promotional tools include displays, at points of purchase (in stores or at retail outlets), of posters, stickers, electronic and neon-sign boards.

Demonstrations are also arranged at the retailing outlets or at super stores and malls to convince the consumers of the performance of the product (e.g. washing/stiching demonstrations.)

(6) **Prizes and Contests:** Some companies induce customers to take part in contests, jingles, quizes and win prizes. Some companies place lottery tickets in packs while some others give it on purchase of their product. The tickets give them a chance to win prizes in the form of cash, merchandise, gift coupons, free air tickets, free trips, two or three nights' stay in a hotel at a hill resort etc. Now many companies like Pepsi Cola, Wills, Coca Cola provide chance to win prizes in contests concerning sports like the "Best Catch of the Series", "The Best Goal in a tournament" etc.

(7) **Gifts:** A number of companies distribute free gifts to their customers through their dealers and distributors. Gifts include pens, calendars, bags, watches, bed sheets, chairs, table decorations. They carry the name of the company.

Recently Newspapers like Hindustan Times and Times of India distributed bags, watches, bed sheets, 8 piece tea set if a new customer paid 6 months' subscription as advance to them. Femina and India today offer 25 per cent discount alongwith a gift if one pays one year subscription in advance.

Then there can be other types of incentives like premiums or extra free merchandise like 20 gm free with 50 gm pack of Coffee, Cash dividends, patrongae awards etc.

B. Trade Promotion Incentives

Some major incentives for retailers, dealers and channel members are—

(1) **Price Discounts** on purchase of product(s) in a particular period. They are accounted for in the bills.

(2) **Bonus Offers:** The retailers or dealers are offered special bonus terms on bulk purchases e.g. 13 to a dozen or 8 items charged as 6 on a minium order of one gross (144) or two gross (288).

(3) **Allowances** for displaying or arranging the company's product in a special way or advertising the company's product. These are called display allowance, advertising allowance.

(4) **Free Goods:** Some companies offer free speciality items like pads, pens, ashtrays, bags to their dealers. Bonus offers as mentioned above are also termed as free goods on purchase of bulk quantity of merchandise.

C. Incentives for Salesforce

These incentives include bonuses, contests, gifts such as appliances, free vacation or trips to persons getting the highest sales volumes etc.

Trade Shows: Manufacturers also organise trade shows, conventions, sales contests and many other shows for select audiences to promote their products. In these shows, products are exhibited and their use demonstrated. Buyers can examine the products themselves. A number of companies book space in 5-star hotels for such displays to attract upper class customers.

In developing a promotion programme, the mangers have to make many decisions such as the size of incentives, timing of promotion, conditions of participation and the sales promotion budget.

Step 4.   Setting Sales Promotion Budget

Market managers who are charged with the responsibility of developing and implementing sales promotion programme must estimate the budgeted cost of incentives for implementing various promotional activities. They must take into the account the *administrative cost* and *incentive cost* of each tool.

There are two approaches to estimating the sales promotion budget, *top-down* and *bottom-up* approach.

(i) Top-down Approach: In this approach, the top management decides the amount of sales promotion budget. The expenditure on promotional activities is to be limited to the authorised budget provision. The main defect of this approach is that decisions made are arbitrary and may be irrational.

(ii) Bottom-up Approach: This approach is a sort of objective task approach. In this approach, the target audiences are identified, promotional objectives established and then activities and incentives are decided to fulfil the set objectives. Then costs to be incurred in implementing each promotional activity are estimated. The total costs for performing all sales promotional activities give the total budget estimates. The estimated budget is discussed with the top management in details and got reviewed and approved.

In fact, sales promotion budget is a part of the promotional budget which again is a component of overall marketing budget. Therefore, the decision on it should be made during the process of setting budget for the overall promotional programme.

Step 5.   Evaluating Sales Promotion Performance

As discussed earlier, evaluation of promotion performance can be made before implementing an activity on a sample basis (i.e. Pretesting) and during and after the implementation of sales promotion activity (i.e. Post-testing). Both involve marketing research studies.

Protesting is a study used to evaluate audience response to a sales promotion activity before a decision is made to implement the activity on a full scale. The incentive is tried on a sample out of the target audience and results recorded.

Post-testing is a research study used to evaluate actual audience response to a sales promotion activity during its implementation or after it has been implemented in full. This study will involve examining sales data, counting number of orders, number of coupons redeemed, number of enquiries/orders received during trade shows etc.

It has to be ensured that the money spent on sales promotion activities contributes to the accomplishment of the marketing objectives set by the company and to its growth and prosperity.

19.3   PUBLIC RELATIONS (PR)/PUBLICITY

Public Relations (PR) is another tool of promotion mix designed "to improve, maintain or project a company or product image."

A.   Public Relations: Concept and Importance

Public Relations is referred to everything that is conducted to improve mutual understanding between an organisation and target groups and all those with whom it comes into contact both within or outside the organisation with the aim of building goodwill and good image. Communication is the root of PR. It is a two-way communication between the organisation and the target group to establish understanding based on truth, knowledge and complete information. The UK Institute of Public Relations has given the following definition.

> *Public Relations is the deliberately planned and sustained effort to establish and maintain mutual understanding between an organisation and its publics.*

The above definition focuses on:
  (i) good relations with the publics and
  (ii) creating goodwill among the organisations' publics.

The *publics* of PR refer to those groups of people with whom the organisation wants to communicate specifically i.e. these refer to those target groups such as customers, trade, media, workforce and other target audiences to whom messages need to be communicated.

So, broadly speaking, public relations (PR) implies the creation and maintenance of a favourable climate of opinion or audience's good-will towards the company to build a good image of the company and it involves performance of a number of activities, namely press relations, publicity, corporate communication, lobbying, counselling.

All these activities do not support the product. We are here concerned with those activities which contribute to marketing effort i.e. our subject of study is marketing-oriented or marketing PR. In this context, publicity and lobbying are worth special consideration.

Many authors use the term "Publicity" instead of term "Public Relations" for this tool of promotion mix. Publicity is a potent tool of communication and is regarded as highly credible and authentic. Publicity is referred to as communicating with an audience by personal or nonpersonal media that are not explicitly paid for delivering the messages. As a promotion tool, publicity is defined in the following way:

Publicity is referred to as nonpersonal stimulation of demand for a product, service or business unit by planting commercially significant news about it in a printed medium or obtaining favourable presentation of it upon radio, TV, or stage that is not paid by the sponsor.

Besides publicity, the modern concept of PR involves undertaking many other below-the-line activities such as sponsoring sports events, sponsoring activities and campaigns for social causes, organising exhibitions, competitions, seminars, cultural activities, reality shows, music and dance contests as important parts of PR programme.

Importance of PR

Public relations is a powerful tool for creating image, building awareness and consumer preference, and establishing useful liaisons with influential groups. It contributes in the following ways:

1. It helps in creating a good and favourable image of the company products in the eyes of consumers.
2. It helps in the introduction of new products by providing information about them to news media by press releases or in any other way.
3. It helps in improving the sagging image of an existing product and in repositioning it in the market.
4. It helps in influencing the national and local governments on decisions affecting the organisations.

5. It helps in influencing specific groups and establishing relationships with them which may be favourable to organisations.

6. Above all, it helps in keeping good and healthy relations with various agencies like the government, trade unions, customers, community and other agencies, which is crucial for the success of the company.

19.4 DEVELOPING A MARKETING PR/(PUBLICITY) PROGRAMME/CAMPAIGN

As stated earlier, the main publics with whom the organisations want to communicate include owners, potential shareholders, customers, potential customers, local community, employees, decision makers, opinion formers like educators, writers, politicians, lobbyists, editorial staff (media), suppliers, channel members, trade unions. Designing and developing a marketing PR programme invovles the following steps:

(1) Determining publicity objectives.
(2) Designing message and selecting PR tools.
(3) Implementing the PR programme.
(4) Evaluating PR effectiveness.

Step 1. Determining PR/Publicity Objectives

Each company will have its own publicity/PR requirements depending upon the nature of its business and type of products. Therefore, every campaign will have different specific objectives. The PR objectives can be:

(1) to create awareness among publics of the company's productive services or performance.

(2) to build company's credibility and image.

(3) to motivate and help distributors and channel members and salesforce sell the product.

(4) to save promotion cost.

(5) to promote a social cause.

Step 2. Designing Messages and Selecting PR Tools

The next step is to find PR messages, rather to create PR messages for communication to various publics. The messages may be in the form of news stories, editorials, arguments, news reports, features, interviews in printed media or on TV, and radio speeches.

PR information to the public or external agencies can be given through press releases, company magazines, special numbers, journals, newsletters, annual reports or through formal and informal talks by the company management or through films and television programmes. External communication i.e. communication with clients, dealers, customers or community people requires particular ability and skill. For example, a PR manager must be competent in conducting meetings, in meeting procedures, in public speaking, in reporting etc. The responsibility of building image of the organisation in public lies on the manager. Some major tools of marketing PR (Publicity) are given below.

(1) Company Publications/Literature/Audio-Visuals: Company publications include company magazines, annual reports, newsletters, brochures, booklets, pamphlets, audio-visual materials like film slides, audio-video cassettes. This literature tells the audiences about the company, its functioning philosophy, its products and how they work, and other public service activities being performed by it.

(2) News and Press-releases: One of the most important tools for PR is the news giving facts and information about the company, about its products and their release to the press for publication. Writing and creating interesting newsworthy, commercially significant, credible stories and articles require special skill on the part of PR personnel. Again, they need to strike special rapport with the media editors and reporters to gain their favour and get news stories, features, articles, editorials accepted for publications.

Press receptions and conferences on special occasions such as on launching of new product, inaguration of an event, a seminar, a contest or show, annual meetings and dinners, go a long way in publicising the company and its products. Many big companies hold such conferences in 5-star hotels and decorated halls, where press people are entertained with refreshments, lunch or dinner like other VIPs and guests.

(3) Events: Companies can organise and sponsor a number of special events such as sports tournaments, music, dance and other arts and cultural programmes, contests and competitions, exhibitions, outings, fairs to capture the attention of the target audiences. The recent examples of event sponsorships are—Wills Cricket World Cup Matches, Godrej's Miss World Beauty Contest, Mahindra's World Billiard Championship, Hero-Honda Motor Cycle Race and the like.

(4) Advertising: Messages involving PR can also be communicated through advertising e.g., that the company is the best, the market leader in a product-market. A number of companies, industries and their associations, and institutions use Ads for publicity and lobbying purposes. A few examples of such advertisements are given in the concluding part of this chapter.

(5) Social Service Activity: Companies can build goodwill by donating or contributing money and efforts to good social or community causes. It can be done through product coupons signifying that a small part of money received from sale of the product would be given for a charitable purpose, say for eradication of Aids, leprosy, for education of orphans etc.

What is important for the PR manager is to make a creative and credible use of these tools and media to the betterment of company's prospects.

Step 3. Implementing the PR Programme

We have already stated that a publicist needs special and creative skill in writing a newsworthy story or message. He must be competent in organising and conducting meetings and should be well skilled in public speaking and reporting. They need to have the ability to maintain personal relationships and rapport with the media correspondents and editors. Now-a-days, PR or publicity work is entrusted to PR agencies. The same criteria are used to select a PR agency as for advertising agency given in chapter 18. Some companies appoint PR consultants to advise on matters concerning public relations and publicity.

Step 4. Evaluating PR Effectiveness

Though evaluating PR results is a difficult and complicated exercise. However a few methods are available to assess its performance.

(a) **Exposures:** It means to measure the coverage that was given to the company and its products in print media or upon radio and TV and number of persons exposed to company related news. This can be measured in terms of column covered, number of paragraphs, air time on radio and TV, estimated audiences of readers, listeners and viewers to assess.

(b) Other methods to assess the performance, namely measuring awareness, attitude change and sales-profit contribution have been explained in chapter 18. The same holds good for evaluating PR results.

19.5 LOBBYING

As referred to above, lobbying is an important activity of PR or publicity. Lobbying is referred to as publicising one's arguments and views in support or against an issue or matter of concern or interest. In marketing context, *lobbying implies dealing or communicating with public, legislators and government officials to promote or defeat legislation and regulation.* The case 19.3 (a), (b) presents a classic example of lobbying.

APPENDIX TO CHAPTER 19*

SOME EXAMPLES AND CASES OF PUBLICITY AND LOBBYING

*(For MBA Students in Professional Institutes)

CASE 19.1	CASE 19.2

CASE 19.1
IMI admission notification

CASE 19.2
DU programme in Radio Management

NEW DELHI: International Management Institute, New Delhi, invites applications for admission to its PhD programmes and Fellow Programme in Management (FPM) approved by AICTE. An applicant possessing any one of the following qualifications shall be eligible to apply for admission to a PhD programme.

A master's degree in Management, Engineering, Technology, Science, Architecture, Humanities, Commerce, Medicine, Law, Education or Pharmacy from a recognised Indian University, or a degree approved by Association of Indian Universities, or any other equivalent qualification to the satisfcation of the Academic Council of the University, in the relevant field with not less than 60 per cent marks in aggregate.

NEW DELHI: Lady Irwin College, Delhi University, has joined hands with South Asia's Academy of Radio Management (ARM) and launched a unique industry-oriented certificate programme in Radio Management. The course would be offered in three modules and the total course duration would be 120 hours. The curriculum includes radio jockeying, advertising, marketing production management, radio programme editing and broadcast management.

The programme aims to create radio professionals rather than just RJs. This would include programming directors, music managers, copywriters, editors, promo producers, marketing professionals and thousands of others who work behind the scenes. This is an effort to bridge the gap between the growing number of options in the industry.

CASE 19.3(A)

Cement Manufacturers' Association issued two advertisements *for the information of the public* in Times of India, January 20 and 28, 1997. We are giving here their major contents (CASE OF LOBBYING)

A SACK FULL OF PROBLEMS

Under the Jute Packaging Act, the Government is considering to order that a certain percentage of cement produced must compulsorily be packed in Jute bags.

Consider the colossus national waste it would result in

Jute bags cause seepage and transit loss of 4-5%.

Rs. 18200 crore worth of cement is produced annually. This seepage would cause a staggering Rs. 900 crore loss to the country, besides posing a serious pollution hazard.

Jute bags can put an additional burden of Rs. 980 crore on customers like you.

Because a jute bag costs Rs. 7 to Rs. 8 more than a plastic one.

Cement packed in jute bags loses strength due to natural humidity. This deterioration in quality is virtually non-existent in plastic bags.

Jute bags are technically unsuitable for packing cement.

Cement, a fine powder, tends to react with the moisture in the air. So, it is technically unfeasible to pack cement in a porous material like jute bags.

Jute bags will be absolutely anti-consumer.

Correct weight, proper quality and easy identification will not be possible with jute bags.

Issued for the information of the public by the Cement Manufacturers', Association.

If cement were packed in jute bags, the money lost could educate 3,60,00,000 children annually...

*(For students in Professional and institutes)

CASE 19.3(B)

As a reaction to the Cement Manufacturers' Association advertisement of January 20,1997, there appeared a newsitem in the Hindustan Times dated January 25, 1997 containing views of jute Industry. This presents a good example of unpaid publicity.

JUTE INDUSTRY TO FIGHT CMA IN COURT
KOLKATA JAN 24

The Jute industry is going to take the Cement Manufacturers' Association (CMA) to court for having launched a massive "disinformation campaign" through countrywide newspaper advertisements against the compulsory use of jute bags for packing cement.

Stung by the advertisement, captioned "sackful of problems", purported to highlight the adverse economics of customer resistance to jute bags, the India Jute Mills Association (IJMA) Chairman Mr. G.M. Singhvi, told newspapers here today that the advertisement had been reissued even after Delhi High Court expressed its stern diapproaval of the first issue on Jan 20. It was on that day that the court had taken up the case on compulsory reservation of 50 per cent for jute bags in the cement industry.

Mr. Singhvi said that the cement industry, in conjuction with the synthetic bag manufacturers, had been trying to scuttle the Government's move to enforce the Jute Packaging Act of 1987, which had been upheld by the Supreme Court in April, 1996. "They are scared because the Jute Commissioner has started issuing show cause notices to them and most of them are facing the prospect of paying penalty for noncompliance of the order, for the last four years."

West Bengal alone had lost Rs. 10,000 crore in this period because neither could the jute industry sell bags to the cement industry nor could the growers and the market for their raw jute, he said. Had the cement industry complied with the mandatory provisions, the monthly offtake of jute bags would have been 25,000 tonnes and the jute industry would have been rid of many of its current problems, he added.

The IJMA Chairman also came down heavily on the Central Government which though officially on the jute industry's side, had been dragging its feet in the submission of its affidavit before the Delhi High Court.

Mr. Singhvi reiterated that the charges of seepage loss, consumer resistance etc., against jute bags were "highly exaggerated and misplaced". He claimed that the seepage loss was only marginal whereas these bags were more robust, easier to handle and not given to bursting. Jute bags being bio-degradable were more eco-friendly, too.

On the question of jute bags being priced higher than plastic bags, he said that this could be offset through the resale of jute bags. Also, there would be hardly any consumer resistance if the cement industry used jute bags of required specifications.

EXERCISES

1. What is meant by sales promotion? Explain its aims.
2. Describe the steps involved in designing a sales promotion programme.
3. Describe some major sales promotion vehicles (incentives).

4. Define the term public relations. What is its importance?

5. Explain the process of developing a publicity programme (or PR programme).

6. Write notes on the following

(i) Lobbying

(ii) Major tools of marketing public relations

(iii) Estimation of sales promotion budget

(iv) Determining publicity objectives.

7. Is sales promotion a substitute for advertising? Can the two promotion tools achieve exactly the same kinds of objectives? Discuss your answer.

8. Discuss sales promotion and the various methods undertaken for it.

Managing the Salesforce

20.1 PERSONAL SELLING: CONCEPT AND IMPORTANCE

Personal selling is a key tool of promotional mix and in modern business it has assumed great importance in firms' marketing strategies. American Marketing Association has given the following definition of personal selling

Personal Selling is oral presentation in a conversation with one or more prospective purchasers for purpose of making sales. It includes in-person sales presentations and telesales, sales meetings, samples. We have already referred to personal selling as a tool of marketing communication.

Personal selling is communicating directly with the target audience through paid personnel of the company or its agents.

This neccessiates hiring and deployment of competent salespeople (salesforce) to engage themselves in face-to-face selling. Thus managing the salesforce effectively is the key to effective marketing. Studies show that annual expenditure on personal selling are generally more than advertising expenditure and substantial percentage of marketing budget in companies goes to personal selling.

As stated earlier, each tool of promotion has its own advantages or trade-offs over others. Advertising and personal selling are two alternative ways to communicating with the audience. A brief comparison between the two is given here.

Comparisons between Advertising and Personal Selling

Advertising is highly popular tool of communication and reaches far larger audience than personal selling. It is an all pervasive method of delivering message and allows the audiences the opportunity of weighing the message with those of the competitors. It permits creativity. It is impersonal in the sense that audiences are not obliged to pay attention or respond to the message. As stated earlier, it is a one-way communication and is an imperfect tool in this sense. Its credibility is low.

Personal Selling on the other hand, is targeted to a particular audience. It is highly personal, makes the audience to listen, pay attention and respond to the message. Salespeople can develop an interchange with existing and potential customers, know their needs and requirements, answer their queries and meet their objections. They can build a credible interactive relationship with the customers. They can demonstrate products to the prospective customers and explain their working and uses. Thus personal selling permits even complicated messages presented in a much better way. Besides, such interaction can be helpful in generating new ideas and requirements for development of new or improved products. The greatest advantage of this mode is that it accords salespeople the opportunity to build rapport with the audience and serves company's long term interests.

20.2 SALES JOB AND CLASSIFICATION OF SALES POSITIONS

(a) Sales Job

Personal selling directly connects company and its products with the target audience or customers. Sales job differs from firm to firm and from product to product. The job of a salesman or saleswoman representing a wholesaler in soaps and detergents will be different from that of a medical representative or an insurance agent or sales engineer. Sales job involves performance of a number of activities including travelling which need financial resources. Therefore, the nature of sales job is largely influenced by the nature and characteristics of the product or service for sale, customers' needs and preferences, and financial resources available for marketing mix.

Typically, sales job consists of the following three tasks

- (i) Generating Sales
- (ii) Providing Market Information
- (iii) Providing Customer Service

(i) *Generating Sales* implies obtaining business and involves activities like locating potential customers, keeping constant touch with them, presenting product or service, developing an interchange, inter-acting with customers, obtaining orders and effecting, sales, developing new customers (i.e. *prospecting*).

(ii) *Providing Information* has two aspects—one that of making sales presentation to the customer i.e. supplying information about the product, its benefits, advantages and uses. The other aspect involves collection of information on competitor's products and their prices, customers' reactions and responses to the company's product, stock levels with channels and delivery and service problems. Some of the information so collected needs to be provided to the company as feedback.

(iii) *Providing Customer Services* include delivery of the consignment, promotional assistance, credit evaluation, after-sales services like repairs and replacements.

(b) Classification of Sales Positions

Relying on the nature and type of the product, customer needs and requirements and financial resources available, a firm decides the tasks to be performed by salespeople, and accordingly designs job positions and job openings in the marketing department to be manned by salesforce. At salesperson level, the salesforce may comprise sales-trainees or apprentices, salespeople, senior salespeople, sales supervisors. As referred to above, salespeople are called by various names — salesmen, sales representatives, medical representatives (MRs), sales engineers, marketing representatives etc.

(i) One way of classifying sales positions is to categorise them on the basis of technical knowledge or distinctive abilities and skills required for performance of sales functions.

(ii) Robert N. McMurry classified sales positions as *deliverers order takers, missionaries* (who provide information about the product, educate the existing and potential customers and create goodwill), *technicians (i.e.* persons with technical knowledge e.g., engineering sales persons), and *demand creators* (persons with distinctive creative skills who create demand of both products and services.)

Various positions signify job descriptions, working conditions, knowledge requirements and distinctive abilities requirements etc.

Chart 20.1, on page 323, shows typically job profiles in sales and marketing department in a company.

20.3 SALESFORCE MANAGEMENT DECISIONS

As stated above, major share of the personnel assigned to the marketing department are engaged in personal selling. In fact, salesforce is an important human resource of a firm and as Gerald J. Carney has observed, "salespeople are being viewed as managers of their assigned market areas". Therefore salesforce management, in fact, implies management of human resources (i.e. personnel management) in the functional area of sales and marketing. In other words, the principles of personnel management are applicable in managing salespeople as well. We can define the salesforce management as follows:

Salesforce management is a distinct process, consisting of planning, organising, staffing, directing and controlling the selling effort by use of salesforce to determine and accomplish the salesforce objectives. In our present context

Planning implies determining salesforce objectives, formulating a strategy and scientific and systematic estimation of salesforce requirements over a period of time. (Salesforce Objectives, Strategy, Size).

Organising means determining the necessary component tasks to achieve salesforce objectives and grouping and assigning responsibilities to them (Deployment Decisions).

Staffing involves recruiting, selecting, training and compensating salespeople and making plans for meeting future staff needs of the organisations.

Directing means guiding, motivating and supervising the sales people (Motivation, Supervision).

Controlling involves those evaluative activities that are essential to ensure that objectives are accomplished as planned (Control and Evaluation).

Besides that, managing salesforce also involves creating cordial social relationships with them (*Interpersonal relations*).

20.4 SALESFORCE MANAGEMENT (SALES MANAGEMENT)

Salesforce management or sales management is the planning, organising, staffing, directing and controlling of the fourth element of promotion mix, the personal selling element which covers field sales and telesales.

1. Planning and Determining Salesforce Objectives

Planning here implies what to be done, how, by whom and by when it is to be done in order to meet overall marketing objectives.

(a) Determining Objectives

The first step is to develop salesforce objectives. The objectives are the goals i.e. the end results to be achieved and need to be based on clearly and accurately established planning premises. In our present context, the salesforce objectives should be set keeping into account the characteristics of the target audience, marketing environment and the company's marketing objectives.

Again, though personal selling is the most effective tool of communication at some stages, yet it is the most costly tool out of promotion mix. Therefore it is also necessary that the objectives developed match with the budget provided for the purposes.

It is the objectives that form the very purpose of an activity and guide the efforts to that direction, so they are to be developed very carefully. Moreover the objectives or end results provide performance criteria for evaluating the performance of salesforce. Therefore, objectives need to be expressed in clear, specific and unambiguous terms and in case of quantitative end results, they should be accurately measured.

Objectives need to be defined at macro level i.e. for the overall period (annual or so) for the total salesforce as a whole and translated to micro level i.e., for shorter time durations (monthly and quarterly) and to the specific goals of individual salesman. This necessitates close cooperation and coordination among the marketing manager, promotion manager and members of salesforce. Individual selling objectives are to be set taking in view the market or territory potential, capability of sales-person, nature and type of product, and past sales-experience.

Some major areas in which objectives are set are:

(i) **Market Performance:** Objectives are set in terms of sales volume, sales targets (productwise), market share.

(ii) **Profits and Cash flows** including selling expense levels — achieving desired proportion of cash and credit sales.

(iii) **Customer Service:** Objectives are set in terms of approaching their existing customers, giving them information about products, finding out their responses, effecting sales, rendering pre-sale and after-sales service including providing technical guidance, arranging finances.

(iv) **New Accounts and Prospecting:** These objectives can be in terms of opening new accounts i.e. convincing and cultivating new customers for company's products and in terms of time to be devoted to prospecting.

(v) **Market Development and Business Growth:** Objectives in terms of expansion of channel members, dealers.

Chart 20.1: Typical Job Profiles in an Engineering Company.

Sl. No.	Functional Area	Brief Description of the job	Working Conditions	Nature of Man Management	Knowledge Requirement	Distinctive Abilities Required
01	After Sale Service	Responsible for setting right the technical faults that occur in the products sold. Collecting date about performance of the products in the market.	Involves manual work at the customer's end. Has to travel extensively and prepare reports.	Has to work by himself and also to guide technicians of the dealers/customers.	Good knowledge of the product from service point of view Knowledge of commercial practices.	(a) Manual skills (b) Diagnostic ability (c) Written communication skills (d) Ability to get along with people.
02	Sales and Marketing	Responsible for and selling various products of the company to a wide range of customers and dealers in a specified geographical area/market. (includes application of engineering products to meet specific requirements of the customers.	Involves a lot of travelling, long hours outside one's residence. Adjusting to varied climatic conditions and outside food.	Has to most often work by himself but be in constant touch with customers/dealers, factory and warehouse personnel, deal with the transport operators, etc.	Sound knowledged of product range, customer's needs technology, application of engineering and commercial practices.	(a) Ability to get along with people, (b) Good communication and selling (c) perseverance and (d) resourcefulness.

(vi) **Training and Development:** Objectives pertaining to training of dealers, sales personnel and customers.

(vii) **Marketing Programme Support:** Objectives relating to collection of information, marketing research, marketing intelligence for the company, contribution to other promotional programme.

After establishing overall objectives and goals for the entire salesforce, they need to be translated into specific objectives and goals for individual salesperson. These are expressed in terms of sales targets or sales quotas, time for existing customers and prospecting, customer service, filling in-call reports, information collection, training dealers and customers. Sales targets can be determined by product, time, with the level of likely accomplishment to be decided on monthly or quarterly basis.

Whereas these objectives help in planning salesforce strategy and effort, they also serve as the criteria to monitor and evaluate the performance of the sales organisation and of each salesperson or sales representative.

(b) Formulating Salesforce Strategy

In modern business where a company confronts strong competitive challenges and customer choice is exercised, approaching the right customer at the right time and in the right way assumes great significance and herein lies the importance of personal selling or what is called *sales presentation.*

A major decision involved in planning salesforce is on the strategy to be devised to approach a customer or a group of customers. Sales people can approach customers in the following ways:

(i) **Sales Presentation - One to One:** A salesperson can approach an existing or prospective customer personally or through a phone call.

(ii) **Sales Presentation to a Group - One to Group:** Here a sales-person makes sales presentation to a group consisting of a number of buyers. Obviously, it is a difficult task to talk to a group of customers. In a group, members have different perceptions about the company and product, different attitudes and different buying behaviours and salespeople need to be specially skilled to deal with them.

(iii) **Sales Presentation - by Sales Team to a Group:** In this approach a sales team consisting of sales person, a technical hand and a company manager meets a group of buyers for sales presentation.

(iv) **Sales Presentation Conferencing:**   Company resource persons confer with a buyer or more to discuss their problems and for sales presentation as in a conference. It is called *conference selling*.

(v) **Sales Presentation - Seminar Selling:**   Here the company organises a seminar to acquaint teams from purchasing firms with advanced technology used in its products. It is called *seminar selling*.

Besides, some big companies producing products having seasonal demand like wool, woollen clothes, etc. invite dealers and wholesalers to meetings or conferences held in 5-star hotels and halls to book orders for their products, four or five months in advance of the commencement of actual season.

(c) Planning the Size of the Sales Force

The estimation of size of salesforce depends on the quantum of work and organisational requirements. We shall discuss it in the following pages.

2. Organising the Salesforce

Organising in our present context is a process of defining and analysing the selling activities of the company to be performed to achieve the salesforce goals, grouping them into distinct areas or departments and establishing the authority relationships among them. Organisation, thus, implies establishing a structural framework within which efforts of individual sales people are coordinated and pooled together in relation to each other to accomplish the salesforce objectives.

(a) Salesforce Structures

In the context of sales, the salesforce can be organised or structured on a number of bases. The structural framework can be of any one of the following types or a combination of them.

(i) Organisation by territory or geographical area

(ii) Organisation by product

(iii) Organisation by customers

(i) **Organisation by Territory (Territorial Structure):**   In case of large companies with a large number of salespeople, the salesforce is organised on the basis of geographical area. In this type of organisation, sales departments are organised region-wise or geographically to serve the particular regions effectively and efficiently with the products and services of same or similar nature. A region may be further organised or sliced into sub-regions, districts or territories depending on the area covered, types of customers, the product or service, selling duties and size of the sales-force. At salesperson level each salesperson is assigned a definite territory or area for personal selling. There is a hierarchical line linking different levels. At the national level, there is Sales Vice-President or General Manager -Sales. At regional level, there may be regional sales manager supervising the working of district or sub-regions or branch sales managers and they in their turn guiding and supervising sales in areas or territory assigned to individual salespersons. Such an organisation is useful in business like soft drinks, bakery products, dairy, soaps and detergents etc.

In a purely territorial structural framework, the salesman handles all the products of the company in the territory or area assigned to him and is responsible for showing the results. He is, in fact, the manager of his assigned market area.

(ii) **Organisation by Product (Product-based Structure):** Another way of organising the sales force is do so on product basis or along product lines. Several large companies in India have product departments supervised by product or brand managers. In this framework, a salesperson handles specific products or brands manufactured by the company for personal selling. Such an organisation is very useful in selling products which need technical knowledge and proficiency to sell them. Thus in product-based structure, a number of salespersons may be operating in the same territory, each selling different products manufactured by the same company. For example, Brooke Bond Lipton owned by Hindustan Unilever has separate salesforce for each of its products like Tea, Dalda, Dairy Products, Jams and Squashes etc.

(iii) **Organisation by Customers (Market-based Structure):** In this structural framework, sales force is organised keeping customers in consideration. In this case, salesforce is organised so as to cater to the specific needs of different classes of customers or target audiences. Separate depots may be organised to meet the needs of children or adults or ladies. Again, different salesmen may be approaching individual buyers, institutional customers, distributors or industrial users. The greatest advantage of this structure is that salespersons acquire specialisation in selling to a particular market or audience.

Large companies producing a large number of products and services for different types of customers and markets spread over a large area often organise their salesforce on a combination of these bases (complex structure) such as salesmen for a particular product in a given territory or catering to a particular audience or market in a specified territory or selling a product to a specific group of customers.

(b) Determining Size of the Salesforce

While determining the size of the salesforce, a number of factors such as marketing opportunities available, competitive challenges and company's sales effort need to be considered. It also needs to be assessed that how much job responsibility can be assigned to a position or a salesperson in terms of geographical area, number of customers or prospects in the time available with him i.e. how much work a salesperson can manage and control efficiently and effectively in a given time, say a year. This is called a *work unit*.

An important approach used in making decisions regarding the size of the salesforce is *workload analysis*.

Workload analysis: In this approach, total workload or sales effort in terms of time requirements to service a firm's customers or prospects is estimated. The total workload is divided by the effort or time an average salesperson can devote i.e. by work unit to arrive at the required salesforce size. The process can be explained in the following steps.

(1) Customers are categorised according to their annual sales volume or potential sales, profitability, customer servicing requirements, as large, medium, or small accounts, or they are grouped into categories A, B, C etc.

(2) The number of accounts in each customer category is determined. This would also give the total number of customers or accounts to be served.

(3) The next step is to determine the sales effort required to properly serve each category of the customers during a year. As a salesperson is to make sales presentation in person or through a call. *Sales effort is measured in terms of desirable call frequencies i.e. number of calls required to be made on an account in a year.*

(4) Then the total work load or sales effort i.e. total number of calls required to be made to serve all the accounts/customers is calculated. It will be the sum of the products of number of accounts in each category and the corresponding call frequencies required.

(5) The next step is to find an estimate of the sales effort or number of calls a typical salesperson can make during the given time i.e. in one year. This is determined keeping in view the factors like

geographical area to be covered, concentration of accounts, average time per call etc.

(6) The total workload is divided by estimated sales effort (number of calls that can be made by an average salesperson) to determine the number of salespersons needed and to be deployed.

The approach is illustrated in the following example

Table 20.1   Determining Salesforce Size (Work Load Approach).

Customer category		Number of Accounts / Customers	Desired Call frequency per year	Total calls required
Large	(A)	100	48	4800
Medium	(B)	200	24	4800
Small	(C)	1700	12	20400
Prospects	(D)	500	6	3000
Total		2500		33000

Total work load or calls needed = 33000

Let us suppose that average person can make 1100 calls per year, then number of salespersons needed = 33000/1100 = 30

3. Salesforce Staffing

Staffing involves recruiting, selecting, training and compensating salespeople and making plans for meeting future staff needs of the organisation. Salesforce staffing activities include:

    (i) Describing job and determining job specifications (job-analysis)
    (ii) Setting selection criteria
    (iii) Recruiting
    (iv) Selecting and hiring suitable people
    (v) Training and Developing Salesforce
    (vi) Deciding about Compensation (Compensating)

A. Job Analysis

A firm hires employees to achieve its objectives. The first prerequisite to providing competent and efficient manpower is to study the jobs and their requirements. It would require description of the job or in other words the nature of work to be performed and its demands of persons who have to perform it. Thus the first step in staffing is job analysis. Job analysis involves gathering information about jobs and job holder characteristics or job specifications. Job analysis information is presented in the form of job description and job specifications.

(a) **Job Description:** Job description is a written statement describing and specifying in detail the duties, objectives, authority vested, responsibilities and results expected of each job or position. It explains the duties, working conditions and other aspects of a specified job.

(b) **Job Specifications:** Job specifications is a profile or list of personal qualifications and personal traits deemed necessary for a job. Job specifications are the requirements in terms of qualifications, experience, technical and human skills and other desirable personal characteristics to perform the work effectively. As we shall see in the following pages, job specifications help the selector in selecting proper people.

In section 20.2 of this chapter, we have defined, in details, the sales job i.e. the job of a salesperson. We have described the tasks expected to be performed by a salesperson. We have also explained the designing of job positions and job openings in a sales department and different ways of classifying job positions. Chart 20.1 shows typical job profiles in a sales and marketing department. The job descriptions, working conditions and job specifications given in the chart are self explanatory. The chart throws light on the characteristics and distinctive abilities desired in prospective salespeople which can serve as criteria for selecting competent salespersons for the company.

B. Setting Selection Criteria

After deciding on the sales positions and having done the necessary job analysis for each position, the next step is to find and hire persons who may have the requisite characteristics, abilities and skills to man the sales job efficiently and effectively. The characteristics or traits required in a salesperson to be a good performer are derived from job description and job specifications. What is important to ensure in this context is that characteristics and skills possessed by a person must match the selling job for which he is required. These characteristics and skills serve as *'criteria for selection'*.

Broadly speaking, criteria are the factors which influence and evaluate the quality of performance of a salesperson. Marketers have drawn up a number of lists of traits and characteristics that a salesperson should possess to be successful. Besides knowledge and skills, these traits include aggressiveness, energy, industriousness, self confidence, empathy, ego drive, interest, urge to solve customer problems, motivation. Factors that have influence on performance are categorised as follows:

(i) **Aptitude and Personality:** Physical appearance, knowledge, mental ability, education, experience, personality and health traits.

(ii) Skill Acquisition: Technical and interpersonal skills acquired for sales job including negotiation skills.

(iii) Motivation: Effort that a salesperson can put in his work.

(iv) Other Factors: Competitor's challenges, quality of management, market opportunities, interests, personal and outdoor.

A number of tests are used for objective evaluation of candidate's knowledge, aptitude, interests, skills and other personality traits. They are used as devices to assess whether or not the applicant matches with the requirements of the job. Some of the commonly used tests are - knowledge tests, intelligence tests, psychological tests (to measure personality or temperament such as self confidence leadership ability, whether extrovert or introvert), performance tests, aptitude tests, interest tests etc. An organisation may administer one or two tests only, depending upon its requirements, or may use a combination of tests. We shall revert to the matter later when we study selection of the sales persons.

C. Recruiting the Salesforce

After having determined the salesforce requirements of the company in terms of job description, job specifications and the number of persons and when they are needed, the next step in staffing function is to find out and search for those capable and suitable persons who may be willing to seek job in the organisation. This will require publicising and dissemination of information about vacancies or jobs in the company and inviting applications from job seekers. It may also involve persuading and inducing suitable persons to apply for and seek job in the company. This process is called *"recruitment"*.

> *Recruitment is the process of searching capable people and inducing them to apply for seeking employment in the organisation.*

The process of recruitment begins with seeking of new recruits and ends once when the candidates submit their applications and seek employment in the organisation. The main purpose is to create a pool of capable and potential candidates and prospective job seekers for selecting suitable persons.

One of the most popular and effective method of recruitment of persons is to advertise the vacancies and jobs in newspapers and journals. An advertisement usually contains the name of the post, identity of the employer, brief job description, qualifications and experience required, compensation package and last date of applications. *Want Ads* is the most familiar way for searching potential candidates. Services of placement agencies are also used to employ salesforce. Most companies require candidates to send their biodata or application forms requiring the candidates to give their personal data such as name, address, date of birth, sex, marital status, physical data, educational qualifications and skills, professional training, job experience, compensation packet, strengths and weaknesses, hobbies, references and other information etc.

D. Selecting the Salesforce

The next step after recruitment is selection of those who are most suitable for the job. In order to find out right persons per needs of the jobs, the manager has to take a number of steps such as screening of the application forms to know the persons who have not got the required abilities, skills or experience and need to be rejected and the persons who appear to meet the requirements, evaluating the prospective suitable candidates with the help of employment tests and interviews, verifying references, ensuring their physical fitness (getting them medically examined) and finally making hiring decisions. All these steps form part of selection process. *The selection process, thus, comprises a number of specific steps taken to decide which recruits should be hired.* The selection process begins with screening of applications and ends with hiring decisions and placement.

In many firms, recruiting and selection are combined into a single function known as the employment function.

Selection process, generally, consists of the following steps:

(i) *Preliminary Screening/interview:* In order to find out suitable persons for a job, the applications received need to be screened and those who are found not to be suitable for the job are rejected and eliminated. Some firms call the applicants for a preliminary interview for purpose of screening and short listing them. This preliminary reception or interview which is also called "courtesy interview" helps the organisation and the applicant to exchange information about each other. It must be borne in mind that selection process is a two way process. It is not only that the organisation has to select a candidate suitable for it, the candidate is also to see whether or not the organisation suits him.

(ii) *Selection Tests:* We have already referred to various tests used to evaluate the suitability of the candidates for the job. These tests may include aptitude tests, knowledge tests, personality tests, skill tests etc. Sometimes the scores achieved by the applicants in these tests are credited to those achieved in the selection interview for deciding merit or final selection.

(iii) *Selection Interview:* Selection interview is a formal face to face conversation or in-depth question-answer session between the selector and the candidates to evaluate the overall suitability of the applicants and for exchange of information. The selection interview aims at assessing whether the applicant has the necessary ability i.e. aptitude, education, experience etc., whether he has the necessary drive, motivation and dynamism to accomplish it, and how does he compare with the applicants or he is above other candidates.

(*iv*) *Verification of References and Medical Examination:* After interview, selectors can get work history and performance verified from the references given by him. The candidate is got medically examined before he is hired.

(*v*) *Hiring Decision:* The selection process ends with hiring decision. It may be made by sales manager or personnel department.

An appointment is an act of faith, therefore the selector should select persons strictly on merit. If the appointments are made by personnel department, it will be desirable that the sales manager and direct sales supervisor are also included in panel of interviewers and are associated with the work because it is they who have to get the work done through the salesperson.

E. Training and Developing Salesforce
(i) Need for Training

After a person is selected and appointed to the position or job for which he was selected, the next task for sales management is to see that he adjusts himself in the job as quickly as possible so that he may be able to perform his job efficiently and successfully. Moreover, a new employee may not be capable of fully performing his duties at the out-set. He may be in the need of some training to sharpen his knowledge and skills. This suggests that the training programme for a new entrant should start with his very induction and placement in the organisation. Even experienced sales persons with a long service need to improve their knowledge and skills so as to perform their job in a more efficient way. They also need to be groomed and developed to meet their future responsibilities. All this will require a consistent and continuous programme of training and development.

(ii) Concepts of Training and Development

(a) *Training* is defined as "the act of increasing the knowledge and skills of an employee for doing a particular job". *Training implies activities that teach employees how to perform their present jobs better.* In short, training prepares people to perform their present jobs more efficiently. Training teaches employees required skills, knowledge or attitudes and helps them in improving their performance by imparting new skills, new techniques of doing the work and by improving their work habits. Training has career-long benefits.

(b) *Development of employees refer to those activities that prepare them for future jobs or positions.* Development represents all those activities or programmes of teaching skills, knowledge or attitudes that increase employees' potential and prepare them for future or higher jobs.

The difference between training and development is primarily one of intent and can be made on two counts, namely contents and the level of employees for which they are directed. Both training and development programmes involve imparting instruction in skills, specific job techniques, principles, concepts, philosophy etc. Each training has a development aspect.

(iii) Training Costs

We have already referred to the need for and importance of training in helping the new entrants in preparing them to perform their job according to the organisation's expectations, in improving the performance of existing employees in enabling them to cope with changes in work techniques and technology. Training costs time and money. A training programme involves substantial expenditures on instructors, materials and venues.

In recent years, there have appeared a number of private Institutes of Sales at different stations that run one year and one and half years courses at different stations and impart training in salesmanship and sales management. Some of them are managed by corporate companies. National Institute of Sales (NIS) is one of such institutes. The approximate fees being charged from a student by these institutions vary from Rs. 30000 p.a. to Rs. 50000 p.a.

Many large Indian and multinational companies (MNCs) spend large amounts of money on training of new entrants (sales persons) in addition to their salaries. The training period for a new entrant ranges from 6 months to 2 years depending upon the nature of task and policy of the company, before they are in the field on actual selling task.

(iv) Importance of Training

As referred to in above, training costs time and money, but it needs to be regarded as organisation's investment in its salesforce (human resources). The investment in employees pays rich dividends in the form of benefits to the organisation so much so that the benefits derived as a result of salesman training far outway the cost. Some of the main benefits of training are given here.

(1) It improves efficiency and profitability. It helps the employee to acquire improved knowledge and skills besides teaching them new

techniques thus increasing their efficiency. It improves their technical capabilities.

(2) It helps them to make better use of resources and reduce cost, time, wastages.

(3) It ensures the morale of employees by inculcating a sense of greater self-confidence.

(4) Trained people require less supervision. This gives them an opportunity to work in freedom and perform their job successfully.

(5) It gives them an opportunity to cultivate better interpersonal and human relations. It improves communication between employees and target audience.

(6) It contributes to stability and growth of the organisation.

(7) It helps the individual in making better decisions and solving problems of the customers effectively.

(v) Training Needs, Objectives and Contents

Training is aimed at improving the performance of people in their present jobs. There can be many objectives in designing a training programme such as to enable new appointees to reach required standards of performance as quickly as possible, to reduce wastage of time or money, to introduce correct methods and innovations to achieve, maintain or improve standards. In short, the main objective of a training programme is to make an employee a better and effective performer.

As stated above, job analysis, especially job descriptions provide good guidance in finding out the training needs of newly appointed salespeople and designing the training programmes for them. Again, existing salespeople also need training to improve their knowledge and skills.

Another way to look at this problem is to enlist the sterling qualities that should be possessed by a salesperson to be a good and effective performer and develop a training programme and its objectives in their light. Such a programme will be really need-based.

Qualities of an Effective Salesperson

An effective salesperson generally possesses the following qualities:

(1) He has a complete knowledge of his company, its history and philosophy, its policies, programmes and procedures, distribution channels.

(2) He knows his product thoroughly, its benefits and added value and how is it different from competitor's products, how it is produced and how it can be used and the techniques involved therein.

(3) He knows the target audience and its characteristics i.e. his customers, their needs and problems, factors that would motivate them.

(4) He has a perfect knowledge of the competitors' challenges.

(5) He has necessary technical capabilities to train the dealers and the customers in the use of the products

(6) He knows his strengths and weaknesses and possesses a thorough knowledge of selling techniques.

(7) He has learnt the principles of personal selling and their application in day-to-day work and acquired the necessary skills needed for his job such as negotiation skills, interpersonal skills.

(8) He contributes to the market effort through cooperation.

(9) He is ever eager to improve his own performance, and participate in self-improvement and self-appraisal.

(10) He achieves the set selling goals/targets.

The list can be expanded. This list suggests that a typical training programme should cover the following areas.

(a) Company history, philosophy, policies, procedures and practices.

(b) Product knowledge and competitors' challenges.

(c) Target audience and its characteristics.

(d) Time and Territory management.

(e) Principles of salesmanship — selling techniques and skill-development.

Depending upon the training needs of sales personnel, the training programmes can be designed and specific objectives set for each such programme such as making salespersons to know and identify with the company's products, its uses and benefits, how to use a product including its repairing and replacement, helping sales representatives to learn the principles of salesmanship and acquiring negotiations and interpersonal skills, improving salesperson's efficiency in time and territory management etc.

As training a salesperson is a costly affair, therefore their selection for participation in a training programme should be done very carefully. Training to be effective should be by and large need-based.

(vi) Training Techniques

Training needs to be linked both with the operational tasks and company's marketing goals. The techniques applied for both training as well as development are mainly common. There are several techniques or methods of imparting training. The technique to be used depends upon the nature of the job, needs of salespeople and objectives of training.

Techniques of training that are generally used to impart training are categorised:

(i) on-the-job techniques and

(ii) off-the-job techniques.

On-the-job techniques/methods include job instruction training, apprenticeship and coaching.

(a) *Job-instruction Training:*   In this method, instruction is imparted directly on the job and salesperson learns how to perform his work while working on his present job. It is in a way learning by doing. The training is imparted by a sales supervisor or senior and experienced salesman at the very work situation. It is rather the most cost effective technique.

(b) *Apprenticeship and Coaching:*   Under this technique, an apprentice or an under-trainee is imparted training in actual working situation. In it, an apprentice gets practical work experience under the supervision of an experienced salesperson. An apprentice is usually paid a paltry pay or stipend in lieu of his services during the period of apprenticeship. Sometimes, this training is supplemented by off-job classroom instruction.

Off-the-job techniques are a number of methods which are used to impart training to the trainees away from the workplace for a specified period. Some important off-the-job techniques are lectures and demonstrations, video and film presentations on salesmanship and company's products, role playing, laboratory training and sensitive training (group training based on sharing of experiences and self-interactions), conferences and seminars, self-study, programmed learning. Now-a-days, computer programmes are used for training which encompass some learning principles.

(vii) Evaluating Training Programmes

A training programme is said to be successful if it fulfils the needs for which training was imparted, if the learning achieved from training is transferred to job and results in improved performance and in positive change in behaviour of the trainees. It is also necessary to ascertain the influences or worthwhileness of the training to find out whether money and time spent on it has been worthwhile and should be continued or whether there is need for modifying the training programme in terms of period, content or methods.

Though it is extremely difficult to evaluate a training programme, yet the criteria for evaluating its effectiveness can be established and it can be measured in terms of its impact on salesforce turnover, sales volume, time spent, calls-to-close ratio, number of new accounts opened per unit time etc. Lower turn over, increase in sales, increase in number of accounts, reduced selling costs, greater calls-to-close ratio etc. provide evidence of effectiveness of the training programmes.

F. Compensating Salespeople

Compensation is the amount received by salespeople in exchange for the work or services performed by them. Compensation in the form of salaries or commissions enables the salespeople to meet their needs and those of their families. Financial compensation is a vital source of satisfaction. A proper compensation structure not only retains competent and efficient workers but also attracts others from outside the organisation.

An appropriate compensation package gives employees a sense of satisfaction and motivates them to strive more and more to achieve the set objectives. It helps in creating an effective workforce and in maintaing and retaining enterprise's human resources. A pay cheque is a symbol of status, a source of self-respect and an avenue of security. An inappropriate compensation adversely affects performance, motivation and satisfaction. Therefore a proper remuneration or compensation package is of vital importance if effective and competent workforce is to be retained and maintained.

Besides a simple, fair and adequate compensation package it should also assure internal as well as external equity.

Internal equity means that similar job gets similar pay and the pay be related to the worth of job within the organisation. There should be no discrimanation and wage differentials should be based on rational and objective basis.

External equity means that employee is paid at least the same compensation which employees belonging to similar positions receive in other organisations. Only an equitable wage structure can retain and attract competent persons.

Compensation Plans

The compensation modes followed by various companies differ from company to company. The compensation plans range from straight salary to a straight commission or a combination of both. A brief explanation of the terms is given here.

(1) Salary: It is a regular fixed amount paid every month as compensation to a salesperson for the services rendered. It ensures the salesforce stability of income and keeps the morale of the sales force. Such a system is easy to understand but the chief weakness of this practice is that it does not make any distinction between an efficient and inefficient salesperson. The system offers no incentive to hard worker or better performer and creates a tendency among salespeople to go slow and work at leisure. It requires more supervision and increases its cost.

(2) Commission: Under this practice, salesperson gets commission at fixed or sliding rate on their sales or profit volume as compensation. This is, in fact, a performance-based system. It has a number of advantages such as it provides incentive to salespeople to work harder and perform better. The rates of commission can be different on sale of different products. The chief weakness of this system is that it provides little stability or security of income and adversely affects the morale. The income is variable.

(3) Bonus: Bonuses are extra payments beyond basic salaries or commissions that are made as incentives to the salespeople for better performance. This is a supplement to salary or commission as a reward for special effort or results. They are decided by the management according to their judgement which is often criticised for lack of objectivity.

Sometimes, special bonus is paid to a group of salespeople for exceeding the targeted sales. These are known as *group incentives*.

(4) Financial Incentives: Incentives are always related to performance. Any reward which is not linked to performance or sales cannot be called incentive.

(5) Fringe Benefits or Perquisites: In addition to salary, commission and financial incentives, the companies provide a number of benefits and services to their employees to satisfy their societal, personal and family needs. These benefits include insurance against illness, accident, paid vacation, leave travel concession, hospitalisation, pensions etc., and provide security and job satisfaction.

(6) Other Costs: They include selling costs like expenses incurred on travel, lodging and boarding, telephone, correspondence, entertainment etc.

Now-a-days, a large number of companies offer a compensation mix comprising a basic salary, plus commission plus bonus plus fringe benefits or perks and selling expenses. Though no reliable survey or study as to compensation plans being followed by Indian companies is available, the findings of a survey made by American Marketing Association in 1980s in USA in this context is worth noting.

Compensation plan	%age of firms
Straight salary	22%
Straight commission	6%
Salary plus commission	28%
Salary plus bonus	30%
Salary plus group bonus	2%
Salary plus commission plus bonus	12%

The above study suggests that only a small fraction of companies (only 6%) in USA were following the practice of paying straight commission as compensation. 94% had salary as a component of their compensation plan. Whereas 72% of the firms adopted salary plus some incentive plan.

Most of large companies in India employ regular salespersons who receive a basic salary with (cost of living) price level adjustments plus some periodic incentive (on quarterly, half yearly or annual basis) plus fringe benefits. Bonus is usually paid after every completed year of service at pro-rata basis e.g., 20 or 25 or 40% of the basic pay or salary.

Components of a Sales Person's/sales Representative's Earnings

The typical components of a salesperson's earnings in Indian companies can be summarised as follows:

(a) *Salary:* Basic pay, personal pay, special pay, deamess allowance, house rent allowance, medical allowance, kit allowance, conveyance or scooter allowance.

(b) *Incentives:* Bonus as percentage of basic pay or salary, performance incentive (on periodic basis; quarterly, half yearly or annually)

(c) *Fringe benefits (Perquisites, perks):* Leave encashment, paid vacation, leave travel allowance, employer contribution to provident fund, family - pension fund/super annuation fund etc.

In addition, expenses incurred on travel, board and lodging, correspondence, entertainment are also reimbursed.

The emoluments of a salesperson in India at present range between Rs. 50,000 to Rs. 100,000 per annum.

4. Directing Salespeople

In sales management, directing is process of guiding and motivating the salespeople towards achieving marketing goals. Sales people in a company should know what they are expected to do. Directing means inspiring the salesforce to work in a desirable direction so as to realise the determined objectives. In nutshell, directing consists of providing instruction, coaching, counselling, guiding, leading, inspiring and motivating salespeople towards attaining marketing objectives with quality results in a given period of time. Effective directing depends on effective communication. Thus the main elements of directing are— supervising, motivation, leadership and communication.

Supervising Salespeople

Supervision means overseeing the performance of work assigned to salespersons at operative level. Viteles has defined supervision as "direct and immediate guidance and control of subordinates in the performance of their tasks. It is the function of assuring that work is being done in accordance with the plan and instructions".

It is evident from the above that effective supervision comprises giving intelligent and precise directions, providing advice and counsel in improving their know-how and skill, motivating the salespeople to reach company goals and coaching and training them. Obviously, effective supervision implies effective sales management and necessitates creation of sales positions at different levels such as manager, assistant manager, sales executives, sales supervisors etc. to oversee and ensure accomplishments of the goals set for the salespeople.

Sales managers or sales supervisors are the main link between the salespeople and higher levels of management. They are, in fact, in direct touch with salespeople and customers. Besides guiding, directing and supervising salespeople, their task also entitles solving salespeople's problems, communicating their grievances and problems which could not be solved at their level to the higher authorities and getting them redressed. The typical functions of sales management or supevisory management are—

- Helping the salespeople in determining their selling objectives and planning.

- Creating a suitable work environment for salespeople so as to enable them to meet deadlines. It would involve maintaining discipline, team spirit among the sales persons, and keeping good personal relations with them to ensure proper functioning.

- Knowing strengths and weaknesses of each of the salespersons and making best use of their abilities.

- Providing on-the-job training to salespeople and ensuring proper operation performance.

- Organising training and developing programmes for salespeople to meet their increasing training needs and help them in improving their knowledge about the company, its procedures and practices, its products, time and territory management and principles of salesmanship.

- Setting realistic and fair standards of performance for salespeople. They may be expressed in number of calls, sales volume, quota or qualitative norms.

- Maintaining communication links and ensuring proper unclogged information flow bothwards i.e. upward and down-ward. It involves providing feedback to the authorities.
- Motivating the salespeople to reach marketing objectives.
- Motivation has strong influence on the effort put in by a salesperson, job satisfaction and hence on his performance. Motivation, thus, leads to a greater effort, improved performance, and higher remuneration and higher satisfaction. There are a number of factors including incentives that affect satisfaction and motivation.
- Monitoring, controlling and evaluating the performance of salesforce.

5. Monitoring, Controlling and Evaluating the Salesforce

Performance Standards/Evaluative Criteria

As referred to above, a primary function of sales management is to monitor, control and evaluate the functioning of the sales department as well as the performance of individual salespersons. The sales management also has the responsibility of setting realistic and fair standards of performance for salespersons in terms of sales effort (number of sales calls), sales quota, sales volume, profit to be contributed, expenses or average cost per sale call, percentage of orders per hundred sales calls, salesforce cost as a percentage of sales-volume, number of new customers, new accounts. The performance is evaluated by making comparisons between their individual efforts or comparing the performance with the performance in the last year or so.

Information Sources/Instruments of Sales Control

Management receives regular information from the salespeople in the form of travel programme, tour report, sales report, expenditures incurred, calls made etc. It also gets information from sales supervisors, executives, managers in the form of inspection reports, field visit reports, and observations regarding meetings with salespeople. They also gather information from dealers, channel members and directly from customers. Customers also make complaints and grievances. These reports and data serve as effective instruments of sales control.

It is also the responsibility of the sales management to see that each salesperson is able to achieve the selling objectives and goals as determined in his individual plan. Monitoring individual efforts and performance is also a critical factor in the direction of accomplishing the overall marketing objectives and goals. In case the results are not upto the expectations, necessary remedial measures like motivating and guidance and training the concerned salesperson should be taken in time, so as to improve his performance.

20.5 SALESMANSHIP AND SELLING PROCESS

We have thus far discussed the role of personal selling as an important tool of promotion mix and management of salesforce to achieve marketing objectives. We shall now briefly explain the tasks of a salesperson i.e. functions of salesmanship and the selling process involved in the sale of a product.

Selling and Customer's Satisfaction

The all important task of a salesperson is to sell the product but he has also to see at the same time that the interests of the customers are also served well. We have already referred to Blake and Mouton Selling Theory in chapter 2, which throws light not only on marketing philosophy but on selling styles as well.

> *According to R.R. Blake and Jane Mouton, two major dimensions of selling style are — (1) concern for the people i.e. concern for customer's needs and satisfaction and (2) concern for the sale. In their view, these two concerns can be combined in different degrees. They suggest that an optimal selling style involves maximising both dimensions viz. concern for sales and concern for the customers.*

Functions of Salesmanship

Personal selling is a creative art. It creates new wants and preferences. It involves establishing a cordial and biding relationship between the company and its customers and is based on mutual trust. Some important functions of salesmanship are:

(1) to introduce products to the customers.

(2) to help the customer to make buying decisions.

(3) to see that customer's needs are transformed into wants and demand.

(4) to negotiate and conduct effective selling at least cost.

(5) to gather information about markets and competitors' products and transmit it to the company.

Performance of these functions is rather the most difficult and most important of the marketing functions. An effective accomplishment of these functions necessitates creation of a direct selling process.

Selling Process

Direct selling is the ultimate offer of the value or benefits to the customer. Besides face-to-face dialogue, it needs creation of an environment where the salesperson is completely free to communicate and negotiate and the customer is utterly receptive to his proposal. In this context, it must also be borne in mind that it only takes very small time to clinch a sale but choosing the right time to contact and call on the customers need careful thinking and planning.

Moreover, clinching of a sale deal requires on the part of the company and its management to empower the salesman with sufficient flexibility to negotiate and deal with customer, accommodate individual needs, and clinch the sale.

An effective selling process consists of the following steps:

(1) Identifying the customer (prospecting and qualifying)

(2) Making appointment with customer (approach)

(3) Choosing suitable time for presentation and demonstration

(4) Ensuring product performance at presentation and demonstration

(5) Handling objections, if any

(6) Closing deal quickly

(7) Following up sale with prompt delivery and service

(8) Ensuring enhancement of customer value.

Clinching of deals and sales sometimes requires protracted negotiations and hard bargaining. The deals involve not only setting price for the product but also many other things concerning quality delivery such as period for completion of delivery, quality of the product offered, volume of the product purchased, financial terms, risk taking and other conditions of sale/delivery. In order to achieve good results a salesman needs to develop negotiation, communication and inter-personal skills and the art of probing.

Negotiation and Bargaining

Negotiation means entering into conversation with the customer with a view to set the price for the product and other terms of exchange in a sale deal through bargaining.

Bargaining denotes the process of mutually agreed agreement about price and other terms between the parties i.e. salesman and customer involving proposals and counter proposals. *Thus negotiation and bargaining are interchangeable terms.*

The above definitions of negotiation and bargaining give the following features of bargain:

(1) It is a *bipartite process* i.e. it is a two way process. Except salesperson and customers, no third party is involved and terms are settled by parties concerned.

(2) *Negotiation is the base.* In bargaining sessions, deliberations, discussions and compromises take place on relevant issues and terms.

(3) *Bargaining is a flexible process.* It is not rigid and there is always a scope for adjustments. Negotiations, proposals, counterproposals, offers, counter-offers go on till a mutually agreed deal is accomplished.

(4) It is *voluntary by nature.* By nature it is an activity leading to voluntary buyer-seller relationships.

Skills of Negotiation

The skills of negotiation to be acquired by a salesperson include:

(1) ability to converse effectively and confidently with members of his company.

(2) ability to plan carefully and knowledge of company's product, its benefits and uses.

(3) good business judgement—an ability to understand the real issues.

(4) ability to tolerate others' views.

(5) ability to listen, to the customer patiently and intently.

(6) self confidence.

(7) willingness to use other company experts.

(8) commitment to honesty, integrity and mutual satisfaction.

(9) ability to manage ambiguity, making sense out of confusion.

We have already explained interpersonal skills required for a marketer in chapter 4.

EXERCISES

1. What is meant by personal selling? Compare it with advertising?

2. What do you mean by salesforce management? Explain the process.

3. Describe sales force staffing activities.

4. What are the functions of salesmanship? Describe the main steps involved in an effective selling process.

5. Write notes on

   (a) Classification of sales positions

   (b) Salesforce structures

   (c) Work load analysis

   (d) Recruitment and selection of salesforce

   (e) Compensation of salesforce.

6. Apart from personal selling skills, what are the other factors which can help and hinder a salesperson's efficiency.

7. Describe various compensation plans employed by companies. What are the situations where straight salary plan is appropriate?

21

Managing Distribution Channels

In chapters 12 to 20, our discussion has been devoted to decision making on three components of the marketing mix, namely product, price and promotion. In this chapter and the next, we shall study the fourth component of the marketing mix *viz.* place (distribution), or in other words, making decisions on how to make the product or service available to the consumer or end user efficiently.

Distribution, in the marketing context, is the process of making the products or services available to the consumer and involves movement of goods and services from manufacturers to the end users.

A manufacturer or supplier may sell or supply the product directly to the consumer or end user or may take the help of intermediaries or middlemen (channels) to make the same available to the end user. Again, distribution implies physical movement of the products from the factory site (or the producer's place) to the place of the consumer or customer which necessitates transportation, inventory management and control, warehousing etc. Therefore, in context of our present study, a marketer has to make two types of decisions–decisions on selection and management of distribution channels i.e. of middlemen through which products are to be marketed and sold, and decisions on physical distribution of the products.

We shall study decision making on distribution channels in this chapter and on physical distribution in chapter 22.

21.1 DISTRIBUTION (MARKETING) CHANNELS

A. Marketing Channels: Definition

The marketing system as a whole consists of organisations, institutions and individuals interconnected by flows of information, products, title (ownership), negotiations, promotion risks etc. The system includes organisations and agencies that perform all the activities that link producers and users and vice-versa to accomplish the marketing task. These organisations are called marketing or distribution channels.

Distribution or marketing channels are means employed by manufacturers and sellers to get their products to the market and into the hands of the users. These are tools hired to do the job of getting goods from factory or place of production into the hands of the ultimate user. They add possession value. It is through intermediaries or middlemen like wholesalers, distributors, dealers, brokers etc. that this purpose is achieved. Marketing intermediaries are organisations that perform the functions that connect products with end users and thus form the distribution or marketing channels. We can define marketing (distribution) channel as below.

Marketing channels or distribution channels refer to an organised network of interconnected organisations and agencies involved in the process of making a product or service available to users or consumers.

As stated above, these are also called *marketing intermediaries or trade channels*. The key functions between two channel members are—information, promotion, negotiation, ordering, financing, risk taking, possession, payment and title.

B. Functions and Importance of Marketing Intermediaries/Channels

Marketing intermediaries work as a link between producers and users and fill the gaps created when the direct exchange between producers and end users is not possible or feasible. They perform a number of important functions such as

(1) Providing *information* about the products to the customers and about customers, competitors and marketing conditions to the producers i.e. providing *feedback and intelligence*.

(2) *Promoting* the marketing offer and sales process.

(3) *Negotiating* with producers as well as with customers to settle the terms of deals w.r.t prices, supply etc.

(4) Helping in actual transfer of *title or ownership* from one organisation or individual to other.

(5) Acquiring stocks, maintaining inventories and catering to even small size requirements of the consumers.

(6) Providing pre-sale and post-sales *service.*

(7) *Risk taking.*

(8) *Financing* including supplying goods on credit besides making deposits with manufacturers in advance. They reduce manufacturers's burden. Film distributors have been providing finances to reputed producers for production of films and taking risk.

(9) Acting as agent of change and transferring technology to users (electronic items, TVs, computers, training in operation of industrial machines)

The above list can be multiplied. Besides the above functions, members of distribution channels also perform other functions like procurement, storage, packaging, transportation etc.

In addition,

(10) Intermediaries reduce the number of buyer-seller transactions thus increasing *transactional efficiency.*

(11) Channel members sell products in assortments.

(12) They deal in bulk purchase, therefore they can help in supplying goods at reasonable prices to the users.

(13) They help in decentralising markets.

Marketing channels have assumed such importance that many companies now employ *channel managers* or executives to coordinate and control channel operations.

21.2 CLASSIFICATION OF INTERMEDIARIES

Marketing intermediaries are classified on the basis of ownership of goods and by their level.

(a) Classification by Ownership of Goods

Intermediaries are categorised as wholesalers/distributors, retailers and agents/brokers.

(i) Wholesalers/Distributors

These are the business organisations or individuals that buy goods and services (i.e. merchandise) in large quantities for resale, mostly to retailers and other traders or to institutions, industries and commercial enterprises for their business use. They exclude retailers-manufacturers, farmers and others engaged in primary production. Generally, they take title to the goods they sell.

They include merchant wholesalers, manufacturers' sales depots and branch offices, commission merchants, C&F agents, and brokers (who do not take title). Some major functions of wholesalers are:

(1) They purchase, sort, assemble, grade and store goods in bulk quantities and resell it for profit.

(2) They extend credit (financing) and provide other services and counselling to retailers and suppliers.

(3) They provide delivery service (transportation) and sales promotion for their customers.

(4) They provide marketing intelligence to their suppliers.

(5) They bear risk including risk due to price changes and of damage, obsolescence etc.

(ii) Retailers

These are the business organisations and individuals that buy goods and services, mostly in small quantities, for resale to final consumers and organisational end users. They may or may not take title to the goods.

They may handle the goods on consignment basis and sell them on commission to end users for their personal non-business use.

They are categorised as store retailers, non-store retailers and retail establishments and include general merchandise stores, shopkeepers, food stores, restaurants, speciality stores, departments stores, super markets, malls, show rooms, petrol stations (pumps), mail order houses, persons conducting saree sales, jean sales etc. Non-store retailing may take the form of direct selling and automatic vending (use of coin-operated machines), consumer cooperatives, corporate chains, franchise stores and organisations. Presently in India, a number of big Corporate Companies and Organisation like Reliance, Subhiksha, More (Aditya Birla group), Big Bazar have engaged themselves in retailing goods.

(iii) Brokers/Agents

Brokers are individuals or business organisations that negotiate purchases or sales or both on behalf of their clients. They do not take title to the goods or maintain any inventory and therefore bear no risk. Agents are representatives of their clients, sellers or buyers, and work on a more permanent basis. They generally get commission or fee for the services rendered. They include brokers, stock brokers, real estate agents, insurance agents, selling agents, manufacturers' agents, purchasing agents, commission merchants. A number of large companies particularly those in automobile industry have a network of main dealers, dealers and subdealers.

(b) Classification of Intermediaries by Levels

Each intermediary in the channel brings the product nearer to the end user or consumer. The number of intermediaries in between the manufacturer and the final customer denotes the *level* of the channel and explains the length of the channel. For example,

  (i) Zero-level Channel: Implies that there is no intermediary or middleman between a manufacturer and the final consumer. In other words, the manufacturer sells the product directly to the final consumer or end user. It is also known as direct marketing and may take the form of door to door selling, mail order, home parties and company's depot. We have already explained direct marketing in chapter 18.

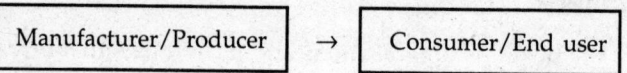

  (ii) One-level Channel: One level channel contains only one intermediary such as retailer, dealer or agent/broker.

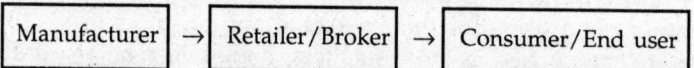

Jenson and Nicholson has one level channel. Retailers purchase paints from their depots.

  (iii) Two-level channel: Two level channel contains two intermediaries. Most often, the intermediaries are whoesalers/distributors and retailers.

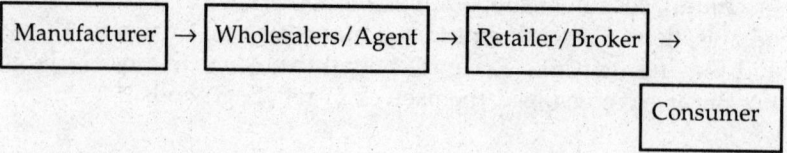

  Wipro Info-tech has two-level channel of 15 C&F agents and 4000 dealers.

  (iv) Three-level channel: This channel consists of three intermediaries, may be agents, wholesalers, retailers or main dealers, dealers and sub-dealers.

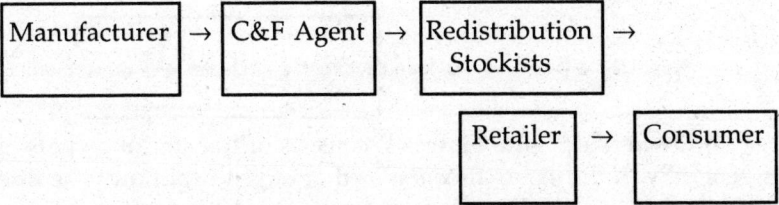

Ashok Leyland Limited has a three level channel distribution net-work consisting of main dealers, dealers and sub-dealers.

Hindustan Unilever Limited has also a three-level channel network comprising Clearing and Forwarding agents, 3000 redistribution stockists and 600000 retail outlets and lakhs of other shops in rural areas.

Similarly, there can be higher than three level channels also but they are found in exceptional circumstances. Channel levels vary from company to company depending upon its objectives.

Besides these channels, each large company establishes its own salesforce consisting of sales managers, sales executives, sales supervisors and salespersons to contact and assist channels and to accomplish its marketing goals.

21.3 MARKETING CHANNEL FLOWS

As stated earlier, distribution involves handling and movement of products and services from the manufacturers to consumers through various intermediaries. In other words, there is a flow of physical products through the marketing channels. Besides physical flow, there are other flows such as information flow, title or transfer of ownership flow, communications/promotion flow, financial transactions, transfer of risk flow etc. These flows connect the channel members and are essential for the efficient performance of these functions. These flows are briefly explained here.

(1) Physical Flow: As the name suggests, this flow takes place between suppliers, manufacturers, intermediaries and consumers. The flow originates from suppliers towards end users with the help of transportation firms. It is often a one-way flow provided that products are not supplied on returnable basis.

$$\text{Suppliers} \xrightarrow[\text{Warehouses}]{\text{Transporters}} \text{Manufacturers} \longrightarrow \text{Dealer} \longrightarrow \text{Consumers}$$

(2) Information Flow: It is communication in all directions between channel members. Communications comprise despatch of catalogues, brochures, change-in-price lists, advertisements, placing of order by mail, phone or fax, grievances and complaints or requests for post-sales service sent by consumers, feedback or marketing intelligence.

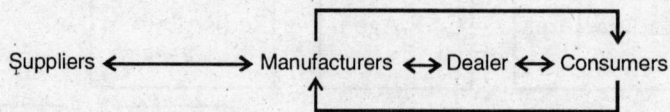

Suppliers ←————————→ Manufacturers ←→ Dealer ←→ Consumers

(3) Ownership Title Flow: It consists of transfer of ownership and is generally in the same direction as that of physical flow. The title rests with the buyer or seller. Some channel members such as brokers and agents do not take title or physical possession of the product as they work only on commission or fee.

(4) Financial/Payment Flow: It is the money flow that moves through the channel. It originates from the consumers and flows upto the supplier comprising payments being made through banks or other financial institutions. Financial payments are also made between channel members and organisations that provide services such as advertising, transportation, warehousing, marketing research etc.

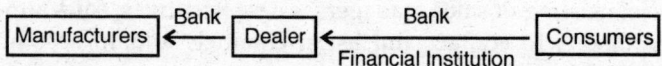

Manufacturers ← Bank ← Dealer ← Bank / Financial Institution ← Consumers

(5) Transfer of Risks: There are many risks involved in transactions i.e. in sale and purchase of products that channel members bear such as price change, damage, spoilage and pilferage, obsolescence and other trading risks. Risk flows almost through the entire channel.

Understanding of these channel flows is essential for performing various channel functions properly and efficiently.

21.4   CONSIDERATIONS AND CONSTRAINTS IN DESIGNING A CHANNEL STRATEGY

The main purpose of all marketing effort is to reach the target markets effectively. Therefore, while designing a channel strategy a manufacturer needs to search for the best means to achieve this aim. A manufacturer has two options to reach the target markets—

The first, option is that the manufacturer should sell its products or services directly to the consumers or end users i.e. it should follow direct distribution approach. *The second option* is that the manufacturer should use the services of intermediaries, the middlemen, to meet the needs of the target audience or it may follow *channel approach.*

We have already explained advantages and limitations of direct marketing and personal selling (selling though sales persons) in chapter 18 and 20. We have also referred to the relative importance of intermediaries in section 21.1 of this chapter. However, in the present context, we must know the pros and cons of the options.

To recall, direct distribution has the following advantages

1. It is the simplest form of channel system.
2. It allows the manufacturers to have a direct control over marketing activities.
3. It facilitates better buyer-seller communication and inter-change.
4. It eliminates intermediaries' profit margins.

The main limitations of this method are—

1. It requires substantial resources to perform channel functions especially to reach target market.
2. It is not suitable when buyers purchase the product in small quantities and frequently.

Which option of the above two is more appropriate depends on a number of factors or considerations such as company capabilities and characteristics, product service characteristics, target-market characteristics, competitors' channel design, marketing programme requirements and channels' characteristics?

(i) *Company capabilities and characteristics:* The decision to go in for direct distribution or utilise channel is affected to a large extent by company capabilities or characteristics such as its long-term objectives, resources, experience, availability of time, marketing strategy. For example, for direct distribution, a company must have resources enough to establish adequate salesforce. It should have an experience of marketing similar products directly to similar target markets and it should have sufficient time to develop a direct marketing programme before competitors enter the field.

(ii) *Product/Service characteristics:* The nature or characteristics of the product or service also influences the decision. For example, perishable goods, complex, non-standardised, high value industrial goods, innovative products need to be sold directly to the customer, whereas for bulky products, products purchased in small quantities, more frequently by different types of customers, and standardised products, intermediaries are necessary.

(iii) *Target market characteristics:* Intermediaries are necessary when target market consists of a large number of customers or customers who are dispersed sparsely over a vast territory or of buyers who do not have a long-term commitment or make large purchases in terms of quantity or unit price or the margin of profit is small. Direct or personal selling will be more appropriate in case when target market consists of few customers, or customers concentrated in a small geographical territory or when profit margins can meet the cost of direct selling or purchasers have long-term commitment to buy the product or brand.

(iv) *Environmental conditions/characteristics:* Decision on a channel design is also influenced by environmental factors like market economic conditions, prosperity or depression, legal regulations and restrictions. For example, the need for reducing physical distribution costs may affect levels of channels.

(v) *Competitors' channel designs:* A company's decision on channels is also influenced by channels used by the competitors. The manufacturers may need to use same channels or outlets or may need to find their intermediaries adjacent to those of the competitors' or may like to avoid them. Sometimes manufacturers have to vie and compete for the same outlets/channels to sell their products. Pepsi Cola and Coca Cola are the examples in this context.

(vi) *Marketing programme requirements:* Decisions on channels are also affected by the company's marketing programme or strategy. For example, the company's marketing strategy may be depending heavily on direct marketing or personal selling or the manufacturer may not be feeling the necessity of intermediaries to pursue its marketing programme and would like to perform these functions on its own or in opposite case the company's marketing programme may be relying mostly on channels to reach the target markets effectively.

(vii) *Channels' characteristics:* Channel decisions are also influenced by the characteristics of various intermediaries or middlemen e.g. by their strengths and weaknesses, their capability to handle storage, promotion, negotiation, local credit, inventory, package, etc.

The above study suggests that marketing managers should be very careful and take all these factors into consideration when designing a channel strategy. It is all the more important because channel decisions involve long-term commitments and relationships with channel members. It takes time, costs and much effort to create and establish a new channel. Again, channel decisions also affect other components of marketing mix e.g. pricing of the product, promotion and advertising strategy. We have already explained *pull and push strategies* in chapter 17. It suggests that management needs different promotional and advertising strategies when channel designs or distribution strategies are different.

21.5 PLANNING AND IMPLEMENTING CHANNEL STRATEGY

Developing a channel system involves the following steps:

1. Establishing channel objectives.
2. Determining type of distribution channels to be used.
3. Identifying channel alternatives.
4. Evaluating alternatives and selecting the channels.
5. Choosing the channel members.
6. Managing channels and evaluating their performance.

Step 1: Establishing Channel Objectives

A company develops a channel system to reach its target markets and accomplish its marketing objectives. Therefore, while determining and establishing channel objectives, a marketer is guided by company's target markets i.e. customer needs and wants and by company's marketing objectives and performance. Channel members not only connect customers with the producer and its product, they are also required to provide a number of services expected and desired by customers. They are expected to provide the right type of product in required quantity, at the needed time and place at a reasonable price and with prompt pre-sale and post-sales service. These services are called as *service outputs*. These service outputs pertain to *quantity, time, space, product variety and service backup.* While selecting channel members, the services expected, by target customers, of the channels and their capability must be kept in view. In short, channel objectives or goals specify what the company expects from the channel with regard to a company's target customers, and firm's marketing support to the channels in pursuance of achieving its marketing objectives. The channel objectives will include:

(1) *Channel performance objectives* to be expressed in terms of service output levels, sales, market share and profit contribution goals, extent of coverage of target market.

(2) *Control objectives* specifying company's role in coordinating channel operations.

(3) *Financial support objectives* specify financial resources to be deployed to establish and support channel operations.

(4) *Operational objectives* specifying the extent of support to be given to channel members in their operations and product servicing. It would involve understanding channel members'/distributors'

needs to ensure increase in customer value and service the target customers better.

Step 2:   Determining Type of Distribution Channels

There are two main types of marketing channels or systems *viz.* conventional marketing channels and vertical marketing systems.

(i) **Conventional Marketing Channels (CMC):**   Conventional marketing channels comprise independent and almost autonomous business entities, each negotiating the terms of marketing or sale agreement through hard bargaining without being dominated or controlled by any other channel member.

In conventional marketing system, channel members *viz.* manufacturers, agents, brokers, wholesalers, retailers, are loosely aligned by mutual agreements, each having its own decision making power and freedom of entry or exit. Each wants to maximise its own profits. No channel member has dominant controlling or coordinating authority or power over other members in the channel. Conventional channels are highly fragmented networks lacking integrated programme approach to channel management. In other words, there is lack of channel coordination. For example, independent and autonomous supermarkets, shoe stores, cloth merchants, retail shops mostly use conventional channels. Small producers especially manufacturers use agents and brokers to reach end users.

(ii) **Vertical Marketing System (VMS):**   Vertical marketing system is a channel system in which a channel member has a dominant or controlling authority over the functioning of other channel members. Vertical market system thus consists of manufactures/producers, wholesalers and retailers in which one channel member, may that be manufacturer, wholesaler or retailer, has the authority to coordinate and integrate the marketing activities of all other channel members or organisations in the channel. The vertical marketing system operates as a cohesive, unified, integrated and centralised network of organisations, professionally managed by one channel member. The dominance in VMS may be secured through ownership of channel members by a single member or through contractual arrangements or through power and influence of the dominant firm. Thus there are three major types of VMSs *viz.* Corporate or Ownership VMS, Contractual VMS, and Administered VMS. A VMS is managed as a coordinated group of channel members. The main characteristic of vertical marketing system (VMS) is that one channel member dominates and coordinates the marketing functions of other channel members to achieve operating economies and maximum marketing impact eliminating conflict and duplicated services of conventional marketing system (CMS).

VMS has proved very appropriate in case of consumer goods. Examples are Bata Shoe Company, DCM, Coca Cola, Pepsi Cola, Godrej, Maruti Udyog, Car manufacturers.

(a) *Corporate or Ownership VMS:* This is a channel system, in which a single channel member, may that be manufacturer, wholesaler, or retailer, owns the other channel organisations. Sometimes there are certain franchisers owning a portion of retail outlets.

(b) *Administered VMS:* In this system, a dominant organisation achieves coordination through its power and influence, integrating production as well as distribution of goods and services. This system is pursued when a manufacturer is a brand leader. Amul (Dairy Products), Cadbury (Choclate), Ashok Leyland (Buses, Heavy Vehicles) are some examples.

(c) *Contractual or Franchising VMS:* In this system, independent organisations in the channel enter into a contract or franchise arrangement involving a written agreement specifying responsibility of participating channel members to achieve operating economies and maximum marketing impact. Usually they consist of wholesaler sponsored chains, retailer cooperative organisations like Khadi Commission (Khadi Fabrics), manufacturer sponsored whoesalers (stockists), franchise system such as Pepsi Cola, Coca Cola, manufacturer-sponsored retailer franchise system, e.g., Maruti Udyog (Maruti Car), Bajaj Scooters, Hero Honda (etc).

There are a number of other variations of marketing systems like horizontal marketing system, multichannel marketing systems etc.

It is after determining the type of the channel system to be utilised that company can take the task of identifying the specific channels or channel alternatives that it should consider.

Step 3: Identifying Channel Alternatives

We have already referred to levels of channels in a marketing system which shows that within each type, conventional or vertical, there are a number of alternatives which can be considered for building a channel system. The present step involves identification of the same.

Typically, there are four criteria that can be used for identifying channel alternatives. They are briefly described here.

(1) Intensity of distribution (Number of middlemen): The first criterion to identify specific channels to be considered (i.e. channel alternatives) is the desired intensity of distribution. *Intensity of distribution refers to the extent to which a particular manufacturer's brand is being distributed through middlemen at a specific channel level in a specific geographic area.* The distribution is said to be saturated when all middlemen marketing a particular product type in that specified area sell the manufacturer's brand. Thus the desired intensity of distribution would mean deciding on the number of middlemen required at a specific channel level (zero 1, 2, 3,) to distribute the manufacturer's brand in the specified geographic area. The alternatives in this context are exclusive distribution, selective distribution and intensive distribution.

(a) *Exclusive distribution:*   In this arrangement, only one firm, wholesaler or retailer, markets or distributes manufacturer's brand in the trading area. The wholesaler, distributor or retailer does not sell competing brands or lines. This arrangement limits the number of dealers handling a manufacturer's brands. For example, cars, trucks, scooters etc.

(b) *Selective distribution:*   It is an arrangement in which more than one outlet or chain are selected to market the manufacturer's product(s). The decision is to be on few selected who would be willing and capable to market the company's products and give it a wide coverage. This arrangement, while covering adequate marketing area, allows better control and is relatively more cost-effective.

(c) *Intensive distribution:*   In this arrangement, a producer requires a large number of intermediaries to market its products. This arrangement is suitable for distributing grocery items, convenience goods and common raw materials. For example, Hindustran Unilever distributes its products through 3300 redistribution stockists and 600000 retail out-lets throughout the length and breadth of the country. Soft drinks, cigarettes, soaps, detergents, chewing gums, toffies and candy are intensively distributed.

(2) Access to end users (types of intermediaries):   The second criterion to identify channel alternatives is their access to end user i.e. to discover those intermediaries that might reach the target market and be accessible to the customers. While identifying the channel alternatives, their financial position, geographical coverage and sales potential should also be considered.

(3) Existing distribution practices: The third criterion to identify channel alternatives is to study the competitors' channel systems, their performance and other distribution's practices prevailing in the market for marketing similar products or services. This will help in identifying the types of intermediaries and functions and responsibilities of various middlemen.

(4) Activities and functions to be performed by channel members: In order to identify channel alternatives, the manufacturers or producers need to determine activities, functions and responsibilities to be performed by the intermediaries in the channel. They also need to assess the channel flows (referred to in section 21.3) required to serve the target market.

They have also to determine distributors' territorial rights, policies, procedures and conditions regarding pricing, sale and payment, mutual services and responsibilities that would rule the trade-relations and conduct of intermediaries.

On the basis of above mentioned criteria, the producer will identify a number of channel alternatives as described in terms of types of middlemen, number of middlemen and functions, responsibilities, terms and conditions of each channel member.

A small firm producing washing soap may have the following channel alternatives.

(1) It may hire five salesmen on a base salary plus commission on regular basis to contact retailers and sell it to them.

(2) It may appoint two wholesale stockists and ten retailers each through them.

(3) It may use a sales agency that has contacts with retailers through its salesforce of 12.

Step 4:   Evaluating Alternatives and Selecting Channels

A.   Evaluating Alternatives

This step involves evaluating each channel alternative and selecting that alternative which the manufacturer feels is the most appropriate one to distribute its products. The criteria to evaluate and assess each channel alternative can be — expected economic performance (sales-cost analysis), time needed for developing a channel, control considerations, legal regulations and restrictions, channel availability.

(1) Economic Criteria (Sales–Cost Analysis):   The most important criterion to evaluate channel alternatives is to see the expected economic performance of each channel alternative. This would involve sales-cost analysis of each alternative and thus determining expected revenues and likely channel costs. Evaluation of each alternative implies estimating expected sales and estimating channel costs at different sales levels as percentage of revenue (sales) for each alternative. With the help of sales-cost analysis (combining sales with channel costs), we can compare the profit margins, and expected economic performance of each alternative. There is every possibility that the selling costs for two channels for the same sales volume may be the same. Prior to reaching this volume, one channel alternative may be more advantageous to use. For example, for smaller firms, employing salesforce may not be economically desirable or feasible.

(2) Time Needed for Developing a Channel: The time that a channel typically takes to develop and mature is also a factor in evaluating channel alternatives. Marketers will favour those channel alternatives that take less time to develop and mature.

(3) Control Criteria: Another criterion to evaluate the desirability of a channel alternative is the extent of control that the company's management can exercise over the alternative channel systems. If the channel members are maximising their profits without any consideration to sale of manufacturer's goods, the alternative may not be a good choice. Again such a company may pose control problem. It is assumed that a greater control in terms of pricing, market coverage and concentration of effort on promotion, sales and service of manufacturer's products will achieve better results for the manufacturer.

(4) Legal Regulations and Restrictions: Legislation measures and legal restrictions including local restrictions imposed on sale and movement of goods also affect the decisions in channel selection.

(5) Availability of Channels: Another criterion to evaluate rather make selection decision, is to know whether the desirable channel, distributors and/or retailers would be available to the company for marketing its products/ brands. Some distributors or retailers may not be willing to take on the work. This would need an in-depth study of this aspect of the channel alternative to be selected.

(b)   Selecting the Channel Alternative

After evaluating all the channel alternatives by applying the above criteria, namely sales-cost analysis, time needed for its development, control considerations, etc., the company can select the most appropriate channel level that would help in achieving its marketing objectives. However, the most important factor to be considered in this context is expected economic performance. But it is open to the management to select more than one channels if it helps in achieving marketing objectives.

Step 5.   Choosing the Channel Members

After having decided on the most appropriate channel alternative or system needed i.e. on the type of channel, number of intermediaries and functions and responsibilities to be performed by them, the next step is

to choose the channel participants or intermediaries who will actually undertake distribution work.

Selecting channel members is, in fact, a two way process. It is not only that a producer or manufacturer selects channel members, the intermediaries have also to decide whether or not a manufacturer or producer suits them. The company needs to match the expected performance of the selected channel with its marketing requirements. On the other hand, channel members (intermediaries) have also to see that their expectations from the producer measure upto their needs and such a relationship will be in mutual interest. Companies use the medium of newspaper and invite organisations/channel members to send their particulars for dealership.

While selecting the channel participants i.e. intermediaries, the producer needs to be sure about intermediary's capability to reach the target markets and ability to deliver sales results . The producer needs to know the financial position, capability, image, past experience in sales and services, managing ability and the like. They need to have good rapport with trade and customers.

On the other hand, the intermediary would also consider standing and image of the producer-company and its products/ brands, support or help to be provided by it, expected profit margins, suitability of the product with their line and how the addition of the principal's products promote and help its growth when entering into distribution relationship with the producer.

Selecting channel members is a difficult task and requires careful handling. Possibly, an intermediary of company's choice may not be willing or ready to market the company's products due to earlier commitments or any other reason. Good middlemen are difficult to find. Therefore building an effective channel demands some sacrifice or give and take on the part of both parties. There should be a feeling of joint partnership and joint responsibility. Anyhow, the intermediary's capability and commitment to deliver goods is the first requirement of this relationship. In case of selection of two channels or so the producer must see that there is no conflict or clash of interests among channel members and there exists a smooth relationship among them.

Step 6:   Managing Channels and Evaluating their Performance

Managing Channels

Managing a channel is a difficult and complex task as it involves a number of organisations functioning independently, structurally different, each with its own objectives and interpersed at different and distant places with no direct control by the manufacturer or producer over them. Effective functioning of channel system mainly depends upon an abiding, enduring and smooth relationship devoid of any conflicts, and demands active support of the producer in matters of pricing, promotion and servicing of the product(s). The employees of channel organisations may also need training in the technical aspects of products and their use. In some cases, producers support them financially as well.

In order to achieve channel objectives, the producer has to ensure that channel members are committed to promote producer's products and show sales results. For this, channel members need to be continuously motivated so that they contribute their best towards accomplishment of producer's marketing objectives.

A. Motivating Channel Participants

Besides suitable terms of agreement that may inspire the channel participants to show good results, channel members also need other support in the form of training, supervision and encouragement. A few measures that can help in motivating channel members to do their best are listed below.

(1) Providing incentives e.g. special rebates, rewards for attaining results exceeding the sales quota.

(2) Regular contact with channel members: Meeting and conferences facilitate two-way information flow — from producer to channel members communicating information about new products, change in policies and other matters and from channel members to the producer concerning their problems, marketing conditions and the like. Prompt service or response to complaints and grievances directly or through channel members is also a great motivator.

It must be kept in mind that channel members are not only the agents or representatives of the firm to sell its products, they are in real sense *the customers of the firms' products* and should be treated as such.

In order to achieve channel objectives, the producer needs to monitor and manage channel properly and effectively. Monitoring and managing a channel will imply:

(1) Understanding channel's needs.

(2) Monitoring inflow and outflow of stocks.

(3) Replenishing stocks.

(4) Ensuring channel's profitability.

(5) Providing incentives and concessions, whenever feasible.

(6) Getting wholesalers/distributors to service retailers.

(7) Monitoring channel's performance.

(8) Servicing select retailers directly.

(9) Keeping communication flows unclogged.

B. Evaluating and Modifying Channels

The achievement of marketing objectives depends, to a large extent, on the performance of channel. Therefore the manufacturer should continuously evaluate and assess the channels' performance against set channel objectives *viz.* services output levels, sales-volume, market share profit contribution, delivery time, market coverage, inventory levels, accessibility, customer service and satisfaction. It would involve gathering and analysing information on sales, costs and marketing research.

In case the performance falls short of the expectations as envisaged, the producer should identify the performance gaps, diagnose the reasons for the same and take up necessary corrective action. Corrective action may comprise providing incentives and concessions to channel members, intensifying promotional efforts or conducting training programmes for dealers etc.

Moreover channel members also need to respond to changing customer needs, varying marketing conditions and new competitive challenges. Marketing managers on their part should continuously monitor

channels' marketing activities, their effectiveness and performance and should provide maximum support to this end.

In case the marketing conditions and firm's requirements call for revision or modification of the prevailing channel system, the producers can go in for a change. This may involve addition or dropping of individual intermediaries or of a type or the entire channel may be replaced by a new arrangement, if the circumstances warrant so.

EXERCISES

1. What is meant by "marketing channel? What are their important functions?
2. How do you classify intermediaries on the basis of ownership of goods?
3. Explain a three level channel of distribution.
4. Describe the factors which influence the decision to use direct distribution or intermediaries.
5. What steps are involved in developing a channel system?
6. "The decision of the type of channel system to be used is critical to corporate success." Comment.
7. Explain various channel flows which contribute to an efficient performance of the channel system.
8. Write notes on:
    (a) Conventional marketing channels
    (b) Vertical marketing system
    (c) Selecting channel members
9. What are channels of distribution and how are they determined?

Managing Physical Distribution Systems

22.1 PHYSICAL DISTRIBUTION: DEFINITION, NATURE AND SCOPE

The ultimate purpose of all marketing effort is to provide products or services to the end users or customers according to their needs and requirements and at the time and place they need them. Distribution is an essential element of marketing effort (marketing mix) and besides calling for the creation of an efficient and effective channel system, involves movement, delivery and customer services of products from the point of production to the point of end use in compliance with the purchase orders. We have referred to the physical flows that move through the channel system in chapter 21. Physical distribution also necessitates building a distribution infrastructure, namely transport organisations, warehouses and stores to facilitate the process. Physical distribution can be defined as below:

Physical distribution refers to a marketing activity concerning handling and movement of materials and finished goods efficiently from the producers to the consumers. It involves planning, implementing and controlling the physical flows of materials and final goods from place of production to the place of end use to satisfy buyers' needs.

Physical Distribution Management

In our present context, physical distribution management refers to management of activities concerning handling and movement of materials, parts, and furnished products from the end of production line to the consumers. These activities include sales forecasting, freight, transportation, warehousing, material handling, protective packaging, inventory control, processing and customers service.

According to Donald C. Bowersox, physical distribution management is "the process of strategically managing the movement and storage of materials, parts and furnished inventory from suppliers, between enterprises facilities, and to customers." *In short, physical distribution management is the management of all activities which facilitate movement and coordination of supply and demand in the creation of time and place utility in goods.*

The above study suggests that there are four comp ments of physical distribution system, namely order processing, transportation, inventory management and control, and warehousing, the decisions on which need to be intergrated in order to create an effective PDS.

Importance of Physical Distribution

The success of marketing effort in general and efficiency and success of channels in particular depend upon the efficiency of the physical distribution system. Its role is crucial, in the sense, that it is this activity which makes the product reach the consumer to satisfy him. It helps the marketing effort in the following ways.

1. It makes the products available to the customer at the place and time corresponding to his needs and thus creates place utility and time utility of the product.

2. It helps in building up a market by providing an efficient and satisfying customer service and earns for the company and its products a good image.

3. An efficient management of physical distribution can help in reducing the costs and in this sense, it helps in saving costs.

22.2 PHYSICAL DISTRIBUTION DECISIONS

We have already referred to various components of physical distribution system, namely order processing, transportation, inventory and warehousing. The designing of an integrated physical distribution system thus necessitates making proper and sound decisions on issues concerning the above mentioned elements of PDS. Decisions pertaining to each component have great impact on costs and on contribution of other components to the system as a whole. We will now examine the decisions that need to be made on each of these components.

A. Order Processing

In business today, goods are supplied by manufacturers, stockists/ distributors or mail order departments on orders received from wholesaler-stockists, retailers or consumers, through mail, fax or phone, on line or as a result of personal visitation of salespersons. The efficiency of marketing effort depends on how quickly and promptly these orders are processed and complied with at the supplier's level.

Moreover, various warehouses or godowns stocking the inventories may be distantly located from the order department or office that processes the received purchase orders. This would pose problems of transportation to be tackled. Again, there may be goods that are not available in stock and need ordering back for production of new stocks. Also the goods delivered out of warehouse stocks must be replenished from time to time so that no break down or unnecessary obstructions occur in smooth functioning of the channels, and delay is avoided.

Specifically, order processing comprises verifying customer's credentials and checking up his credit status, knowing stock position in various warehouses and godowns, preparing invoices in multiple copies and sending the advice to various departments, issuing order to transport the required quantity to a given address, preparing bills for customers, updating stocks and sending a production order for new stocks and sending the advice to the customer or the salesperson on despatch of the consignment. Order processing thus forms what is called *order-shipping-billing cycle.* It will be in the interest of all concerned parties and agencies that this cycle is completed as quickly as possible. Now, all large companies in the country have installed computers to expedite order processing. Computers are used to verify customer's credit standing, check up inventory position and prepare various bills, despatch advices and documents simultaneously and thus to eliminate any delay whatsoever.

It may be borne in mind that timely handling of orders, despatching of documents and advices quickly and correct flow of information through the distribution networks reduce the costs of other distribution activities like transportation, inventory control and wareshousing. Accurate and quick processing of order helps in transportation scheduling and improving inventory management and delivery system besides helping the organisation to achieve its marketing goals.

B. Transportation

All movement of materials and goods outside the undertaking is called transportation. It provides a systematic movement of products from one place to another and is an essential function of marketing. Specifically, it refers to physical movement of goods from the points of production/ supply to the points of use, resale or consumption.

A good transportation system is reflected in its efficient customer service performance and cost effectiveness or in other words, on-time delivery of goods in proper condition and quantity at feasible cost. Therefore, some major operating characteristics of an efficient transportation system are cost, speed, availability, dependability, capability and frequency. These characteristics serve as criteria to make decisions on transportation system as well as to asses carrier performance.

Transportation Decisions

Transportation-related decisions would include designing and developing a cost-effective transportation, selecting transportation modes, and decision on hiring or owning transportation services. Transportation management also involves evaluation of transportation performance.

(1) Developing Cost Effective Transportation

It would mean devising a transportation system that may provide efficient customer service at least cost. It should be kept in mind that a transportation system with low cost that compromises with quality of service or causes delay in delivery of goods or that may adversely affect the condition of the goods during transit would not be suitable or desirable. The system should be such as may provide cost-effective service as well as excellent customer service.

(2) Choosing Transportation Modes

There are five basic transportation modes to ship goods to warehouses, stockists, dealers and customers and a company has to select the mode or modes that may offer quality customer service at low cost. These modes are rail, road (truck), air, water and pipeline. We have already referred to various criteria that can be used to asses the suitability or otherwise of a particular transportation mode for shipping purposes of the company. These criteria are–cost, speed, availability, dependability, capability and frequency. We shall examine here the relative strengths and weaknesses of these modes.

(a) Rail: Railways are the country's largest transportation carriers. This mode of shipping is easily available and capable enough of shipping bulk quantities over long distances. It is rather the most cost-effective mode but is relatively slow in speed. Its dependability is moderate and frequency very low. A number of improvements have been made in rail roads in recent years to increase speed and customer oriented services. There are graded rate costs, the wagon load being charged at the lowest rate.

(b) Road (Highway Trucks): This is increasingly becoming a more popular and widely used mode of transportation of goods. It is the easiest available mode ensuring speedily movement of goods with high degree of capability and dependability at moderate costs. It has high frequency. This mode is suitable for shipping goods over small and middle distances. In India, role of bullock and camel carts and tractor— trollies in transportation cannot be underestimated. Now-a-days, tempos are used to carry and deliver goods and products (like domestic gas cylinders) to retailers as a mode of local or short-distance distribution.

(c) Water: Water transport is the cheapest mode for shipping bulk quantities of goods by ships and barges on coastal and inland waterways but this mode has extremly low speed, low dependability, low availability and lowest frequency.

(d) Pipeline: This mode is used only for petroleum and gas. It is a cheap mode, least available and capable but is highly dependable.

(e) Air: It is the most expensive and speediest mode of shipping goods with relatively moderate availability and frequency but least dependable.

After asessing the strengths and weaknesses of these modes, the company has to decide and identify the most appropriate transportation mode or combination of two modes for efficient shipping of goods and hence to select the transportation supplier or carriers that may provide the needed service at lowest cost.

(3) Hiring or Owning Transportation Services

An important decision to be made in this context is whether the company should provide its own shipping services or hire these services from outside suppliers or transporters. This involves cost-benefit analysis in both cases. However, a company that owns a fleet of trucks has more flexibility and control over shipping services. Now-a-days, even container services are available on hire.

(4) Evaluating Transportation Mode Performance

As stated above, monitoring and controlling mode performance would involve its evaluation in terms of speed, size of order, time of delivery, transit time taken, damage, cost etc. In case the performance is not as desired, necessary corrective action will have to be taken to set the things right.

C. Inventory Management and Control

Inventories are goods held in stock by manufacturer or channel members for future use or sales. Inventories comprise raw materials, partly manufactured goods, general stores and supplies, machinery, spare parts, components, purchased or manufactured for stock, and finished goods. In simple words inventory is a stock of goods procured to meet future demands.

Manufacturers and channel members must maintain and carry enough stock of goods to meet customers' orders whenever they are received and hence to achieve desired level of customers' satisfaction. Maintenance of inventory or stock of goods requires substantial funding and warehousing facility which put added financial burden on the company or channel members. The levels of inventories need to be managed and controlled in such a way that resources are not unnecessarily blocked on these thus incurring huge cost on the concerned organisation. They need to be planned very carefully. Decisions on inventories concern when to order and how much to order so that carrying that much inventory incurs the least possible cost.

Inventory Control

Inventory control is a planned method of determining what to indent, when to indent, how much to indent and how much to stock so that purchasing and storing costs are lowest possible without affecting production and sales. It is a method to maintain inventories at the optimum level keeping in view the operational requirements and financial resources of the business. The basic assumption is free availability of materials and goods as and when required in any quantity.

The manufacturer and each channel member, wholesaler or retailer, have to carry a stock, *a normal amount of inventory* with them, say a two weeks' stock, a month's stock or two months' stock, depending upon their resources, to satisfy their customers. Ordering for goods takes place accordingly. The total sales volume in a year is many a times of sale of the normal inventory amount. If it is ten times, then we say that inventory turnover is 10.

Inventory Turnover means the number of times the normal volume of inventory carried by a firm is sold. Obviously, it is influenced by the speed and cost of transportation, location of warehousing facilities, efficiency of communication and handling and storage requirements. This suggests that change in inventory turnover and hence in inventory carrying costs will affect costs of other activities like transportation, warehousing and order processing etc.

As stated earlier, decisions on inventories concern how much inventory to be carried by the company, when to order and how much to order to carry the inventory at least cost and yet maintain a high level of customer satisfaction. In order to reduce inventory costs, the best way for the firm will be to reduce its carrying costs and buying goods in economical lots. It must be kept in mind that reduction in carrying costs as a result of lower level of inventory may lead to increase in transportation costs due to frequent movement of goods in smaller lots. There are a number of decision models that can be used such as fixed order time at regular intervals or economic order quantity model.

Economic Order Quantity (EOQ) Model

This model has been suggested by Douglas M. Lambert and James R. Stock. *Economic Order Quantity is the quantity to buy or order for at one time that will achieve the lowest unit cost. It refers to size of order of an item at one time for which inventory costs (carrying costs, ordering costs and usage) are minimum.* It is also called *least cost quantity.*

Determination of this quantity involves determination of acquisition or ordering cost, carrying cost and quantity discount available.

(i) Acquisition or Ordering Cost: This is the cost incurred in placing an order. This is the cost incurred on indenting, calling for quotations, making a comparative statement, getting sanction of the proper authority, issuing and posting the order, receiving, inspection and storage, verifying bills and invoices, passing and making payment. All these costs make up the ordering cost, also known as *setup cost.* This cost decreases with the inventory.

(ii) Carrying or Possession Cost: Carrying cost includes interest charges on the cost of inventory, storage and handling costs, insurance, taxes, physical deterioration and obsolescence. Carrying cost increases with the inventory.

The above two type of costs, *viz.* the ordering cost and the carrying cost are opposing costs.

The EOQ or the least cost quantity is the quantity at which acquistion or ordering cost equals carrying cost.

Ordering costs = Costs of carrying the item

The EOQ is thus the order quantity for which total cost of inventory ordering and carrying is minimum.

The formula for calculating EOQ is

$$EOQ = \sqrt{2PD/CV}$$

where   P = ordering cost (rupees per order),
                 D = annual demand in units,
                 C = annnual inventory carrying cost as a pecrentage of product cost and
                 V = average cost of one unit of inventory.

This estimation is based on the assumption that demand is known.

In case of uncertain demand, other models are used. For example, fixed orders (orders for fixed quantity) may be placed when inventory reaches a predetermined minimum level or replacement order may be placed at fixed intervals (fixed order interval). Recently, there have been new approaches to inventory control. Just in time is one such approach.

Just-in-time (JIT) Management: This is an approach to inventory management designed to minimise the inventory cost in a way that not only cuts the inventory cost but also boosts product quality. It means providing inventory (goods) exactly when needed and no sooner. In this case there is no inventory that stands waiting in storage. The effectiveness of this practice depends upon dependability of supplies.

D. The Warehousing Function

All firms, manufacturers, stockists, retailers engaged in providing goods have to store goods and maintain stocks to satisfy needs of customers when they need them.

Storage and warehousing is the process of holding and preserving goods between the time of their purchase/production and the time of their sale/resale. *Warehousing is a marketing function which involves retaining the goods for a sufficient period of time so that commodities (products) are available at proper time when they are needed without any deterioration in quality.* It becomes necessary at every stage of marketing process.

Building warehouses involves huge capital investment in construction of buildings and equipping them with other fixtures including machinery and cooling facilities. A company may build and own its private warehouses or rent space in warehouses built by outside organisations for the purpose. Many companies providing warehousing facilities have installed automated systems in their warehouses and cold storages. The warehousing decisions concern the size, number and location of these facilities. These decisions affect the costs of other distribution facilities also.

Another important aspect is the need for efficient handling and storage of inventories. Proper handling and storage together with right type of packaging and advanced methods of movement can result in cost reductions and cost savings.

A recent development in this field has been the movement of large assortments of products from the product line directly to the retail outlets without using warehousing facilities. However, it has not yet become popular practice.

An organisation needs to coordinate all the decisions concerning various components of PDS to design an intergrated distribution system.

22.3   DEVELOPING THE PDS PROCESS

Management of physical distribution system involves many functions like inventory management and control, transportation, warehousing and order processing. The PDS process consists of the following steps.

(1) Determining PDS objectives
(2) Measuring customer service level
(3) Determining PDS components combination and cost trade offs.
(4) Monitoring and controlling physical distribution.

Step 1.   Determining PDS Objectives

We have already referred to the main purpose of developing PDS *viz.* getting the right goods reach the customer at the time and place needed and at the least cost. However, this objective does not throw any light on the level of the customer service to be perormed. Therefore, the objectives need to be expressed in terms of level of customer service performance.

Customer Service Level refers to the extent the customer service function is being performed. A 100 per cent customer service level means that all customers are fully satisfied with the product availability or in other words that the products are made available at the right place and time in adequate quantities to satisfy needs of all customers who were willing and able to purchase them. As it is not possible because it will incur exorbitant costs, therefore what is required is to determine a level of customer service less than 100 per cent. We have already referred to the factors that affect customer service namely time, dependability, communication, availability, convenience. The objectives thus may be determined as 90 per cent customer service level or customer satisfaction, 95 per cent customer satisfaction and so on. Thus one PDS objective can be - *to make goods available to at least 95 per cent customers within three days, or 24 hours* (95 percent customer level or satisfaction level). There will be a set of such objectives.

Again, level of customer service has relationship with sales and the cost of physical distribution. A minimum cost PDS is a system that provides a specified level of customer service at the lowest cost. It has been observed that higher the level of customer service, higher the sales and higher the distribution costs. Thus the major task for market managers is to determine the minimum level of satisfaction (customer service) required to meet the marketing objectives. In other words, objectives will be expressed in terms of guaranteed minimum level of customer service keeping the costs of PDS as low as possible. In determining these levels, the marketer has to keep the following in view-

- the requirements/desires of customers and channel members
- customer service level of competitors (competitors' practices)
- specific requirements of some specific organisation, if any
- extent of possible flexibility
- the opinion of salespersons

First, the marketer has to determine the range of service levels that is necessary to achieve the marketing objectives and then he has to identify that service level within the range that gives the maximum excess of sales over physical distribution costs.

Step 2. Measuring Customer Service Performance

Lambert and Stock have enlisted some important elements that can be used to measure or assess customer service performance. These are categorised as *pre-transaction elements* that determine service capability before it is provided to the customers. These determinants are — availability of inventory and target delivery dates or planned delivery time. *The transaction elements* are factors that assess customer service that is being transacted or performed by PD System. These determinants include order status, order tracing, back order status, shipment shortages, shipment delays, routing change. The *post-transaction elements* that measure customer service as actualy rendered include actual dates of delivery, returns, adjustments due to damage or loss during transit. This suggests that service level is not only to be expressed in terms of percentage, it can be assessed in terms of lost orders, delays, stock outs.

Step 3.   Determining PDS Components Combination Vis-à-vis Cost Trade-offs

We have seen that the physical distribution activities, namely order processing, transportation, inventory control, warehousing are inter-connected and interrelated. A marketer must evaluate the costs of each component of PDS (trade-off analysis) so as *to arrive at a combination of components that provides a specified customer service level at least total cost.*

In order to achieve specified service level objectives, we can design a number of PDS alternatives (combinations of PDS components) that may be acceptable to the company. As stated earlier, the marketing manager or distribution manager has to select that alternative that yields the highest profit contribution implying higher difference of estimated sales over physical distribution cost.

It is now clear that PDS costs consist of transportation costs (freight costs : T), inventory costs (I), warehousing costs (fixed and variable : W), order processing costs (O) and cost of loss due to delay and damage in delivery (S).

$$\text{PDS Costs} = T + I + W + O + S$$

The objective is to minimise the total costs of physical distribution to provide the specified level of customer service.

The company has to design and select the PDS system that provides the service level at least costs and yields best results but at the same time it should be flexible with a scope to be improved upon. This would mean improving customer service level at no extra cost or the same service level at reduced total physical distribution costs.

Large companies spend usually from 10 to 12 per cent of sales on physical distribution activities.

Step 4:   Monitoring and Controlling PDS

The main purpose of designing a PDS is to work out a unified framework that may coordinate and integrate decisions on various components of physical distribution in such a way as may provide a desired level of customer service at least total cost. After such a system has been designed and implemented, the next step is to monitor it and evaluate its performance. This would necessitate setting of some organisational set up to perform this responsibility and with the authority to modify the system, if need be. This responsibility can be entrusted:

(1) to a task force or a permanent committee consisting of managers concerned with PD activities.

or (2) to an existing functional department executive like channel manager, transportation manager, marketing manager, planning manager etc.

or (3) to a newly created department or Vice-president of physical distribution.

Controlling implies evaluation of the PDS performance. It would mean collecting correct information on actual results and comparing them with desired results against norms or standards as expressed in PDS objectives. Besides, those entrusted with this responsibility are also to find out the problems, reasons for deficiencies in performance and take necessary corrective action in order to enable it to contribute its best towards accomplishing marketing objectives.

AN IMPECCABLE FOLLOW-THROUGH

As a business house, you know it takes time to execute an order, when finished goods have to be sent to a stockist. It also means added tension and late payments. We, at SAFEXPRESS, take over the task of distribution. Delivering your high-value items safely, in exact quantity and on time. That's what we call **Distribution Redefined.**

We, at SAFEXPRESS, always select the right route to ensure that your product reaches its destination, not only on time, but even meets any critical deadline, without the interference of the usual channels.

EXERCISES

1. Define the term "physical distribution". What is its significance?
2. What are the major decisions involved in an efficient physical distribution system?
3. What steps are involved in developing the physical distribution system.
4. Describe components of physical distribution system.
5. What is meant by customer service level? Suggest possible ways of measuring customer satisfaction for a company.
6. Write notes on
   (a) Choosing transportation modes
   (b) Inventory control
   (c) Economic order quantity model
   (d) Monitoring and controlling PDS
7. How can a PDS be designed to attain customer service objectives as well as physical distribution cost objectives?

23

Controlling Marketing Performance

23.1 MARKETING CONTROL: MEANING AND NATURE

Controlling is an important element of management. Controlling function of a manager consists of assessing the work in progress, measuring it against the predetermined yardsticks or performance standard, determining the deviations if any and to take timely corrective action to set the things right so as to accomplish organisational goals. It is the work a manager performs to assess and regulate results or performance. Thus controlling is a function of management which makes it sure that objectives are being achieved, according to the plan. It sees to it that activities proceed as planned and results are achieved as predetermined in the plans, that policies and programmes are carried out in accordance with the plan. It is evident that the process of controlling has planning as its basis. Planning is a primary function of management and controlling is an essential counterpart to planning. In a nutshell, controlling is the process of monitoring, measuring and comparing operating results with the plans and taking corrective action when results deviate from plans.

Controlling a marketing programme means monitoring and evaluation of marketing activities and providing necessary feedback. It will include a systematic and comprehensive audit of marketing activities, periodical reassessment of marketing effectiveness through instruments of marketing audit, marketing profitability, and marketing efficiency studies.

Again, controlling marketing performance does not only imply monitoring and analysing performance with a view to finding deviations or problems and taking necessary action, it also involves finding out new opportunities that might emerge during implementation of marketing programme and adopting necessary measures to take advantages of those opportunities. We can thus define marketing control as follows:

> *Marketing Control is the task of monitoring and evaluating the performance of marketing programme or marketing department of an organisation, finding out performance problems or opportunities, taking corrective actions to resolve problems or measures to take advantages of the opportunities.*

We have already explained the methods to evaluate the performance of various marketing activities/programmes pertaining to different components of marketing mix in chapters 12 to 22. In this chapter, we shall examine the analysis and evaluation of the marketing programme or the organisation as a whole.

In an organisation, the work of analysing and evaluating the marketing performance may be entrusted to the marketing department or a controller or a manager with a number of such positions at lower levels to monitor financial inputs and their impacts.

In large organisations, the responsibility of controlling is entrusted to the managers of marketing activity centres. There is a line hierarchy or hierarchy of levels starting from lower level managers going upward. *Marketing activity centre is a component of marketing programme entrusted to a manager for achieving marketing (sales) objectives for that component.* Thus performance of marketing programme starting from small territories to district, state and national levels can be monitored and controlled by *marketing activity centres.*

23.2   THE CONTROLLING PROCESS

The work of a manager in controlling involves four specific steps:

(1) Setting performance standards or criteria.

(2) Measuring performance and comparing it with standards.

(3) Finding serious performance deviations (problems) or emerging opportunities and searching for causes.

(4) Taking corrective action to fill in gaps between standards (goals) and performance.

Step 1.   Setting Performance Standards/Criteria

The first step in the direction of performing the controlling function is to set standards or criteria against which to judge results or one's performance. For example, the criterion to measure performance might be sales figures, profits earned, readership of an advertisement or any other measure of accomplishments.

We have seen that while planning any marketing activity, a marketing manager is required to determine some specific objectives or goals. These goals are expressed in terms of quality, quantities, sales, costs, time schedules or desired results etc. These objectives and goals set up in planning marketing programmes become standards for evaluating performance in the control process. Some common types of objectives are:

(a) Sales-oriented objectives

(b) Profit margins

(c) Customer-attitude objectives (market support objectives)

(a) Sales-oriented Criteria of Performance: According to sales-oriented criteria, performance is measured and evaluated in terms of actual sales and market share of the company achieved as a result of implementation of marketing programme against the goals set in the planned programme. It involves sales-analysis. Market share may be expressed in terms of percentage of total sales or as a percentage of sales achieved by a group of competitors. The higher market share shows gains over competition.

(b) Profit Margins: Now-a-days, objectives or goals are not set only in terms of sales volume or market share, profit contributions attributed to marketing activities also measure their performance. The measures used to assess performance are profit margin and return on assets. This involves financial analysis

$$\text{return on assets} = \text{Profit margin} \times \text{asset turnover}$$
$$= (\text{operating income/sales}) \times (\text{sales/assets})$$
$$= \text{operating income/assets}.$$

(c) Budgeted Costs: Budgets are the monetary evaluation of resources, i.e. of time, men, material, money and other units that are required to perform marketing activities and to achieve marketing objectives. A budget is a plan which starts with setting and projecting both long-term and short-term goals and specifies the expected money costs of resources required to achieve these goals and end results. In order to maximise return on the capital employed, it is necessary to ensure that expenses are kept within the limits specified in the budgeted plan and at the same time end results are achieved as planned. Budgets can thus serve as effective tools of controlling. This will involve *marketing expense-to-sales analysis.* Budgets are the measures of cost performance. We can find the relationships of budget allocations to various marketing activities like advertising, sales promotions, personal selling, physical distribution etc. to the sales-volume to assess their performance. Efficient utilisation of resources to achieve desired results helps in reducing costs and increasing profit margins.

(d) Customer-attitude Objectives (market support objectives): In chapters on marketing communication and advertising, we have explained the objectives seeking qualitative change in attitudes, feelings and preferences of target audiences such as creating awareness, arousing interest, stirring up desire, effecting change in attitudes and purchase intentions. All these objectives support company's marketing efforts. These can serve as qualitative criteria to monitor attitudes of customers, dealers and other parties that take part in marketing operations.

The use of above criteria to measure marketing performance would involve collection and analysis of financial and accounting data and results and the concerned qualitative information. It would also involve conducting marketing surveys and marketing research. It may require what controllers call *modular accounting system.*

Modular accounting system is a system that collects and stores data on sales and costs pertaining to marketing activities, programmes or departments and is responsible for preparing reports on their performance.

Step 2. Measuring and Comparing Performance to Standards

In order to find out serious problems warranting immediate attention (or opportunities of which advantage could be taken of), the data on actual performance and information so collected are compared against the standards set to be sought. This gives an idea on how far the actual performance deviates from the preset standards or marketing programme's objectives and goals. In case the actual performance achieved by the marketing department/centre suggests that goals set in the plan were overambitious, then the management might have to modify the marketing objectives or goals.

Step 3. Finding Performance Deviations and their Causes

The comparison of actual performance to the set or designated standards shows its deviation from the plan's goals and objectives. The main task of evaluator or controller (or marketing activity centre) here is to search for the deviations and identify causes for the same. This will necessitate examining and evaluating performance in each marketing activity centre. Searching for problems involves a number of control comparisons and may take the following forms:

(i) Search for deviations of actual performance from set standards during a specified period.

(ii) Search for deviations of actual performance in a specified period from the performance achieved in past e.g. in previous year.

(iii) Comparing performances achieved in sequential time periods e.g. quarterwise or monthwise comparisons.

(iv) Search for problem may involve a *brand switching study.*

Brand-switching analysis requires measuring consumers' actual purchases of a particular brand within a product line over several time periods. This will need marketing research and involves gathering data and information on the brands bought by customers from period to period.

Identifying Causes

The above comparisons only discover performance problems or serious deviations. One important controlling function is to identify causes of perforamnce problems. These problems may arise due to inadequacy or deficiencies of marketing efforts or on account of external environment.

Usually, causes of performance problems can be identified from the existing information and evaluation reports, however, in some situations marketing research may be necessary to find out the causes.

Step 4. Taking Corrective Action

After having analysed the actual performance vis-à-vis goals and discovered the problems and their causes, the next task for the manager is to undertake corrective action to resolve those problems. It must be kept in view that it needs some *lead time* before corrective action can be introduced into the marketing efforts. Sometimes, the situation warrants immediate action as the lead time for taking action may be small.

Generally, corrective actions are directed toward boosting sales and controlling costs to improve performance and fill in the gaps. Other areas that may be needing improvement can be—profitability, efficiency and marketing effectiveness. For example, in case of efficiency control, corrective action may consist of measures aiming at more efficient management of salesforce, advertising, sales promotion, physical distribution. We have already discussed these measures in chapters dealing specifically with management of various elements of marketing mix.

23.3   THE MARKETING AUDIT

Monitoring and controlling various marketing activities, operations and programmes are continuous and regular functions of marketing managers or departments. Companies also undertake studies to gauge marketing effectiveness of its programmes and operations. The attributes that are examined in a marketing effectiveness study are company's customer philosophy, marketing organisation, availability of adequate marketing information and nature and extent of marketing research activities, form and quality of marketing planning and strategy and efficiency of marketing operations and activities. Such a study is helpful in discovering the strengths and weaknesses of the company's marketing programmes and functions.

In case a company finds that its marketing programme is not as effective as it should be or problems occur more often than occasionally, then it should undertake a more intensive and in-depth analysis of its marketing functions or what is called *marketing audit*. We can define marketing audit as follows:

Marketing audit is a systematic, in-depth analysis and assessment of marketing function (or department) i.e. of company's mission, marketing strategy, plan and organisation, market targets, objectives, programmes and activities in fields pertaining to components of its marketing mix with a view to identify problem areas and formulate a plan to resolve them and thus improve company's marketing effectiveness and performance.

In other words, a market audit includes comprehensive analysis of the commitment of management to the marketing function (mission), marketing environment, company's marketing planning approach, organisation of the marketing function and contribution of various elements of marketing mix to the company.

Marketing Audit Procedures

Obviously, it requires much effort, time, money and expertise to conduct such a study. A marketing audit can prove useful only if it is conducted objectively by persons who are well aware of company's business and marketing environment and have requisite expertise in the concerned fields of study. This responsibility may be entrusted to staff members or managers but their findings and recommendations may not stand the test of objectivity. The company can create a marketing audit department. This work can be entrusted to outside consultants or auditors who have requisite expertise and knowledge of the company's business environment. Many large companies follow combination approach entrusting the work to a committee or task force (auditing team), consisting of outside consultants or auditors and company personnel to arrive at the findings and conclusions and suggest their recommendations objectively and independently.

Designing the Audit Programme

This task involves preparing a detailed outline identifying areas to be studied, assigning responsibilities to team members, determining a time schedule, enlisting information to be gathered and sources of such information and deciding on procedures to be used in conducting the audit. The information to be collected usually pertains to sales-volume, costs, other financial statements, corporate, business and marketing plans and strategies, description of product lines and products, marketing opportunities available, managers' attitude, various marketing activities or operations etc. The information may be available from company's records, or as a result of interviews with concerned people.

Coverage of a Marketing Audit

As referred to in above discussion, a marketing audit requires analysing the entire marketing function of the company and consists of examining various management and marketing components, namely corporate mission and vision, marketing environment, corporate, business and marketing strategies, marketing plans, organisation and programme, marketing operational activities concerning different elements of marketing mix and implementation and controlling measures. The audit calls for answers to a list of auditing questions that are framed to gather information which forms basis for drawing out conclusions and making recommendations. The audit may take the form of some critical questions like the following. A sample of some critical questions is given. The number can be multiplied depending upon time, money, men available for the purpose.

A.   Area: Vision and Mission of the Company

   1. Do the "vision" and "mission" statements of the company specifically mention its philosophy in market-oriented terms?

2. Does the mission statement clearly specify the product-lines of interest to it?

3. How big is customer value in company's vision statements?

4. Have the corporate as well as marketing objectives been set by the company?

5. How far have these objectives been achieved?

6. How far have the objectives been appropriate?

7. Has the corporate strategy been able to meet its objectives?

8. Which problems (or opportunities) have come to the notice of company that require action or some change in marketing strategy? etc.

B. Area: Marketing Environment

Here information is sought on important aspects, developments and changes in demographic, economic, ecological, legal, technological, political and cultural conditions and environment. For example:

1. What have been the major developments or trends in demographic environment?

2. How would changes in income, prices, costs, credit affect the company and its strategy?

3. What changes are discernible in the attitude, tastes or life styles of the customers and how will they affect the company's progress?

4. What developments have taken place in governmental regulations that have bearing on the functioning or plans of the company?

C. Area: Marketing Strategy

1. Does the corporate strategy clearly mention the role and responsibility of marketing in its scheme of things?

2. Who has been enstrusted the responsibility and authority of overseeing the marketing function?

3. How far has the company's marketing strategy been effective in accomplishing company's objective's?

4. What are the problems being faced or likely to be faced due to changes in marketing environment and how would they affect marketing strategy? etc.

D. Area: Marketing Plan, Organisation and Programme

1. Did the company develop annual and long-term marketing plans and make them applicable?

2. Have the responsibilities been specifically and clearly assigned to various functional areas/units in the marketing organisation?

3. What are the strengths and limitations of personnel engaged in various positions in marketing organisation?

4. How far is the organisational structure, set up for marketing appropriate to implement the plan effectively?

5. Have the marketing managers been delegated adequate authority and responsibility to implement the plan effectively and to customers' satisfaction?

6. Has the target market (product-market) been clearly defined and specified?

7. Have the size, growth, potential and strategic significance of target market been considered and what are the estimates?

8. What methods have been used in segmentation of the market?

9. Has the marketing department identified the characteristics of the target audience (e.g. their life style etc.)

    (a) How far is the marketing department familiar with the target segments brand usage?

    (b) How much customer's usage pattern is known to marketing people?

    (c) How does the company track changes in consumer preferences?

    (d) How is the marketing research used by the organisation to track the customer?

10. Has the demand of the target market been estimated and what is the extent of competitive threat/challenge?

11. Does the plan set specific objectives and goals for each target market? Are the objectives realistic? Are they in consonance with corporate objectives?

12. Has the company adequate resources to implement the marketing programme effectively and efficiently?

13. What are the allocations for various components of marketing mix? Are they adequate for achieving set goals and objectives?

14. Is the marketing programme being evaluated regularly? etc.

E. Area: Marketing Activities (Components of Marketing Mix)

(i) Product

1. Has the marketing people understood and evaluated customer's needs?
2. Does the product mix offered by the company cater to customers' needs.
3. Is the product performance evaluated regularly?
4. How does the company get new ideas for new products?
5. How does the company introduce its new products in the market and how does it familiarise channel members with new product?
6. How does the company meet competitive challenges i.e. threat of competitors offering new products?
7. What is the brand image of products manufactured by the company?
8. How are brands differentiated and improved upon?
9. What is the role of packaging in positioning company's brands?
10. How far is the packaging reinforcing brand image and personality?

(ii) Pricing

1. How does the company take its pricing decisions?
2. How does the company respond to rising costs?
3. How much is the target market responsive to change in company's prices?
4. How does company's pricing strategy compare with competitors' pricing strategies?
5. Does pricing strategy need any change?

(iii) Promotion (Advertising, Salespromotion, Publicity)

1. Has the company defined the role of advertising and established advertising objectives?
2. What is the procedure for deciding the creative content of advertising message?
3. How are media budgets determined and analysed?
4. Is budget allocation to promotional mix adequate?
5. How much is the advertising programme effective in meeting the advertising objectives?

(iv) Salesforce (Personal Selling)

1. Have the role and objectives of personal selling been clearly specified in the promotional mix?
2. Is the size of salesforce adequate enough to achieve marketing objectives?
3. Do the salespersons possess proper qualifications to perform their assigned responsibilities?
4. How do salespersons measure to company's expectations?
5. What is the compensation package that a company offers to its salesforce?
6. Are compensation levels equitable and do they match competitors' levels?
7. Are training programmes and facilities for the salesforce adequate to meet their training requirements? etc.

(v) Channels Management and Distribution System

1. How strong is the company's distribution reach?
2. How does the company respond to distributors' and retailers' needs and requirements?
3. What are the post-sales service arrangements and how far are they effective?
4. How does the company build customer relationships?
5. How does the company ensure control over distribution channels and physical distribution vehicles? etc.

F. Area: Implementation and Control

1. Have the problems concerning all aspects of marketing function been identified and causes diagnosed?
2. Are the activities being implemented effectively as planned?
3. Are control procedures adequate and effective to evaluate the performance?
4. Does the marketing plan need any revision or modification?
5. Is there any need for studying further into some aspects of marketing functions before action is taken?

Findings and Implementing Recommendations

After gathering and analysing the complete information on above aspects and undertaking this stupendous exercise, the audit team formulates its findings and inferences in its report and suggests concrete measures and recommendations that the company should implement to improve its marketing operations and also to take advantages of the emerging opportunities. In fact, this has been the theme of our study throughout the book.

EXERCISES

1. What do you mean by marketing control? Explain the controlling process.

2. What do you mean by marketing audit? Draw a draft to analyse various management and marketing components of marketing function.

3. What are the similarities and dissimilarities in the planning and control of marketing strategy.

4. What are the major problems that can come in the way of evaluating marketing activities?

5. Marketing control requires evaluation of company's performance against a set standard or criteria of performance. Suggest how these standards are set?

International Marketing

As referred to in chapter 1, in modern times, the number of goods and services to be produced is very large and varied. The recent advancements in the fields of science and technology, transport and communication together with innovations and inventions and entry of multi-nationals as a result of liberalisation and globalisation of economies have widened the scope of business activities. With expansion in their scales of production by many large companies to reap the advantages of economies of scale, the need to enter the foreign markets and extend marketing activities beyond the boundaries of domestic markets has become all the more important and urgent. Obviously, the task of marketing management in this context has become more varied and more challenging. It offers both opportunities and challenges. Now the management is required to keep pace with the new developments and rapidly changing environment both at home and in foreign markets. It goes without saying that present day marketing management must have a mind set to *think globally* and the confidence in its ability to make its business globally competitive. In context of India, globalisation is becoming almost a reality. Thanks to liberalisation, privatisation, and dynamic business activity, some multinational companies have started overtaking national companies. Some national companies are merging to gain strength and some have got entry into foreign countries and are producing and marketing goods there. No country or no business enterprise can be sheltered from the winds of change and global competition and the business organisations will have to cope with these challenges and adapt and adjust accordingly.

24.1 INTERNATIONAL AND GLOBAL MARKETING: DEFINITIONS

As the name suggests, international marketing means marketing in which marketing activities are directed to more than one country or nation. We can define international marketing as follows:

International marketing (management) is the process of designing, planning and executing marketing strategies to achieve marketing objectives in the markets of other countries or nations.

Internationalisation of business operations would require reorganisation of the company into a multinational or transnational one.

Multinational company (MNC) is referred to as a company which carries on its business operations (production, sales, distribution etc.) not only in the country of its incorporation but also in the number of other countries. It is also called transnational company. In simple words, it is a business firm doing business in two or more countries. Most of the multinational companies or corporations make direct investments in a number of countries and conduct their marketing operations.

International marketing thus involves marketing in domestic country as well as marketing across political frontiers. Therefore, the marketing management functions and processes are almost the same in international marketing as in domestic marketing and so are the marketing mix variables. What differentiates international marketing from domestic marketing is the difference and diversity of marketing environments i.e. cultural, political, economic and competitive environments as they obtain in different countries and companies have to devise their marketing strategies and decide on marketing mix elements keeping in view the macro and task environments as they prevail in each concerned country and make necessary adaptations or adjustments so as to achieve their marketing objectives.

A large number of multinational companies (MNCs) throughout the world have devised and developed standardised marketing strategies including the very products they produce and market, and conduct uniform ad programmes in markets all over the world. In other words, a large number of MNCs are engaged now-a-days in what is called '*global marketing.*' The examples are Procter and Gamble (P&G), Unilever (Hindustan Unilever), Philips, Pepsi Cola, Coca Cola, Suzuki, Panasonic National and a host of others. Most of MNCs manufacturing and marketing similar products or productlines are commercial rivals and pose considerable competitive challenge. Many of them have built a global image or reputation that places them in much more advantageous position as respect to domestic or host country competitors. We can define global marketing in the following way:

Global marketing means using a standardised marketing strategy and uniform marketing programmes of product development, production, advertising and distribution etc. in worldwide markets to achieve marketing objectives.

Obviously, internationalisation of marketing operations have led to the development of global industries. Michael E. Porter has defined global industry as "an industry in which the strategic positions of competitors in major geographic or national markets are fundamentally affected by their overall global positions". Doubtlessly, they have great bearing on the fortunes of domestic firms in such industries.

Whatever may be the compulsions of global marketing, catering to different customer needs and wants in the face of diverse marketing environments across different political frontiers or nations may necessitate use of different marketing strategies and tactics to attain marketing objectives.

24.2 FRAMEWORK FOR INTERNATIONAL MARKETING

As stated above, the process of designing and developing of marketing strategy for international marketing is similar to that for domestic marketing and involves the following steps:

1. Analysing global marketing environment.
2. Determining which markets to enter.
3. Deciding entry strategies.
4. Designing the marketing programme.

Step 1. Analysing Global Marketing Environment (Macroenvironment)

Each country or nation has its unique characteristics which are reflected in its cultural, political, legal, economic and social conditions and environments, the worldwide diversity of which we have already referred to. While designing the marketing strategy for a foreign country, the macroenvironment existing therein needs to be examined and analysed carefully to understand its unique features, marketing opportunities available therein and its attractiveness as a market for the company's products.

(A) Cultural Environment

According to Louis Allen, "cultural environment refers to the traditions, laws, rules and beliefs". Each nation has its own culture, values, customs, attitudes, faiths, habits, taboos, languages, social organisations, classes and ethnic groups. Each of these elements varies from country to country and affects lifestyles and consumption patterns of the people therein and hence their needs, wants and requirements, tastes and preferences. The marketers while designing a strategy for international marketing must take into consideration these cultural elements existing in the market and plan according to the requirements of the situation. Language differences would necessitate communications taking place in the language of the customers and not marketers. A study of social structures will help in market segmentation and targeting. Products to be developed and presented must suit their life styles. The marketer must check before hand what products the people of a particular country buy, when they buy and how much do they buy. The marketing offer must suit and fit customers' culture i.e. foreign cultural environment. Thus marketing programme for international marketing is to be developed keeping in consideration what is called *"marketing relativism"*. The cultural differences pose a great challenge for the marketer and necessary adjustments must be made to cope with these differences.

(B) Political and Legal Environment

Political environment includes political atmosphere and stability, political parties and their philosophies, government administration and policies concerning business and international policies of the government, legal environment, laws, both central or state. These have direct and immediate impact on marketing and seller-buyer relationships.

The marketing managers must take these political and legal factors into consideration. Particularly, the aspects to be considered are—the political stability of the host country, their attitude toward foreign business firms and investments, importance of the company's product to the host nation, monetary regulations, currency convertibility, custom clearance procedures, price controls, efficiency of administrative system, nature of procedures concerned with imports, legal laws and restrictions pertaining to marketing mix decisions etc.

(C) Economic and Competitive Environment

As discussed in chapter 3, economic environment is filled with factors like general economic conditions, market conditions, industrial structure, competitors and nature of competition, economic system, fiscal and monetary policies, financial facilities and constraints, level of economic development. The major economic factors that need to be considered in context of international marketing are:

(i) size of the population of the foreign or host country.
(ii) income statistics—GNP and distribution of income among people, high income population, medium income population, low income population, needs and wants of people, per capita income.
(iii) Industrial structure and stage of economic development- developed economy, subsistence economy, underdeveloped economy or developing economy. This may be reflected in number of automobiles, TVs, telephones per 1000 population.
(iv) importing patterns of the foreign countries.
(v) nature of competition and major competitors, both international and indigenous, existing presently and likely to emerge in near future. The risks involved should be considered thread bare. Investments in medium car industry in India is a case in example.
(vi) Economic infrastructure available in foreign country. Economic infrastructure refers to facilities, activities and services which support operation and development of other sectors of economy. They include (a) sources of energy, electricity, fuel, gas, (b) transport services, (c) communication facilities especially telephone facilities to take advantage of marketing opportunities. Rural markets in underdeveloped and developing counties cannot be reached due to lack of infrastructure.

(D) Business Environment

Due to cultural differences and what Richard Lewis, the author of the book observes in his book, "When Cultures Collide", business norms, behaviours, traditions and procedures differ from country to country and the marketing managers must be well aware of these characteristics of the people while entering into negotiations with them and doing business abroad. For example, British businessmen are chatty and love jokes, anecdotes and telling stories, Germans are morose and do not like jokes or stories, Japanese are sober and humourless, may smile occasionally, Fins are cold, speak little while Spaniards are very touchy. Again, they may have their own reservations about dealing with some particular countries. Marketing managers need to deal with them accordingly.

(E) Other Environment Factors

Other factors to be studied are geographical differences (e.g. terrain, climate, natural resources), distribution system and level of technology within the foreign country.

The above factors, namely product, economic conditions, political and legal atmosphere, available infrastructure, geographical factors-all influence marketing attractiveness of host country.

It is after appraising, understanding and analysing the international marketing environment that the company has to decide whether or not to do business abroad and if the answer to this question is in affirmative, then they have to make decision on which markets to enter.

Step 2. Determining which Markets to Enter

Deciding which markets to enter requires matching company's capabilities and strengths with host country's needs, marketing environment, potential and opportunities. This involves two steps or phases:

(i) On one hand, it would require analysing company's characteristics, home country's constraints, host country's environment and constraints, and determining how much adaptation would be required in the company's marketing mix to enter the market in the host country.

(ii) On the other hand, the marketing managers will have to set a screening criteria in the light of its marketing objectives to select a particular market/markets to enter.

In the first phase, the marketing manager is required to match company's strengths and weaknesses, philosophy, marketing objectives, resources, organisational structure, financial position and constraints, products, home country's macroenvironment with host country's macroenvironment, namely economic conditions, political stability, geographical environment, level of technology, constraints and market potential and opportunities available in the international market.

Again, the company has also to gauge how much adaptation would be needed in the company's marketing mix i.e. product characteristics, pricing, promotional efforts and distribution to market its products in a particular country.

Screening Criteria

After undertaking the above preliminary analysis and matching the company's capabilities, strengths and weaknesses with the foreign markets' needs, the next phase is to establish screening criteria to select which particular market/markets to enter out of the studied markets in foreign countries. Obviously the screening criteria will be:

- market potential
- market attractiveness
- expected profits

- expected return on investment
- competitive advantages
- political stability
- trade restrictions.
- available infrastructure
- transportation costs

We have already referred to some of these criteria in earlier discussion. This phase would require estimation of current market potential in terms of total industry sales in each market, forecasts of future market potential, sales potential, expected market share, forecast of costs and profits and estimates of rate of return on investments and above all the risks involved are assessed and evaluated.

It must be borne in mind that international marketing i.e. doing business abroad involves high risks. It is particularly so in case of marketing in underdeveloped and developing countries. Despite of a large market potential and market attractiveness existing therein, some of the causes that hamper marketing in these counties are—political instability, high foreign indebtedness, lack of exchange stability, tariffs and trade restrictions, corruption, technological piracy, inadequate infrastructure, high cost of product adaptation, inefficient, cumbersome and time consuming bureaucratic procedures etc.

Thus before selecting a foreign market to enter, each market needs to be judged in the light of the above screening criteria. The markets and countries that do not meet the minimum acceptable standards are eliminated and dropped. The markets that meet the set criteria are selected for entering into. The next question is how to enter the market.

Step 3. Deciding Entry Strategies

After having decided the international market or markets in which the company is going to enter, the next step is to design and develop a marketing plan and devise an appropriate entry strategy. As in case of domestic marketing, planning process involves analysing marketing situation or environment in the target market, setting marketing objectives and goals, devising marketing strategy and tactics, allocating budgets and charting out a plan of action. But all this will depend on what mode or strategy the company decides to adopt to enter the market. There are a number of entry strategies. Each has its own advantages and limitations. The costs and risks involved also differ from strategy to strategy.

Broadly speaking, a company can use any of the following strategies to enter a foreign market:

1. Exporting.

2. Joint Venture Arrangements.
3. Investing directly in a wholly owned subsidiary in the host country.

1. Exporting

Exporting means producing products in the domestic country and sending and distributing them to other counties. The products that are exported are a part of domestic product. Exporting may be direct or indirect. The costs and risks involved in indirect exporting are less than those involved in direct exporting.

(a) Indirect Exporting: Indirect exporting means using the help of independent middlemen and sales intermediaries that take the responsibility of sending the products to foreign countries. There are many types of such intermediary firms or middlemen available to the companies. The greatest advantage of indirect exporting is that even the small firms can be able to distribute their products to one or more other counties through middlemen. Some major types of intermediaries are:

(i) Commission agents who are based in home country and represent foreign firms to purchase goods. They are paid commission by foreign firms.

(ii) Domestic based export merchants or export trade companies (ETCs): These firms buy goods from domestic companies to export them to foreign countries.

(iii) Buying or purchasing agents: These are companies that are authorised to purchase goods on behalf of their foreign clients.

(iv) Export agents: These are the firms that export products on behalf of manufacturers on a fee or commission. It is the manufacturer who bears the risk.

(v) Export management companies (EMCs): These are the firms or departments that manage exports on behalf of manufacturers. They also finance export sales and sometimes provide immediate payment to the producers.

(vi) Cooperative organisations: These are organisations that export products on behalf of several domestic firms or producers.

(b) Direct Exporting: In this mode, the domestic firm itself undertakes selling its products overseas and is responsible for dealing with foreign firms directly and hence for shipping the products to them. A firm may carry on direct exporting by any one of the following modes:

(i) By establishing company's own corporate export division or department that may be entrusted responsibility to seek foreign clients and export goods to them.
(ii) By appointing foreign sales representative and agents.
(iii) Through foreign based distributors and retailers.

(iv) Through foreign based state trading corporations.

(v) Through overseas sales branches: Many large companies have established sales offices or branches in one or more other countries to deal with the foreign customers directly.

2. Joint Venture Arrangements

Another strategy that a company can adopt to enter a foreign market is to enter into some sort of joint venture with firm/firms in target markets in foreign countries. The arrangement may take any one of the following forms.

(i) **Joint Ownership and Control:** In this arrangement, foreign companies join with local companies or parties to establish a local business in which foreign company or government shares ownership, management and control of the joint venture business with local principals i.e. party or government. Generally, in this arrangement, foreign investor joins with local investor to build a new manufacturing facility or buys an interest in the local company or a local business. Suzuki - Maruti, Modi Xerox, Nepal Unilever Limited, DCM Daewoo etc. are some of the joint ventures.

The greatest advantage of establishing a joint venture is that the local party is well acquainted with local markets, competitors and the political and legal environment and the foreign investor has to invest less. Again, local party will also have to spend less resources and it can have the advantages of the latest technology and managerial expertise of foreign company. The disadvantage of joint venture is that there may be a clash of interests due to differences in their management philosophies, goals, global marketing policies and aspirations. This may lead to management conflicts.

(ii) **Licensing/Franchising:** In this arrangement, a company enters the foreign market by allowing a manufacturer in the host country a license or the right to use its manufacturing processes, patent or trade-marks, and provides technical know-how and guidance to produce and market the product in that market under agreed conditions. The licencee pays a fee or royalty to the licensor. A large-number of pharmaceutical companies in India manufacture medicines under license from foreign companies. A few examples are Shriram Honda Gensets, Reynolds pens, Godrej Pacific Fax machines, BPL Sanyo etc.

Franchising is a kind of licensing through which a company enters foreign markets. For example, Coca Cola has given franchise to bottlers all over the world to produce its brand. Some companies like Coca Cola provide some ingredients like powder and syrup to franchising bottlers so as to maintain dependency of the foreign party on the franchiser.

Licensing or franchising is a strategy of entering a market rapidly with low investment or cost. The limitations of this arrangement are—lack of control over the licensee and apprehension of competition from the licensee after the contract ends.

(iii) Management Contracting: This arrangement involves participation and management control of an organisation in s foreign market without making any capital investment. This means providing services rather than products. The management contractors receive some share of profits or fees for rendering management services. For example, a company may get a contract to manage and run a hotel, airport, hospital, telephone services for a fee or share. For example, a number of foreign companies (MNCs) have entered into contract with Indian Firms to manage and undertake marketing of Indian products. This is low-risk strategy to enter foreign market. An example is Kwality Wall's Ice Cream.

(iv) Contract Manufacturing: Under this arrangement, a foreign firm enters into contract with manufacturing firms in the host country for production of its products according to its specifications but itself manages and controls marketing in the host market. This arrangement allows the firm an easy entry with low risk into the foreign market. In this approach, profit on manufacturing cannot be reaped by the foreign company. Coca Cola, Pepsi Cola, Procter and Gamble, Reckit and Colman, Kellogg, are some of the companies that have adopted this approach.

3. Investing Directly in a Wholly Owned Subsidiary

The third strategy to enter foreign market for a company is to invest directly in the host country and create a manufacturing or production facilities therein. It is done by floating a wholly owned subsidiary company in the foreign country that carries on the manufacturing and marketing functions. The subsidiary is thus located and operates within the host country over a long time. It almost becomes domesticated. Hindustan Unilever, Bata Shoe Company, Philips are a few examples.

Advantages

1. The parent company has complete control over production and marketing operations in the foreign market.
2. It ensures flexibility in operations.
3. In case of underdeveloped and developing countries, the parent company can take advantages of cheaper labour and raw materials and effect other production economies.

4. This strategy ensures protection of company's trade mark, patent and copyright in the host country.

5. Host counties offer a number of incentives to foreign investors like tax exemptions or tax free production and profits for a specific period, freight savings and other concessions.

6. Direct investment leads to creation of more jobs and hence employment in the host country.

7. It enables the parent company to adapt its products better to the local market environment and requirements.

8. It helps the company in safeguarding its long-term interests in international markets and in achieving long-term objectives.

9. Such creation of manufacturing facilities helps in the development of the economy of the host country.

10. For host country, it is the cheapest way to take advantage of latest technology.

Disadvantages

1. It entails high financial risks due to fluctuating rates of currencies like frequent devaluations or deteriorating marketing conditions.

2. It is vulnerable to local political pressures and environment.

3. It is difficult and expensive to close down the operations and exit.

4. It deprives the company of the entrepreneurial and marketing experience of local partnerships as the company has to go alone.

The examples that can be cited are—Palmolive-Colgate, Oberoi Hotels, Tektronics (Laser class colour pritners), Carrier (airconditioners), Nestle, Cadbury, Whirlpool, Pond's, Hongkong Bank, Standard Chartered Bank, Marico Limited.

Step 4.   Designing the Marketing Programme

As stated earlier, developing a marketing plan for a foreign market would require an in-depth analysis of situation or marketing environment in the host country, determination of marketing objectives and goals, formulating a marketing strategy to achieve those objectives, allocating necessary budgets, and charting out the action programme. The most important decision to be made in this context is to design and develop the marketing mix that may suit the target markets and achieve the set objectives and goals. *In other words, the company has to decide how much adaptation is required in the marketing mix to effectively achieve the goals.*

Global Standardised Marketing Mix

A number of companies use what is called *'global standardised marketing mix' across countries*. It involves standardisation of various elements of marketing mix i.e. of the product, packaging, promotion, advertising, distribution channels which require no further change or adaptation. Global standardisation of the marketing mix has many advantages.

- It ensures lower production costs.
- It takes less time to manage marketing operations.
- Monitoring and controlling of international marketing becomes easier and more effective.

Adaptation

However many companies require changes in the elements of marketing mix i.e. in the existing product, pricing, advertising and distribution variables to fit the local conditions and hence fit the target market. The marketing managers have to decide on the extent of adaptation that should be made in the marketing mix variables to cater to the needs of the target market. We shall briefly examine what changes can be introduced in various elements of the marketing mix to meet requirements of the foreign markets.

A. Product Decisions

An effective marketing programme calls for matching the product to the foreign market and its environment. We have already referred to the religious, cultural, climatic, lingustic and geographical diversities from country to country. People in different countries have different tastes, preferences and requirements in respect of quality, colour, size, packaging of the products that are offered. There are three strategies that are generally used to introduce the product in the foreign market.

1. Straight Product Extension: In this strategy, product is introduced into the foreign market without effecting any change in the product. The product is presented to the customers in the foreign market as it is manufactured. Large multinational companies producing watches, cameras, machine tools, electronic items use the strategy of simple product extension.

2. Product Adaptation: This strategy involves effecting changes in the product to meet the needs and wants of customers in the foreign market. Alterations are made in the quality, size, brand, packaging, pricing of the product to meet local conditions.

3. Product Invention: Under this strategy the company specifically creates a brand new product for the new market in the foreign country according the target market requirements. Most often it costs less in case of underdeveloped and developing countries.

B. Pricing

As stated in chapter 16, pricing of a product depends on costs, demand and competition. There are differences in incomes of people and demand elasticities in different countries. Pricing of products is based on various costs of production and other costs that are incurred in a foreign market. These costs concern fuel costs, transportation and communication costs, middlemen's margins, taxes, tariffs etc. Pricing decisions are also affected by monetary policies and currency rates prevailing in the foreign market. A number of large companies pursue the policy of price differentiation depending on economic conditions of different countries. All these factors need to be considered while arriving at appropriate prices in a particular foreign market.

C. Promotion

A company may adopt the same promotion strategy all over the world or may effect a change in it to meet the requirements of the local markets. Most of multinational companies use the same promotion strategy in different countries. For example, Unilever, the parent company of Hindustan Unilever, Ponds, Brooke Bond Lipton follow global promotion especially in English speaking world. Take the case of Lux Soap, the message runs "96% of Movie Stars use Lux" or a beauty hint/tip straight from lovely Leelachitnis or Sri Devi or Madhuri Dikshit or the Lux ad will contain the reproduction of a film star or "Where there's LIFE-BUOY, there's Health". The only difference they make is to change the language, or colour. Now publicity films and docmentaries are being dubbed in local languages for advertising their products.

An example in this context is that of Telebrand Teleshopping. The company shows in India American Ad Video films with American faces but dubbed in Hindi or Tamil or Telugu.

However, there are many barriers that hamper the use of same advertising. These barriers/problems are:

1. Semantic or language differences.
2. Differences in tastes and preferences.
3. Government rules and regulations.
4. Availability of media in the host country and media ad rates.
5. Availability of agency in the host country.
6. Differences in levels of economic development.

Then there are bilingual and multilingual countries and the advertisement will have to be done in several languages. Different words mean differently in different languages. Therefore the same message may convey different meanings in different countries. Same colours signify different associations. Thus name, language, colour, message will need change and have to be adapted fitting the local conditions, mores or beliefs. Again the promotion mix and promotion budget will have to be planned keeping in view the factors mentioned above.

D. Distribution

Though the channel decisions to be made for international marketing are similar to those that are made for domestic markets but they are more complex. Distribution channels function differently in different countries as they are subject to different types of governmental control, rules and regulations. They work in different environmental conditions and even the size and character of channel members vary from country to country depending upon the requirements of customers or end users. A typical distribution arrangement for international marketing can be:

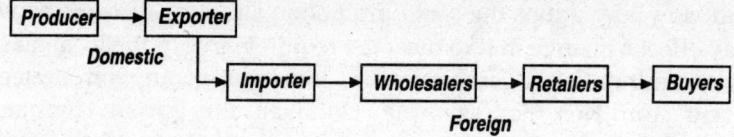

24.3 IMPLEMENTATION, EVALUATION AND CONTROL

This step involves implementation of the marketing programme and the assessment of its performance to find out whether the objectives set for international marketing have been achieved or not and to what extent they have been achieved, what have been deficiencies, what are the causes and what corrective action needs to be taken to attain the objectives. There is thus need for developing control and feedback systems.

Case 24.1   GM India creates new marketing entity

General Motor India, the wholly owned subsidary of General Motors Corporation, announced the creation of its sales and distribution company entitled 'General Motors (India) Marketing Private Limited'.

As a part of the restructuring and realignment of its corporate structure, GM India is tranferring its sales, marketing and distribution activities to the new entity, while all manufacturing activities will remain with General Motors (India) Private Limited.

Speaking about this realignment, P Balendran, vice president, GM India said, "The new company will be operational soon. With GM's operations in India expanding significantly in size and scale, it has become necessary to bring greater focus and efficiency to selling and distributing our world-class products and services."

"This move will help us improve planning, inventory management, transport management, timely delivery and thus, maximize cost synergies." he added.

Formation of new company follows GM India's state-of-the-art Greenfield plant at Talegaon.

The restructuring exercise, which is expected to result in significant cost efficiencies, will also involve the transfer of some employees who are currently enagaged in sales, marketing and distribution activities to the new company.

Source: **Pioneer, September 11, 2008.**

Case 24.2   Marico Limited, International Connection.

List of subsidiary companies

Name	Country of incorporarion interest	Percentage of ownership
Marico Bangladesh Limited	Bangladesh	100. (100)
MBL Industries Limited (Through Marico Bangladesh Limited)	Bangladesh	100 (100)
Kaya Limited (Erstwhile Kaya Skin Care Limited)	India	100 (100)
Marico Middle East FZE	UAE	100 (100)
Kaya Middle East FZE (Through Marico Middle East FZE)	UAE	100 (100)
MEL Consumer Care SAE (Through Marico Middle East FZE)	Egypt	100 (100)
EAIIDC (Through Marico Middle East FZE)	Egypt	100 (Nil)
Sundari LLC	USA	100 (75.5)
MSACC	South Africa	100 (Nil)
MSA (Through Marico South Africa Consumer Care)	South Africa	100 (Nil)
CPF International (Through Marico South Africa)	South Africa	100 (Nil)

EXERCISES

1. What is meant by the terms "international marketing" and "global marketing". Explain the concepts.

2. Describe the steps involved in designing a marketing strategy for international marketing.

3. Describe some major entry strategies in context of international marketing.

4. Identify the environmental factors which need to be considered before entering an international market.

5. Write short notes on
   (a) Joint Venture Arrangement.
   (b) Advantages and disadvantages of a wholly owned subsidiary.
   (c) Global standardised marketing mix.
   (d) Franchising.

Bibliography

Abell, F. and Hammond John S; *Strategic Marketing Planning*, Prentice Hall, Englewood Cliffs NJ 1979

Alien, Louis A; *Management and Organization*, McGraw Hill Book, Company Inc. New York, 1958.

Barnard Chester I; *The Functions of the Executive*, Harvard University Press, Cambridge, 1938.

Blake Robert R. and Mouton, Jane S; *The Grid for Sales Excellence Bench Marks for Effective Salesmanship*, McGraw Hill Book Co. New York 1970.

Blum, M.L. & Naylor, J.C.; *Industrial Psychology*, (Indian Edition) CBS Publishers, New Delhi 1984.

Boyd Harper W; *Marketing Management*, Harcourt Brace Jovonovich, New York 1972.

Burton, Dawn, *Cross-Cultural Marketing*, Routledge, March, 2008

Crane, Andrew, Maatten Dirk and Spence Laura; *Coporate Social Responsibility*, Routledge, January, 2008

Cravens, Hills, Woodruff; *Marketing Management*, (Indian Reprint) A.I.T.B.S. Publishers & Distributors, New Delhi, 1996.

Cravens David W; *Strategic Marketing*, Richard D. Irwin, Homewood, III, 1982.

Cundiff, E.W.; et al, *Fundamentals of Modern Marketing*, Prentice Hall of India Pvt. Ltd., New Delhi, 1977.

Davis, Keith; *Human Relations in Business*, McGraw Hill Book Co., Inc. New York 1957.

Dawson, John; et al edited, *The Retailing Reader*, Routledge, March, 2007

Drucker, Peter F; *Management: Tasks, Responsibilities, Practices*, Harper & Row, New York, 1973.

Engel, Kollat and Blackwell; *Consumer Behaviour*, Dryden Press Illinois, 1973.

Fayol, Henri; *General and Industrial Management*, Translated Constance Stores, Sir Issac Pitman & Sons London, 1949.

Hess John, M. and Cateora, Philip R; *International Marketing* Richard D. Irwin Inc. Illinois, 1966.

Howard, John A; and Sheth, Jagdish N.; *The Theory of Buyer Behaviour*, John Wiley and Sons, Inc., New York, 1969.

Kapur, S.K.; *Elements of Practical Statistics*, Oxford & IBH Publishing Co. Pvt. Ltd., New Delhi, 2008.

Kotler, Philip; *Marketing Management, Analysis, Planning and Control*, Prentice Hall of India Pvt. Ltd, New Delhi, 1995.

Kotler, Philip; *Marketing Management, the Millennium Edition*, Prentice Hall of India Pvt. Ltd, New Delhi, 2001

Lambert D.M. and Stock J.R.; *Strategic Distribution Management*, Richard D. Irwin, Homewood III, 1982.

Maslow, Abraham H; *Motivation and Personality*, Harper & Brothers, New York 1994.

McCarthy E. Jerome and Perreault W.D.; *Basic Marketing*, Richard D. Irwin, Homewood Illinois, 1984.

Newman W.H. and Warren E.K.; *The Process of Management*, Prentice Hall of India Pvt. Ltd, New Delhi, 1979.

Nicosia P.M.; *Consumer Decision Process*, Prentice Hall Englewood Cliffs, N.J., 1966.

Porter Michael E.; *Competitive Strategy, Techniques for Analysing Industries and Compititors*, The Free Press, New York 1980.

Robinson, Paris, Wind; *Industrial Buying and Creative Marketing*, Allyn & Bacon, Boston, 1967.

Stanely, Richard E; *Promotion*, Prentice Hall Inc. NJ, 1977.

Sissors, J.Z and Petray, R.E.; *Advertising Media Planning*, Crain Books Chicago, 1976.

Stanton, William J.; *Fundamentals of Marketing*, McGraw Hill Books Company, New York, 1967.

Vroom, V.H.; *Work and Motivation*, John Wiley & Sons, New York, 1964.

Webster, F.E. Jr; *Marketing for Managers*, Harper & Row Publishers. New York, 1974.

Werther, W.B. Jr & Davis Keith; *Personnel Management & Human Resources*, McGraw Hill Inc. New York, 1981.

Index